T0192334

Exposure–Response Modeling

Methods and Practical Implementation

Chapman & Hall/CRC Biostatistics Series

Editor-in-Chief

Shein-Chung Chow, Ph.D., Professor, Department of Biostatistics and Bioinformatics,
Duke University School of Medicine, Durham, North Carolina

Series Editors

Byron Jones, Biometrical Fellow, Statistical Methodology, Integrated Information Sciences,
Novartis Pharma AG, Basel, Switzerland

Jen-pei Liu, Professor, Division of Biometry, Department of Agronomy,
National Taiwan University, Taipei, Taiwan

Karl E. Peace, Georgia Cancer Coalition, Distinguished Cancer Scholar, Senior Research Scientist
and Professor of Biostatistics, Jiann-Ping Hsu College of Public Health,
Georgia Southern University, Statesboro, Georgia

Bruce W. Turnbull, Professor, School of Operations Research and Industrial Engineering,
Cornell University, Ithaca, New York

Published Titles

Adaptive Design Methods in Clinical Trials, Second Edition
Shein-Chung Chow and Mark Chang

Adaptive Designs for Sequential Treatment Allocation
Alessandro Baldi Antognini and Alessandra Giovagnoli

Adaptive Design Theory and Implementation Using SAS and R, Second Edition
Mark Chang

Advanced Bayesian Methods for Medical Test Accuracy
Lyle D. Broemeling

Advances in Clinical Trial Biostatistics
Nancy L. Geller

Applied Meta-Analysis with R
Ding-Geng (Din) Chen and Karl E. Peace

Basic Statistics and Pharmaceutical Statistical Applications, Second Edition
James E. De Muth

Bayesian Adaptive Methods for Clinical Trials
Scott M. Berry, Bradley P. Carlin, J. Jack Lee, and Peter Muller

Bayesian Analysis Made Simple: An Excel GUI for WinBUGS
Phil Woodward

Bayesian Methods for Measures of Agreement
Lyle D. Broemeling

Bayesian Methods in Epidemiology
Lyle D. Broemeling

Bayesian Methods in Health Economics
Gianluca Baio

Bayesian Missing Data Problems: EM, Data Augmentation and Noniterative Computation
Ming T. Tan, Guo-Liang Tian, and Kai Wang Ng

Bayesian Modeling in Bioinformatics
Dipak K. Dey, Samiran Ghosh, and Bani K. Mallick

Benefit-Risk Assessment in Pharmaceutical Research and Development
Andreas Sashegyi, James Felli, and Rebecca Noel

Biosimilars: Design and Analysis of Follow-on Biologics
Shein-Chung Chow

Biostatistics: A Computing Approach
Stewart J. Anderson

Causal Analysis in Biomedicine and Epidemiology: Based on Minimal Sufficient Causation
Mikel Aickin

Clinical and Statistical Considerations in Personalized Medicine
Claudio Carini, Sandeep Menon, and Mark Chang

Chapman & Hall/CRC Biostatistics Series

Exposure–Response Modeling
Methods and Practical Implementation

Jixian Wang

Celgene International

Switzerland

CRC Press
Taylor & Francis Group
Boca Raton London New York

CRC Press is an imprint of the
Taylor & Francis Group, an **informa** business

A CHAPMAN & HALL BOOK

CRC Press
Taylor & Francis Group
6000 Broken Sound Parkway NW, Suite 300
Boca Raton, FL 33487-2742

First issued in paperback 2020

© 2016 by Taylor & Francis Group, LLC
CRC Press is an imprint of Taylor & Francis Group, an Informa business

No claim to original U.S. Government works

ISBN-13: 978-1-4665-7320-8 (hbk)
ISBN-13: 978-0-367-73804-4 (pbk)

To my parents,
my wife, and sons

Contents

Symbol Description

$E(y\|x)$	Expectation of y conditional on x.
$E_x(f(x))$	Expectation of $f(x)$ over x.
$E^g(X)$	Geometric mean of $X : \exp(E(\log(X)))$.
$\int_x f(x)dG(x)$	Expectation of $f(x)$ over $x \sim G(x)$, where the distribution $G(x)$ is specified explicitly.
$y \sim F(.)$	y follows distribution $F(.)$.
$y\|x \sim F(.)$	y follows distribution $F(.)$ conditional on x.
$y_n \to y$	y_n tends or converges to y (all types).
$y \equiv x$	y defined as x.
$y \approx x$	y and x are approximately equal.
$\mathbf{X}^T$	Transposition of matrix or vector $\mathbf{X}$.
$diag(a_1, ..., a_k)$	A diagonal matrix with elements $a_1, ..., a_k$.
$\mathbf{I}$	An identity matrix of a conformable dimension (if not specified).
$Bin(p)$	Binary distribution with parameter p.
$IG(a, b)$	Inverse gamma distribution with parameters a and b.
$D(a_1, ..., a_k)$	Dirichlet distribution with parameters $a_1, ..., a_k$.
$IW(\rho, \mathbf{R})$	Inverse Wishart distribution with degree of freedom ρ and variance–covariance matrix $\mathbf{R}$.
$\chi_k^2(h)$	The χ^2 distribution with k degrees of freedom and non-central parameter h (may be omitted if $h = 0$).
$g^{-1}(y)$	The inverse function of $g(x)$.
$dy(x)/dx$	The derivative of $y(x)$ with respect to x.
$\partial y(x, z)/\partial x$	The partial derivative of $y(x, z)$ with respect to x.
$N(\mu, \sigma^2)$	Normal distribution with mean μ and variance σ^2.
$\Phi(x)$	The standard normal distribution function.
$\phi(x)$	The standard normal density function.
$N(\boldsymbol{\mu}, \boldsymbol{\Sigma})$	Multivariate normal distribution with mean $\boldsymbol{\mu}$ and variance–covariance matrix $\boldsymbol{\Sigma}$.
$U(a, b)$	The uniform distribution within the range (a, b).
$o(a)$	A higher order term of a.
$x \perp y\|z$	x and y are independent conditional on z.
$x \propto y$	x is proportional to y.
I_A or $I(A)$	$I_A = 1$ if A is true, and $I_A = 0$ otherwise.
$f(x)\|_{x=x_0}$	$f(x)$ evaluated as x_0.

modeling aspects. Although this is an application-oriented book, I have not shied away from using a large number of formulae, as they are an important part of applied statistics. In some situations, the derivations for some formulae are given to illustrate the concept and how a method is developed. However, theoretical details, e.g., technical conditions for some asymptotic properties of parameter estimates, are omitted.

The book emphasizes a number of important aspects: 1) causal inference in exposure–response modeling, 2) sequential modeling in the view of measurement error models, 3) dose–adjustment and treatment adaptation based on dynamic exposure–response models, and 4) model-based decision analysis linking exposure–response modeling to decision making. It tries to bridge gaps between difference research areas to allow borrowing well-developed approaches from each other. This is an application book, but it does not stop at using simple models and approaches. It goes much further to recent developments in a number of areas and describes implementation, methodologies and interpretation of fitted models and statistical inference based on them. Although the focus is on recent developments, no intensive knowledge on ER modeling is needed to read the book and implement the methods. There are models given in general forms with matrix notations. This may not be necessary when a model is explicitly specified, but is useful to understand the concept of some advanced approaches particularly when using software with models specified in general forms such as R and SAS. The contents are arranged to allow the reader to skip these formulae yet still be able to implement the approaches.

A large number of practical and numerical examples can be found in this book. Some illustrate how to solve practical problems with the approaches described, while some others are designed to help with understanding concepts and evaluating the performance of new methods. In particular, several examples in clinical pharmacology are included. However, to apply approaches to a real problem, it is crucially important to consult the literature and seek advice from pharmacologists. A large number of SAS and R codes are included for the reader to run and to explore in their own scenarios. Applied statisticians and modelers can find details on how to implement new approaches. Researchers and research students may find topics for, or applications of, their research. It may also be used to illustrate how complex methodology and models can be applied and implemented for very practical situations in relevant courses.

The book benefits from the numerous people who helped me or supported me one way or another. First I would like to thank Byron Jones, my PhD supervisor and former colleague for many years, for helping me with the planning and writing of the book from the very beginning and for his advice and friendship starting from 20 years ago. During my early career development, professors Jiqian Fang, Robin Prescott and James Linsey, and Dr Nick Longford gave me enormous help. The book reflects some early work I did with Byron and James, and my former colleagues professors Tom MacDonald and Peter Donnan at Medicines Monitoring Unit, University of Dundee, and some more recent research when I was working at Novartis. I would like to thank my former and current colleagues Wing Cheung, Lily Zhao, Cheng Zheng, Wenqing Li, Ai Li, Wei Zhang, Roland Fisch, Amy Racine-Poon, Frank Bretz, Tony Rossini, Christelle Darstein, Sebastien Lorenzo, Venkat Sethuraman, Marie-Laure Bravo, Arlene Swern, Bruce Dornseif and especially Kai Grosch and Emmanuel Bouillaud for their excellent team management and kind help which made project work we shared enjoyable. I gained much knowledge and experience in clinical pharmacology from the collaboration with Wing for several years. Prof. Nick Holford kindly allowed me to use the Theophylline data, and Novartis RAD001 and clinical pharmacology teams kindly allowed the use of moxifloxacin data. My thanks are due to John Kimmel, the executive editor for the CRC biostatistics series, and Karen Simon, project manager, for this book for valuable help and advice, and anonymous reviewers for their very useful comments and suggestions. Finally I would like to thank my wife Sharon, without her support and understanding the work would be impossible.

1

Introduction

1.1 Multifaceted exposure–response relationships

The exposure–response (ER) relationship is a very general concept, since it may refer to different types of relationships between different types of exposures and their responses. This book is concerned with modeling quantitative relationships between drug or chemical exposures and their responses, with an emphasis on modeling ER relationships in the pharmaceutical environment. However, the approaches described in this book are readily applicable to a wider range of topics such as environmetric modeling and areas in biostatistics where the source of exposure is not drugs or chemicals.

In many biostatistics-related areas, ER relationships carry important information about how different types of exposures influence outcomes of interest. For example, the drug exposure in pharmaceutics refers to the situation where a patient is under the influence of a drug. A quantitative measure of the exposure is the key information for ER modeling. It may be measured by drug concentration, dose or even treatment compliance. The exposure could be well controlled, such as the dose level in a randomized clinical trial, could only be observed, such as drug concentrations measured during a clinical trial, or could be partially controlled, e.g., drug concentrations in a randomized concentration controlled trial in which patients are randomized into a number of concentration ranges and the concentration for an individual patient is controlled by dose titration to achieve the target range. Exposures to toxic agents in toxicology experiments are apparently similar to drug exposures in humans, but may have different characteristics in *in vitro* and animal experiments. In environmetrics exposures could be air or water pollution, radiation, or exposure to risk factors in an industry environment, where often we (as the analyst of the exposure–response relationship) have no control over the exposure. Similar situations can be found in epidemiology where the concerned exposure may be a natural cause or an unnatural cause such as a drug prescription to a patient.

There are also different types of ER relationships. They can represent simply an association that the exposure and response apparently occur together (in this context, the word *response* is used in a loose sense), or a causal relationship in which the exposure is the true cause of the response. It may be a steady state relationship between constant exposure and response but may also represent a dynamic relationship between time-varying exposure and response. The relationship may refer to the exposure effect in a population, for example, how much is the risk of adverse events (AE) increased on average if a drug is prescribed for an indication. ER relationships may also be individual, e.g., as a measure of AE risk change related to a dose adjustment for individual patients.

1.2   Practical scenarios in ER modeling

We will use a number of real and hypothetical examples to illustrate practical aspects of ER modeling as well as a practical background for methodological development. Even with a number of examples, we cannot cover all aspects of ER modeling in practice. Therefore, it is important to identify the similarity and difference between approaches and between models applied to seemingly different practical scenarios so that appropriate approaches can be selected and adapted. This point will be emphasized from time to time in this book wherever appropriate. The following are four examples of different types from the area of pharmaceutics.

1.2.1   Moxifloxacin exposure and QT prolongation

The QT interval in electrocardiography (ECG) is an important measure for the time from heart depolarization to repolarization. QT interval prolongation (QT prolongation hereafter) is an often used measure to assess cardiac safety of new non-cardiovascular drugs. Since the interval depends on the heart rate (or equivalently the RR interval, i.e., the interval between two consecutive R-waves in the same ECG), the QT interval should be corrected by the heart rate to eliminate the impact of the RR difference among people with very different heart rates. A commonly used corrected QT interval is $QTcF = QT/RR^{1/3}$, known as the Fridericia correction. QT prolongation is believed to suggest an increasing risk of cardiac arrest and sudden death. A 10 millisecond (ms) prolongation in QTcF has been used by drug regulators as a threshold to indicate increased cardiac risk. Another approach called categorical analysis calculates frequencies of QTc higher than certain thresholds (e.g., 30 ms and 60 ms). A thorough QT(TQT) study, which includes a placebo, a positive control, and two doses of a test drug, all with baseline QT measurements, is used to show that a drug has no QT effect and normally does not need complex ER modeling. However, modeling approaches can be very useful in situations where the TQT study fails to demonstrate no QT effects at the current doses.

Moxifloxacin is known to have a stable QT prolongation effect and is safe to use as a positive control in TQT trials. We use the moxifloxacin data in a

TQT trial as an example to show (i) potential uses of modeling approaches, (ii) the need for advanced modeling techniques and (iii) some potential problems leading to further development of models and model fitting techniques. We will revisit this example in later chapters when a model or a method can be applied to it. The data consists of ECG and moxifloxacin concentration data from a cross-over trial in which a total of 61 subjects all received a single dose of placebo, moxifloxacin or the test drug in each period, and concentration and ECG measures were taken at a number of time points. The ECG parameters were corrected by subtracting the baseline value and the (post-baseline – baseline) difference in the placebo period measured at the same time (known as the $\Delta\Delta$ parameters). Figure 1.1 plots $\Delta\Delta QTcF$s vs the corresponding concentrations at all time points. There appears to be a linear trend of increasing QTcF along with exposure increase. However, since the data contain multiple ECG measures from individual patients, one cannot simply use a least squares (LS) method to fit a linear regression model. A correct approach is to use an appropriate repeated measurements model, as described in Chapter 3. One may be tempted to use a very intuitive approach: first fit the slopes for individual subjects, then analyze the slopes by taking the mean and standard deviation (SD) of them. This approach is, in fact, one type of two-stage approach for fitting a repeated measurement model. Figure 1.2 shows the distribution of individual LS slope estimates together with estimates based on a linear mixed model, assuming the slope for each subject follows a normal distribution. One interesting point is that the individual estimates are highly variable and the distribution shows heavy tails, particularly at the right hand side. It seems the mixed model estimates are shrunken toward the center. Details of linear mixed models can be found in Chapter 3.

Apart from the obvious need to deal with repeated measures, other characteristics in this dataset are also worthwhile exploring. First the concentration–QT relationship may not be linear. Although the trend in Figure 1.1 seems linear, it is not sufficient to conclude that using a linear repeated measurement model is sufficient. The relationship may not be instantaneous since the drug effect may accumulate and cause a delay (known as hysteresis) of the response to the exposure.

All the three features are important when using modeling approaches to help further develop a drug that shows some QT prolongation. For example, if a tested dose showed unacceptable QT prolongation, it would be useful to find the maximum dose that has an acceptable QT effect and test if it also delivers satisfactory efficacy. A valid exposure–response relationship allows calibrating that dose by taking model fitting uncertainty and inter- and intra-subject variabilities into account. Another approach to reduce the QT effects may be to change the drug formulation to reduce the peak concentration, while keeping a comparable overall exposure. Two key factors to be considered here are exposure accumulation and correct estimation of the upper percentiles of subjects with very long prolongation, e.g., defined for the category analysis. If exposure accumulates over time, the effect of cutting peak concentration on

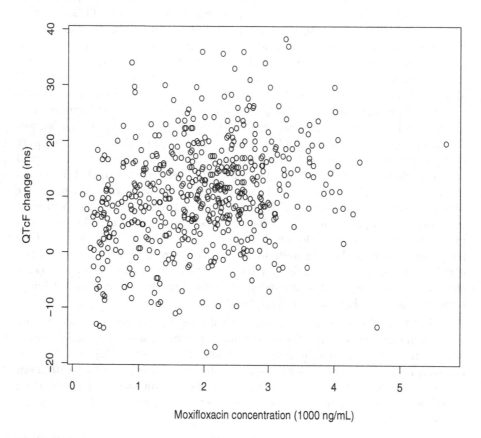

FIGURE 1.1
Pooled $\Delta\Delta QTcF$s vs. moxifloxacin concentrations from 61 subjects measured at multiple time points.

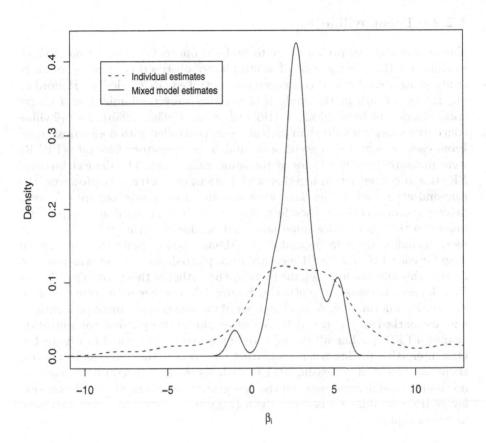

FIGURE 1.2
Individual parameter estimates by linear least squares and linear mixed model estimates for the slope of $\Delta\Delta QTcF$ and moxifloxacin concentration (1000ng/mL).

reducing QT prolongation may be much lower than estimated from a model assuming an instantaneous exposure effect. Appropriate assessment of the effect requires developing a dynamic model describing the temporal relationship between the pharmacokinetics (PK) time profile and the QT time profile. This type of model and related analytical methods will be described in Chapter 6.

1.2.2   Theophylline

This example comes from a study to find the optimal concentration of theophylline for the management of acute airway obstruction. This is an early study using a randomized concentration controlled (RCC) design (Holford et al., 1993a, 1993b). In the study, 174 patients were randomized to a target concentration of 10 or 20 mg/L (Holford et al., 1993a, 1993b). Theophylline concentrations from individual patients were controlled with dose adjustment. From each patient, the concentration and peak expiratory flow rate (PEFR) were measured multiple times at the same time point. The dataset includes PEFR and concentration together with patient characteristics. However, the randomization and dosing data were lost due to a power outage (Halford, private communication). Therefore, one cannot fit a population PK (popPK) model to the data, hence some non- and semiparametric approaches were used, including those in Holford et al. (1993a, 1993b). Since the randomized dose for each PK and PEFR measure were plotted, an attempt was made to recover this information from the plots by the author of this book. The PK and PEFR measurements are plotted in Figure 1.3 together with recovered randomization information. A small number of measurements cannot be identified and are marked as + in the plots. A scatter plot of theophylline concentration against PEFR, pooling all the repeated measures, is presented in Figure 1.4. One interesting finding when comparing the plots is that, although it seems there is a trend of increasing PFER with increased concentration, there is no significant difference between the two groups, while the observed exposure levels are quite different between them (Figure 1.3). We will revisit this issue in later chapters.

1.2.3   Argatroban and activated partial thromboplastin time

Argatroban is an anticoagulant for multiple indications including prophylaxis or treatment of thrombosis in some groups of patients. It is given by intravenous infusion, and drug plasma concentrations reach steady state in a few hours of the infusion. Its anticoagulant effect was measured by the activated partial thromboplastin time (APTT), a measure for the speed of blood clotting, and it was expected that increasing exposure would increase APTT values. The dataset in Davidian and Giltinan (1995) contains 219 measures from 37 patients with argatroban given as a 4-hour infusion, with repeated

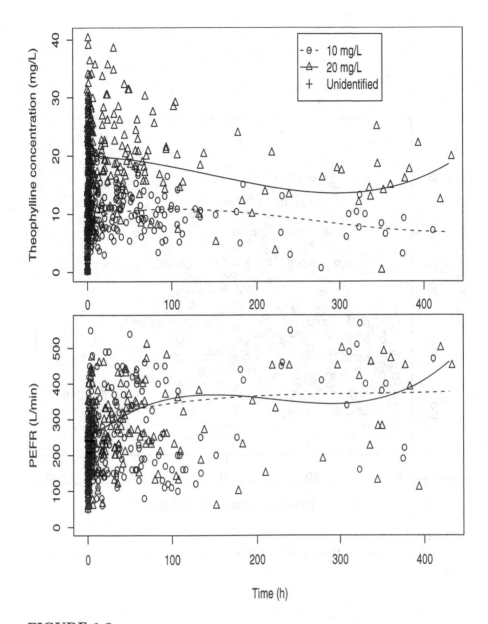

FIGURE 1.3
Theophylline and PFER time profiles with nonparametric curves fitted by spline functions via linear mixed models for both dose groups.

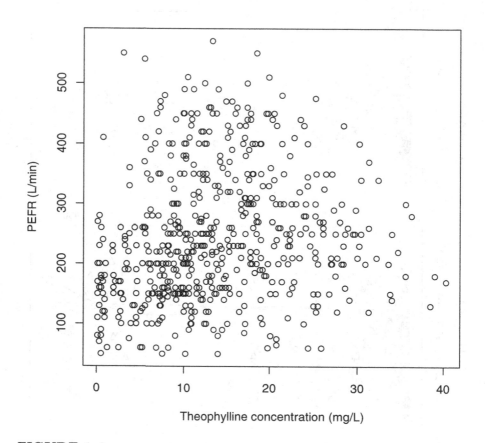

FIGURE 1.4
Theophylline concentrations and PEFR values pooling all the repeated measures.

concentration measures taken up to 6 hours after start of the infusion. Figure 1.5 gives individual APTT time profiles for these patients.

The APTT values were not all measured at the same time as the PK samples were measured. So, to look into the ER relationship, we need to predict the concentration at the time when each APTT is measured. This can be done using a nonlinear mixed model to fit the PK data and using the fitted model to predict the APTT at a given time. The APTT values and predicted argatroban concentrations show a strong linear trend, as shown in Figure 1.6. However, appropriate modeling approaches dealing with potential repeated measures, nonlinearity and hysteresis are needed to quantify the relationship between the exposure and APTT.

1.3   Models and modeling in exposure–response analysis

There are many different types of ER model. However, exposure models describing the relationship between exposure and factors such as dose also have an important role in ER modeling. Two types of model are used in ER modeling. Mechanistic models are based on a pharmacological mechanism for the action of drugs. Pharmacometrics is the main area in which mechanism models are widely used for ER relationships. PK modeling has a long history in describing PK characteristics of drugs and the fitting of drug concentration data (Bonate, 2006). Compartmental models derived from physical models of drug disposition have been well developed and implemented in software for PK and pharmacokinetic–pharmacodynamic (PKPD) modeling such as NONMEM. Relatively recently, mechanistic models for exposure–response relationships have also been developed. Empirical models describe the data in a quantitative manner, taking potential factors affecting the exposure and ER relationship into account.

Dose-response and ER models of both types may be written in the following abstract form.

```
Exposure=h(Dose, factors,Parameters)+Error
Response=g(Exposure, factors,Parameters)+Error
```

where g and h describe the relationships between dose and exposure, and exposure and response, and both are measured with errors. The tasks of modeling include fitting the two models to estimate the parameters, assessing the models as well as quantifying the error terms, and variability and uncertainty in the fitted model and estimated parameters. The models can be used to determine the exposure level corresponding to the desired response level, and how to achieve the exposure by adjusting the dose (or in other ways to control the exposure). The importance of quantifying the variability and uncertainty may be less obvious than estimating the parameters, but it is an important

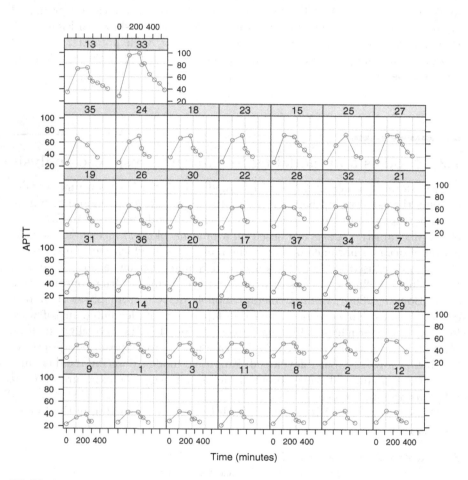

FIGURE 1.5

Individual APTT time profiles of 37 patients after a 4-hour infusion.

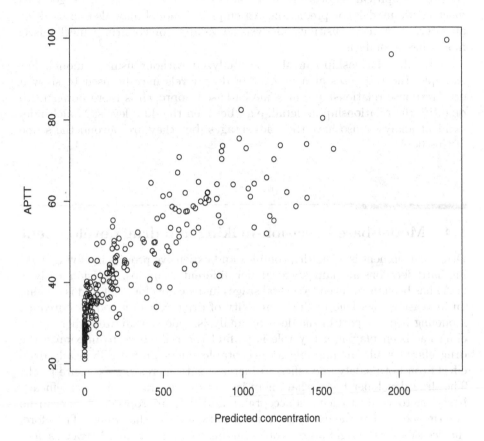

FIGURE 1.6
Observed APTT and predicted argatroban concentration by a nonlinear mixed model.

aspect of statistical modeling and provides important information for the assessment of an ER relationship without considering the mechanism behind it. There are also models that are partly mechanistic and partly empirical.

A mechanistic model has obvious advantages over an empirical model since it may support empirical evidence with a quantified pharmacological relationship. It may also indicate further investigation in specific directions. In contrast, empirical models are more flexible and can often be as good as mechanistic models for prediction. An empirical model may also be used for an alternative or a sensitivity analysis to validate or confirm a mechanistic model-based analysis.

The ER relationship can also be analyzed without using a model. For example, the responses at a number of dose levels may be used to show a dose-response relationship, but a model-based approach is more powerful to quantify the relationship in detail, e.g., between the dose levels. Model independent analyses also have their advantages, but they are beyond the scope of this book.

1.4   Model-based decision-making and drug development

Drug development is a lengthy, complex and expensive process involving making hard decisions at many stages under different scenarios. Decision analysis (DA) has been introduced to several stages in drug development and is playing an increasing role. Due to the complexity of drug exposure and its outcome, modeling is a key part of the decision analysis. One typical area where modeling has been playing a key role is to find the right dose to maximize the drug effects while maintaining an acceptable safety profile . Typically drug effects (on both safety and efficacy) increase with increasing exposure levels. The drug developer hopes that there is an exposure range in which efficacy is satisfactory and toxicity is acceptable and that appropriate doses can be determined so that the exposure in most patients is in that range. Therefore, the decision for the right dose involves finding the right balance between safety and efficacy. ER modeling provides key information for making the decision.

Often a critical question at a certain stage is whether a trial to obtain further information about safety and efficacy is needed, given the trial is expensive and takes a long time to run. The information may be to find a more accurate dose, or to ascertain the efficacy and safety profile of an existing dose. Decision analysis calculates the value of information (VoI) to assess if the benefit of running such a study is worth the costs (both money and possible delay of the development process). Decision analysis offers tools to evaluate the benefit in order to compare it with the costs. For this ER models are needed to provide information about efficacy and safety at a given or a range of doses. On a larger scale than the decision on a single trial, the whole drug development process involving a large number of trials can be considered as a complex process and decision makers make multiple decisions step by step, taking up-to-date information into consideration.

Decision analysis is increasingly used to optimize individual treatment for personalized medicine . For example, if a drug has highly variable exposure among patients, therapeutic dose monitoring, which monitors the exposure periodically and adjusts the dose if necessary, may be needed to ensure individual patients are under the right exposure level. Taking the measure has a considerable cost and inconvenience to the patient. How long do we have to measure before we stop? In fact, the question should concern when collecting further information does not justify the costs and inconvenience. Recent development of dynamic treatment regimes also brings decision analysis to the frontline of developing optimal treatment regimes that maximize the benefit of drugs to individual patients by response dependent treatment adaptation .

Most of these decision making problems require making optimal decisions sequentially. The decision process itself needs a complex model to evaluate. The decision tree is a common tool to solve some simple problems. For complex problems, a dynamic programming approach is needed. For both, ER models have to be integrated into this decision model.

1.5 Drug regulatory guidance for analysis of exposure–response relationship

Modeling exposure–response relationships is now playing an increasing role in drug development, as well as in drug regulation. Modeling approaches are now frequently used to support the submission of a new drug application (NDA). The U.S. Food and Drug Administration (FDA) published a guidance on the analysis of exposure–response relationships (FDA, 2003), in which exposure may refer to dose or PK concentration. The guidance suggests that determining the exposure–response relationship can play a role in drug development to

- support the drug discovery and development processes

- support a determination of safety and efficacy

- support new target populations and adjustment of dosages and dosing regimens.

The importance of the PKPD relationship, compared with the dose-PD relationship was recognized: " ... concentration-response relationships in the same individual over time are especially informative because they are not potentially confounded by doses selection/titration phenomena and inter-individual PK variability."

The guidance considered confounding bias in exposure–response analysis an important issue. It states that if one simply uses the observed drug concentration in a PKPD analysis "...potential confounding of the concentration-response relationship can occur and an observed concentration-response relationship may not be credible evidence of an exposure–response relationship". The randomized concentration controlled (RCC) design, which uses individual dose adjustment to achieve exposure levels randomly assigned to patients, was recommended. Although the RCC design has not been widely used, there is a trend of increasing RCC trials in the last 10 years.

The guidance also gives some details on specific situations when PK/PD modeling may provide important information: "Where effectiveness is immediate and is readily measured repeatedly ..., it is possible to relate clinical response to blood concentrations over time, which can provide critical information for choosing a dose and dosing interval." The guidance suggests that the modeling strategy be described clearly and include a statement of the problem, assumptions used, model selection and validation, and for a PK/PD analysis for a submission, this information should also be presented in the exposure–response study report.

Since the guidance was published ten years ago, it does not reflect recent developments. These include developments in confounding bias and causal effect estimation, modeling temporal exposure–response relationships and exposure-safety modeling. The reader can find details of these in later chapters.

Other health authorities have not published guidance specifically for exposure–response relationships. But using modeling and simulation for drug registration has been encouraged, particularly in some indications or populations such as pediatrics (Committee for Medicinal Products for Human Use, 2006). We expect further increases in the application of ER modeling in drug development and post-marketing monitoring.

1.6   Examples and modeling software

As this is a book about applied statistics, a large number of numerical examples are included. Some of them are illustrations of real examples such as the PK-QT and the PK-PEFR trials. We also use some simulation approaches under practical scenarios to assess the performance of procedures, or simply to demonstrate some specific properties of them. Using real data is the best way to demonstrate the implementation of an approach in practice, but a simulation has the advantage of focusing on the key issues and knowing the correct answers to specific questions (e.g., which is the best estimate for a parameter). The selection of numerical examples is a compromise between using real examples only and, at the other extreme, using simulated data only.

For all simulated data and examples, the codes for simulation and analysis are available such that the reader can not only reproduce the results, but also can explore further in the direction in which he or she is interested.

In the last 20 years we have seen a dramatic increase in computing power. Its impact on modeling is the rapid development of software that can fit very complex models. Most applied statistics texts now include a substantial amount of content dedicated to using software to solve specific problems and implementing approaches. This book is not an exception, since modern exposure–response modeling is heavily dependent on using software. We will use mainly three type of software: SAS (SAS, 2011), NONMEM (Beal, et al., 2006) and R (R Development Core Team, 2008) but will not use them to repeat the same task. The choice of the three is a compromise between software used by modelers (NONMEM and R) and statisticians (SAS and R) in the pharmaceutical industry. Researchers in academic institutes often have a wide range of software to choose from but these three are among the most popular ones. In this book implementation of all three may be given if they are straightforward, and sometimes the most convenient one is recommended. In particular, NONMEM is not a good platform for complex calculation such as matrix operations based on a fitted model. Our view is that these types of calculation should be left for R since NONMEM outputs are easily accessible in R. One may also question the merits of doing all the modeling in NONMEM while only taking data to R for extra calculation. We leave this for the reader to choose since it again depends on individual preference.

Although there is no space for a detailed comparison between the three, a few words may help to make a selection if the reader has a choice, particularly between SAS and NONMEM. SAS is a universal software package that provides a large number of statistical tools. NONMEM was developed for modeling and simulation using population PK and PKPD models. The core part is a nonlinear mixed model fitting algorithm, very similar to that of SAS proc NLMIXED, and the other parts allow easily implementation, e.g., of a standard compartmental PK model. Since NONMEM lacks other facilities such as graphic functions, a partner software is needed and the common selection is R. For those with no access to commercial software (SAS and NONMEM among the three), R alone is sufficient in most cases.

2

Basic exposure and exposure–response models

This chapter introduces simple models for modeling exposures and exposure–responses relationship. It serves as the basis of more complex models for longitudinal or repeated measurement data. Modeling exposure is often an integrated part of ER modeling, hence exposure models are also introduced. Some modeling tools such as transformation are also introduced here.

2.1 Models based on pharmacological mechanisms

2.1.1 Example of a PKPD model

Some PKPD models are derived from drug pharmacological mechanisms, although often based on some simplifying assumptions. Csajka and Verotta (2006) give an excellent review of this type of PKPD model. A typical example of such a model is derived from the drug receptor occupancy theory. According to the theory, the drug needs to bind to its receptor to form a drug–receptor complex to produce its effects. Often there is only a limited number of receptors. There are two processes occurring simultaneously: free receptors and drug molecules bind to form the complex and the complex disassociates back to free receptors and drug molecules. Let $R_c(t)$ be the amount of the complex and $C(t)$ be the drug concentration at time t. We assume that the change of $R_c(t)$ follows

$$\frac{dR_c(t)}{dt} = k_{on}(R_{max} - R_c(t))C(t) - k_{off}R_c(t) \tag{2.1}$$

where k_{in} and k_{off} are, respectively, the rates of forming the complex and disassociating back to free drug molecules and receptors and R_{max} is the total amount of receptors. This is a dynamic model and $R_c(t)$ is a function of time. But if $C(t) = C$ is constant, at the steady state (i.e., when association-disassociation reaches equilibrium so that $dR_c(t)/dt \approx 0, t > t_s$, with a reasonably large t_s), we can solve $R_c = R_c(t_s)$ easily and get

$$R_c = \frac{R_{max}C}{K_d + C}, \tag{2.2}$$

where $K_d = k_{off}/k_{on}$. When the response Y is proportional to R_c, this theory leads to

$$Y = \frac{E_{max}C}{K_d + C},$$ (2.3)

where E_{max} is the maximum effect when all R_{max} receptors have become R_c, when the drug concentration tends to infinity. K_d is also known as the concentration that gives the $E_{max}/2$ effect and is denoted as EC_{50} and the model is known as the E_{max} model. In the following, we often write this model as $Y = E_{max}/(K_d/C + 1)$ to shorten formulas, which should read as (2.3), in particular, $Y = 0$ when $C = 0$. The basic form has a number of variants. The reader is referred to Csajka and Verotta (2006) or a specialized text for details. This model is the most commonly used PKPD model. One may also use this model for the dose–response relationship by replacing C with dose for a drug with linear PK, i.e., the concentration is proportional to dose: $C = KDose$. In this case, $EC_{50} = K_d/K$ is the dose level that gives the $E_{max}/2$ effect.

There is a rich literature on PKPD models and mostly these models are dynamic and used to describe temporal ER relationships. Some typical models will be introduced in Chapter 6.

2.1.2   Compartmental models for drug–exposure modeling

Compartmental models are commonly used for drug concentrations, and sometimes for general exposures . A compartmental model describes drug absorption, disposition and elimination in and out of different body compartments (although some are hypothetical) (Gibaldi and Perrier, 1982). For example, the following are differential equations for an open one-compartment model with first-order absorption.

$$\frac{dA_1}{dt} = -K_a A_1$$

$$\frac{dA_2}{dt} = K_a A_1 - K_e A_2$$ (2.4)

where A_1 and A_2 are, respectively, the amount of drug at the absorption site and in the blood circulation (the central compartment), and K_a and K_e, respectively, are the rates of absorption and elimination. After a dose D at time $t = 0$, $A_1(t) = DF\exp(-K_a t)$, where F is called the bioavailability. Taking it into the second equation we can solve $A_2(t)$:

$$A_2(t) = \frac{DK_a F}{(K_a - K_e)}(\exp(-K_e t) - \exp(-K_a t)),$$ (2.5)

which can be converted to drug concentration in the central compartment as $c(t) = A_2(t)/V$, where V is the volume of distribution, an important PK parameter. K_e is often parameterized as $K_e = Cl/V$ in terms of drug clearance Cl.

If the drug is given by a bolus injection that delivers the drug into the central compartment instantly, the model is simplified to

$$c(t) = \frac{DF}{V} \exp(-K_e t). \tag{2.6}$$

This model may also be considered as an approximation to model (2.5) when the absorption is very quick. This can be verified by letting $K_a \to \infty$ in model (2.5). This model may be generalized to allow parameters depending on covariates. For example, V often depends on weight or body surface area. Some PK parameters such as the area under the curve defined as $AUC = \int_0^\infty c(t)dt$ and the maximum concentration $C_{max} = \max_t(c(t))$ are also important exposure measures in PKPD modeling. In the last example, we have $AUC = DF/(VK_e)$ and $C_{max} = DF/V$.

2.2 Statistical models

The models introduced in the last section describe theoretical situations, while almost always there are random variations in drug exposure and response data. Statistical models can take random variations in the response, exposure and model parameters into account, but we will introduce them by steps. Here, two error terms, representing the variations in the response and the exposure, respectively, are introduced, assuming other variables in the models are known. Variations in model parameters will be introduced in the next chapter and measurement errors in the exposure in ER models will be introduced in Chapter 4. A statistical model may be based on a mechanistic or an empirical model. An empirical ER model describes the relationship between exposures and responses but is not derived from the biological or medical mechanism between them. Drug exposure and its relationship with dose and other factors may also be modeled by an empirical model. Sometimes a mechanistic model may take a simple form. For example, taking a log-transformation of model (2.6) we obtain

$$\log(c(t)) = \log(D) - \log(V) - K_e t. \tag{2.7}$$

Therefore, a simple linear empirical model for log-concentration may also have a mechanistic interpretation. One such model includes log-dose and other factors such as age and weight as covariates, known as the power model, and is frequently used as an empirical alternative to the compartmental model. See the example in the next section

In this book we focus on cardinal (rather than ordinal) exposure measures. Obviously drug concentration is a cardinal measure, and so is the percentage of compliance to a treatment. When the drug dose is the exposure of concern, there are often only a few dose levels available. Indeed, one can consider dose level as an ordinal categorical variable. But we will treat the dose as a cardinal measure, since this way allows, for example, calibration at any dose level based on a fitted model. Therefore, we only consider ER models with a cardinal measure for the exposure, although covariates of other types may be included.

The ER model for a particular analysis mainly depends on the type of response. When the response is a continuous variable, the following regression models are the main candidates. Let y_i be the response, c_i be the exposure (which may not always be drug concentration) and $\mathbf{X}_i$ be a set of covariates of subject i. A simple linear model is

$$y_i = \beta_c c_i + \mathbf{X}_i^T \boldsymbol{\beta} + \varepsilon_i \tag{2.8}$$

where ε_i is a random variable with zero mean. Here the key parameter is β_c since it measures how y_i changes with c_i. To fit the model, the simple least squares (LS) procedure is often sufficient, based on a key assumption that c_i and ε_i are independent. This assumption obviously holds when the c_is are randomized dose levels. If c_i is observed, this might not be true. This issue will be left for later chapters.

Often PKPD models are nonlinear; one example is the E_{max} model (2.3). In general, a nonlinear model with an additive random error can be written as

$$y_i = g(c_i, \mathbf{X}_i, \boldsymbol{\beta}) + \varepsilon_i. \tag{2.9}$$

Note that here we classify a model as nonlinear when the relationship between y_i and the model parameters, not that between y_i and c_i, is nonlinear, since it is this nonlinearity that has a substantial impact on the statistical and model fitting aspects. For example, a model $y_i = \beta_0 + \beta_c \log(c_i) + \varepsilon_i$ is considered a linear model, and statistical inference and model fitting approaches for linear models can be applied. One special form of model (2.9) is

$$y_i = g(\beta_c, c_i) + \mathbf{X}_i^T \boldsymbol{\beta} + \varepsilon_i \tag{2.10}$$

in which the $\mathbf{X}_i^T \boldsymbol{\beta}$ part is linear. This model is often referred to as a partial linear model, and the partial linear structure, specifically the linearity between y_i and $\boldsymbol{\beta}$, may be used to facilitate model fitting and statistical inference based on the model.

Often the variance of ε_i in (2.9) may not be constant, a situation called variance heterogeneity. $\mathrm{var}(\varepsilon_i)$ may be a function of the mean of y_i, e.g., $\mathrm{var}(\varepsilon_i) = \sigma_0^2 + a g^b(c_i, \mathbf{X}_i, \boldsymbol{\beta})$, where σ_0^2 is the constant component, and a and b are parameters determining how $\mathrm{var}(\varepsilon_i)$ changes with $g(c_i, \mathbf{X}_i, \boldsymbol{\beta})$. This occurs when y_i is a nonnegative measure such as a biomarker measure and its variation or measurement error may increase with its mean.

In model (2.9) the error term is additive to $g(c_i, \mathbf{X}_i, \boldsymbol{\beta})$. Sometimes a model with multiplicative error term

$$y_i = g(c_i, \mathbf{X}_i, \boldsymbol{\beta})\varepsilon_i \tag{2.11}$$

may be needed. This model is particularly useful when the value range of the response may be limited to be positive, e.g., the level of a biomarker, drug concentration, or within a range, e.g., the percentage reduction in tumor size from the baseline size. In this case, one may assume that $\log(\varepsilon) \sim N(0, \sigma^2)$, hence y_i is always positive as long as $g(c_i, \mathbf{X}_i, \boldsymbol{\beta}) > 0$. In contrast, the additive model with $E(\varepsilon_i) = 0$ cannot guarantee this property. The PK model (2.6) is also a multiplicative model.

Both (2.8) and (2.11) assume that y_i is continuous, hence they cannot describe some types of response measures such as the status of dead or alive or the number of epileptic events within a day. A class of models known as generalized linear models (GLM) provides tools to model such types of responses (McCullagh and Nelder, 1989). A GLM assumes that the distribution of y_i belongs to the exponential family, and the mean of the response $E(y_i)$ is linked to a linear structure with a link function $g(.)$ in the form

$$g(E(y_i)) = \mathbf{X}_i^T \boldsymbol{\beta} + \beta_c c_i \qquad (2.12)$$

or equivalently $E(y_i) = g^{-1}(\mathbf{X}_i^T \boldsymbol{\beta} + \beta_c c_i)$. This model preserves the linear structure in model (2.8), but allows nonlinearity in the link function. The linear part $\mathbf{X}_i^T \boldsymbol{\beta}$ is known as the linear predictor. GLMs can be used to model a range of outcome types such as continuous variables with a positive distribution, binary, count and categorical variables, since the exponential family contains a wide range of distributions. Two commonly used GLMs are the Poisson and logistic regression models. The former is used for count data, e.g. $y_i \sim \text{Poisson}(\lambda_i)$ may be the count of AEs on patient i with the log-link function

$$\log(\lambda_i) = \mathbf{X}_i^T \boldsymbol{\beta} + \beta_c c_i, \qquad (2.13)$$

where the right hand side is the linear predictor and β_c is log-risk ratio (log-RR) for a 1 unit increase in c_i. Logistic regression models for binomial outcomes $y_i \sim \text{Binomial}(p_i, n_i)$ where n_i is the denominator are also GLMs. For example, when y_i is a binary variable (e.g., $y_i = 1$ if the patient had an AE and $y_i = 0$ otherwise), then y_i may follow the logistic model

$$\text{logit}(P(y_i = 1)) = \mathbf{X}_i^T \boldsymbol{\beta} + \beta_c c_i \qquad (2.14)$$

where $\text{logit}(P) = \log(P/(1 - P))$ and and β_c is log-odds ratio (log-OR) for a 1 unit increase in c_i. For fitting GLMs using different approaches, see the appendix for details.

Now we apply the logistic regression model to the QT prolongation data in the example in Chapter 1. Since the mean PK concentration had the maximum at 2 hours, we explore the relationship between the concentration and probability of QTcF prolongations being more than 20 ms around 2 hours. Using the 20 ms rather than the 30 ms threshold used by drug regulators is due to the small number of patients with QTcF > 30 ms in this dataset. Since QTcF was also measured during the placebo period, we can treat these measures as taken under 0 concentration. Fitting the logistic model to the data of both placebo and moxifloxacin periods, we get logOR = 0.468 (SE

= 0.215) for a 1000 ng/mL concentration increase, which shows a significant increase of the risk of 20 ms prolongation at the 5% level. However, when fitting the model to the moxifloxacin data only, the log-OR becomes -0.104 (SE = 0.557), indicating no ER relationship. In fact, this situation is not uncommon in exposure–response modeling when control (i.e., no exposure) data are available, since fitting the same model with and without control data may lead to completely different results. The difference often suggests that the dose–response relationship in the whole exposure range is complex and may need a more complex model. A simple model is likely to be misspecified at some exposure ranges. Using such a model may give very misleading predictions at these ranges.

A number of special situations need careful consideration when using a GLM model. For example, if y_i is the number of asthma attacks of patient i within a week, one may be attempted to use a Poisson regression model to model y_i since it is a count of events. However, y_i may have a higher variance than $E(y_i) = \lambda_i$: the variance of y_i if it is Poisson distributed. This situation is known as over-dispersion, often caused by correlation between the events on the same patient. One way to accommodate over-dispersion is to assume that λ_i is a random variable following a gamma distribution. This leads to a negative binomial distribution for y_i with var$(y_i) = \lambda_i + \phi\lambda_i^2$ where ϕ is an over-dispersion parameter which can be estimated from the data (Cameron and Trivedi, 1998).

Over-dispersion may also be found in the distribution of the number of failures (or responders) y_i among n_i subjects. In toxicology experiments, often the responses of animals in the same litter are correlated. Therefore, in litter i the number of deaths y_i may not follow a binomial distribution and var$(y_i) \neq n_i p_i (1 - p_i)$, where p_i is the probability of an event. The effect leading to this correlation is known as the litter effect, but the effect can be found in other areas under similar as well as different situations. A common model to take positive correlation into account is to assume that p_i follows a beta distribution. The resulting distribution for y_i is the beta-binomial distribution with var$(y_i) = n_i p_i (1 - p_i)(1 + (n_i - 1)\phi)$ where ϕ is the over-dispersion parameter.

Over-dispersion may not always occur even when p_i is a random variable. For the beta-binomial distribution, when $n_i = 1$, ϕ has no effect and the variance is var$(y_i) = p_i(1 - p_i)$. Hence there is no over-dispersion in a binary response variable. Intuitively, in an animal toxicology experiment, if there is only one animal in a litter then there is no over-dispersion due to variation between animals. It is easy to verify that if $y_i \sim Bin(p_i)$ and $p_i \sim F(p)$ where $F(p)$ is any distribution within (0,1), the marginal distribution y_i is always binary.

Although one may use a maximum likelihood estimate (MLE) approach with an appropriate distribution such as negative- or beta-binomials for fitting over-dispersed data, in practice, an empirical approach is often sufficient and more convenient. The approach uses a robust estimate, rather than its parametric form (e.g., $\text{var}(y_i) = n_i p_i (1 - p_i)$ for the binomial distribution) to estimate $\text{var}(y_i)$, then uses the estimate to calculate the SEs of $\hat{\beta}$ so that the Wald test-based statistical inference is valid under over-dispersion. In the SAS proc GENMOD, the option "PSCALE" in the "model" statement asks to estimate ϕ by the Pearson χ^2 statistic based on the empirical variance estimate $\sum_{i=1}^{n}(y_i - E(y_i))^2/(n - q)$ where q is the number of parameters in the model. Another option "DSCALE" uses the deviance to calculate ϕ. Often the resulting difference in $\text{var}(\hat{\beta})$ by using the two options is small. Although ϕ is introduced by a mixture of distributions that only allows for over-dispersion, the robust approach also works for under-dispersion, i.e., when $\text{var}(y_i) < \lambda_i$ in the count model example. In this case, it is often worthwhile to consider if under-dispersion is likely before deciding if one of the options should be used. NONMEM is a likelihood based software, so there is no readily used option to deal with over-dispersion. One may use either a full likelihood approach with the likelihood function for the beta-binomial or negative binomial distribution, or use the so called quasi-likelihood approach. See Chapter 3.

Sometimes extra variations occur at a particular value. The most common scenario can be found in data with more zero counts than there should be if a Poisson or even a negative binomial distribution is assumed. This scenario is known as zero-inflation and has often been seen in biomedical outcomes, e.g., in the counts of asthma or epilepsy attacks, or adverse events under drug exposure. To describe this type of data, zero-inflated Poisson (ZIP) has been introduced as follows. A ZIP distribution is denoted as $ZIP(\lambda_i, \rho_i)$ and defined as

$$P(Y_i = y) = \begin{cases} \rho_i + (1 - \rho_i)\exp(\lambda_i) & y = 0 \\ (1 - \rho_i)\frac{\lambda_i^{y_i}\exp(\lambda_i)}{y_i!} & y > 0 \end{cases} \quad (2.15)$$

where ρ_i is the probability of Y_i being always zero. Therefore, we have $E(Y_i) = (1 - \rho_i)\lambda_i$. As both λ_i and ρ_i may depend on exposure and covariates, they may be fitted in separate models. These models are also known as two-part models.

In a similar way, one can define a zero-inflated negative binomial (NB) distribution $Y_i \sim ZINB(\lambda_i, \phi, \rho_i)$ and Y_i has $1 - \rho_i$ chance to follow the NB distribution and ρ_i chance being always 0. Zero-inflated models have been widely used for medical decision-making, in which some outcomes such as the number of hospital visits are often zero-inflated. An ER model may help to assess the impact of dosing and dose adjustment on costs due to adverse events leading to hospitalization.

To link the two distributions to exposure and covariates, one may use a GLM structure for both λ_i and ρ_i:

$$\begin{aligned} \text{logit}(\rho_i) &= \mathbf{X}_i^T \boldsymbol{\beta}_a + c_i \gamma_a \\ \log(\lambda_i) &= \mathbf{X}_i^T \boldsymbol{\beta}_b + c_i \gamma_b \end{aligned} \quad (2.16)$$

where each model may only use a part of the covariates in $\mathbf{X}_i$. The likelihood function of the ZIP can be written based on its distribution (2.15), and can be maximized numerically to obtain estimates for parameters in both parts. Fitting some standard zero-inflated models is easy with software such as SAS proc GENMOD. Alternatively, proc NLMIXED can fit a wide range of models that consists of two parts, with a regression model for each.

A similar issue may arise when the outcome is either zero or a continuous variable. One example is the duration of hospital stay for a patient population. The duration may be modeled by a log-normal or a gamma distribution, but for those without any hospitalization, the duration is zero. The similar approach can be applied. But as a positive continuous distribution does not contain zero value, it has no contribution to the zero count, as we have seen $(1 - \rho_i) \exp(\lambda_i)$ in the ZIP distribution (2.15). Hence, the models for the probability of having zero value and the non-zero values can be fitted separately.

The logistic model can be extended to model ordered categorical response. Examples may include clinical outcomes such as a three-level category of responding to treatment, stable disease and disease progression. One may model such an outcome with a general polynomial distribution $Poly(p_1, ..., p_k)$, with $p_k = P(y = k)$ being the probability of outcome y in the kth category, and $\sum_{j=1}^k p_k = 1$. Although each p_i may be linked to exposure and covariates, to exploit the order of categories, one may use a model to represent a trend between, e.g., exposure and the level of outcome. For example, a higher exposure may increase the odds of being in a better category, which is the basic assumption of the proportional odds model with

$$log(P(y > k)/P(y \leq k)) = \beta_c c_i + \mathbf{X}_i^T \boldsymbol{\beta} \tag{2.17}$$

where $\beta_c i$ s the common log-OR for a 1 unit increase in c_i. An alternative is to use a reference category, e.g., level 1, so that the model assumes

$$log(P(y = k)/P(y = 1)) = \beta_c c_i + \mathbf{X}_i^T \boldsymbol{\beta}. \tag{2.18}$$

Both models include the logistic model a special case when the category has two levels. Fitting such a model is straightforward with current software for GLMs such as proc GENMOD.

2.3   Transformations

Transformation is a useful approach in modeling. It can be used on exposure and/or response variables as well as parameters. We will consider the first case and leave transformation of parameters (also known as re-parameterizations) to a later part where we treat parameters as random variables. The log-transformation is the most commonly used transformation and plays a central role in ER modeling. Often a transformation should be applied to both the variable and the model for it. This is known as the transform-on-both-sides approach (TBS, Carroll and Ruppert, 1988). Typically a log TBS is a bridge between additive and multiplicative models. One reason for using a transformation from a statistical aspect is to make the distribution of the response easy to handle with a simple model. The log-transformation is the most common one to use. The transformation also converts a multiplicative model to an additive one, and the latter is often much easier to fit. Some transformations proposed from purely statistical aspects, particularly those depending on extra-parameters, such as the Box–Cox transformation, are used less frequently than the log-transformation in ER modeling. Here we will mainly focus on the log-transformation.

Consider the theophylline concentration data. Since there is no dosing history data, we model the concentration data before any dose adjustment is made. Let c_i be the concentration from patient i at 0.01 h. We fitted a linear model for log-transformed c_i and covariates:

$$\log(c_i) = \mathbf{X}_i^T \beta + e_i, \tag{2.19}$$

where $e_i \sim N(0, \sigma_e^2)$. Note that this model might be considered as an approximation to model (2.6). The data were fitted by the linear regression function lm(.) in R. After fitting models with different combinations of covariates in $\mathbf{X}_i$, we found that only age is related to c_i (denoted as THEO in the outputs) and the fitted model is

```
lm(formula = log(THEO) ~ log(AGE), data = short[short$TIME ==0.01, ])
```

```
Coefficients:
            Estimate Std. Error t value Pr(>|t|)
(Intercept)  -0.8623     0.9418  -0.916   0.3618
log(AGE)      0.5699     0.2655   2.147   0.0339
```

The other factors included in the dataset might have been sufficiently adjusted when planning the initial dose. Transforming back to the original scale, we get the geometric mean of c_i as

$$\exp(\mathbf{X}_i^T \hat{\beta}) = \exp(-0.862)\text{AGE}^{0.570} \tag{2.20}$$

which is in the form of the power model. The key feature is that the contributions of the factors are multiplicative to each other. Note that $E^g(y_i) = \exp(\mathbf{X}_i^T \beta)$ is the geometric mean of y_i and $E(y_i) = \exp(\mathbf{X}_i^T \beta)E(\exp(e_i)) > \exp(\mathbf{X}_i^T \beta)$. Geometric means are commonly used for modeling exposure data, but may not always be appropriate for PKPD modeling.

Transformations are also used for nonlinear models. The error structure of the model is a key factor to consider for selecting an appropriate transformation. We take the Emax model (2.3) as an example to show how to select the transformation. Suppose that y_i is positive (e.g., the tumor size or a positive valued biomarker) and its relationship with c_i has a sigmoid-like shape; then the following model may be considered:

$$\begin{aligned}
\log(y_i) &= \log(E_{max}/(1 + EC_{50}/c_i)\varepsilon_i) \\
&= \log(E_{max}) - \log(1 + EC_{50}/c_i) + \log(\varepsilon_i) \tag{2.21}
\end{aligned}$$

where $\log(\varepsilon_i) \sim N(0, \sigma_\varepsilon^2)$ is an additive error term. This is a direct application of TBS to the Emax model with a multiplicative error term. The assumption for multiplicative error is reasonable when all $y_i > 0$. Consider an additive Emax model

$$y_i = E_{max}/(1 + EC_{50}/c_i) + \varepsilon_i^*. \qquad (2.22)$$

Under constraint $y_i > 0$, the error term has to satisfy the condition $E_{max}/(1 + EC_{50}/c_i) > \varepsilon_i^*$, which makes the distribution of ε_i^* depend on $E(y_i)$ in an awkward way. Model (2.21) is a partial linear model since the parameter E_{max} has been separated from the nonlinear part.

From model (2.21) we can derive

$$E(y_i) = E_{max}/(1 + EC_{50}/c_i) \exp(\sigma_\varepsilon^2/2). \qquad (2.23)$$

Note that the relationship between $E(y_i)$ and c_i is still described by the Emax model if σ_ε^2 is constant. E_{max} and EC_{50} are still the maximum effect and the concentration for achieving 50% of the maximum effect. The absolute change in $E(y_i)$ due to c_i is, however, dependent on σ_ε^2. A constant σ_ε^2 is an important condition for the Emax relationship between $E(y_i)$ and c_i, as the geometric mean $E^g(y_i) = E_{max}/(1 + EC_{50}/c_i)$ is proportional to $E(y_i)$ when σ_ε^2 is constant. But the two may be different functions if σ_ε^2 depends on c_i directly or indirectly.

We note that, with this model, $E(y_i) = 0$ when $c_i = 0$, and $E(y_i) = E_{max}$ when $c_i \to \infty$. $E(y_i) = 0$ may be an unreasonable constraint since often the baseline of response y_{i0} (i.e., the outcome when $c_i = 0$) may not be zero. Therefore, we may want to model the percent change from baseline or the ratio of the response to the baseline. Multiplicative models are commonly used for this purpose. For example, let y_i be the tumor size or a biomarker measure and y_{i0} the baseline value. We are interested in the treatment effect on y_i/y_{i0} or more specifically the reduction of tumor size and/or biomarker measure from the baseline. In this case we want the geometric mean $E^g(y_i) = y_{i0}$ when $c_i = 0$. For this situation a variant of model (2.21) is

$$y_i/y_{i0} = (1 - E_{max}/(1 + EC_{50}/c_i))\varepsilon_i \qquad (2.24)$$

where y_{i0} is the baseline. This model leads to $E^g(y_i|y_{i0}, c_i = 0) = y_{i0}$, as we desire. Note that often a constraint $E_{max} \leq 1$ should be applied as the tumor or biomarker values should be nonnegative. A further extension may allow fitting y_{i0} in the model as a covariate:

$$y_i = y_{i0}^\alpha (1 - E_{max}/(1 + EC_{50}/c_i))\varepsilon_i \qquad (2.25)$$

where α is a new parameter representing the impact of y_{i0} on y_i. With a log-transformation on both sides, the model

$$\log(y_i) = \alpha \log(y_{i0}) + \log(1 - E_{max}/(1 + EC_{50}/c_i)) + \log(\varepsilon_i) \qquad (2.26)$$

can be fitted as a partial linear model. A numerical problem may occur when

$1-E_{max}/(1+EC_{50}/c_i)$ is too close to zero. Often a very small positive number can be added or E_{max} is bounded lower than 1 to avoid this problem. E_{max} can be negative if the exposure effect is to increase y_i from y_{i0}.

Finally, it is important to note that dealing with zero exposures (e.g., exposure of the control group) is difficult for log-transformation, e.g., $\log(y_i) = \beta \log(c_i) + \varepsilon_i$. The ad hoc approach of replacing the zero value with a small value may work fine. But sensitivity of the model fitting to the imputed value should be examined. If the purpose is to extrapolate the exposure–response relationship to a lower exposure range than the observed exposure range, one should ensure the model fits well in the extrapolation range, as the imputed value may not affect the overall model fitting, but have a strong impact on prediction for the lower exposure range.

2.4   Semiparametric and nonparametric models

Often models for a PKPD relationship are empirical in nature (in contrast to popPK models derived from a pharmacological mechanism). Therefore, the modeler may be less certain of the correctness of the model and may not be able to take advantage of using a model with clear pharmacological meaning. Sometimes parametric models may not fit the data well, hence more flexible models are useful alternatives. There are many non and semiparametric approaches described in the literature. We will concentrate on one approach, spline functions (de Boor, 1978). This approach has a number of advantages over the other approaches, the most important one being its ease of use with implementations in R and SAS.

Spline functions have been widely used to approximate a function with known or unknown analytical form in computational mathematics, and algorithms for calculating them are well developed (de Boor, 1978). Without going into details, we simply state that spline functions are piecewise polynomial functions connected at a number of points (knots) where the value and derivatives up to a certain order are continuous, so that the curve they create is sufficiently smooth. The most commonly used are cubic spline functions consisting of cubic polynomials with continuous values and up to the second order derivatives continuous. Although some software fits data with spline functions automatically, sometimes one needs to construct spline functions to fit complex models with software without built-in spline functions. The approach of using spline functions with a certain smoothness controlled by the number knots and grade of splines is called regression splines and is a common approach.

Using regression splines to fit models is straightforward. We use B-splines as an example to introduce this approach. Suppose we would like to construct a smoothed ER curve in the exposure range $(0, 1)$. The simplest B-splines are those with knots at 0 and 1 only (known as boundary knots). In this case, the B-splines are terms of $C_n^k x^n (1-x)^k, k = 0, ..., n$, with C_n^k the number of combinations of taking k balls from a total of n. With only boundary knots, the cubic B-spline has $n = 3 + 1 = 4$ such terms. Using function bs() in R library splines, we can construct

```
> x=(0:10)/10
> cbind(x,bs(x,degree = 3,intercept=T))
           x     1     2     3     4
 [1,] 0.0 1.000 0.000 0.000 0.000
 [2,] 0.1 0.729 0.243 0.027 0.001
 [3,] 0.2 0.512 0.384 0.096 0.008
 [4,] 0.3 0.343 0.441 0.189 0.027
 [5,] 0.4 0.216 0.432 0.288 0.064
 [6,] 0.5 0.125 0.375 0.375 0.125
 [7,] 0.6 0.064 0.288 0.432 0.216
 [8,] 0.7 0.027 0.189 0.441 0.343
 [9,] 0.8 0.008 0.096 0.384 0.512
[10,] 0.9 0.001 0.027 0.243 0.729
[11,] 1.0 0.000 0.000 0.000 1.000
```

gives the function values at different x. With one interior knot at 0.5, there is one more term

```
> bs(x,degree = 3,knots=0.5,intercept=T)
          1     2     3     4     5
 [1,] 1.000 0.000 0.000 0.000 0.000
 [2,] 0.512 0.434 0.052 0.002 0.000
 [3,] 0.216 0.592 0.176 0.016 0.000
 [4,] 0.064 0.558 0.324 0.054 0.000
 [5,] 0.008 0.416 0.448 0.128 0.000
 [6,] 0.000 0.250 0.500 0.250 0.000
 [7,] 0.000 0.128 0.448 0.416 0.008
 [8,] 0.000 0.054 0.324 0.558 0.064
 [9,] 0.000 0.016 0.176 0.592 0.216
[10,] 0.000 0.002 0.052 0.434 0.512
[11,] 0.000 0.000 0.000 0.000 1.000
attr(,"degree")
[1] 3
attr(,"knots")
[1] 0.5
attr(,"Boundary.knots")
[1] 0 1
```

However, one of the terms is redundant as it is a linear combination of the others. This can be seen from that $C_n^k x^n (1-x)^k$s are the probability of having k events in n trials when x is the probability, hence $\sum_{k=1}^{n} C_n^k x^n (1-x)^k = 1$. Therefore, to fit B-splines to, e.g., $\sqrt{x}$, one should use the bs() function without the option intercept= T, which is the default, of not produce the first term:

```
lm(sqrt(x)~bs(x,degree=3))
```

Otherwise, the intercept cannot be fitted. The algorithm of constructing B-splines (de Boor, 1978) is very efficient, but will not be discussed here. The coefficient C_n^i is not needed from a practical aspect, as for each term, a parameter will be fitted. Therefore, we may just note each term as $B_k(x)$. There are other types of spline functions; among them the natural spline functions are also commonly used.

When using spline functions in an ER regression model, one can simply include them as covariates:

$$y_i = \sum_{k=1}^{K} B_k(c_i)\eta_k + \varepsilon_i \qquad (2.27)$$

where $B_k(c_i)$ is evaluated at exposure level c_i. Construction of B-spline functions is very efficient with software such as the R-library splines(.). For example, using the argatroban data at the 6-hour time point and predicted exposures, which will be discussed in Chapter 4, we can use function bs() together with function $lm()$ to fit the spline functions to log(APTT):

```
fittedModel1=lm(log(resp)~bs(pred,3),data=data240),
```

in which data240 is the 6-hour dataset, pred and resp are the predicted exposure and log(APTT), respectively, and $bs(pred, 3)$ produces a matrix of B-spline functions. It produces the model fit

```
lm(formula = log(resp) ~ bs(pred, 3), data = data240)

Coefficients:
 (Intercept)  bs(pred, 3)1  bs(pred, 3)2  bs(pred, 3)3
     3.7845        0.4710        0.2907        0.8023
```

One can also generate the B-spline functions outside of the model fitting function lm(), and then use the matrix as covariates. The following function call gives exactly the same fit.

```
Bspline=bs(data240$pred,3)
fittedModel1=lm(log(data240$resp)~Bspline)
```

Spline functions can be used to construct semiparametric models such as

$$y_i = \mathbf{X}_i^T \beta + \sum_{k=1}^{K} B_k(c_i)\eta_k + \varepsilon_i \qquad (2.28)$$

where $\mathbf{X}_i^T \beta$ is the parametric part. For example, in the argatroban analysis we may adjust for log-baseline values ("base") in the model

```
fittedModel2=lm(log(resp)~log(base)+bs(pred,3),data=data240)
```

which gives

```
lm(formula = log(resp) ~ log(base) + bs(pred, 3), data = data240)
```

```
Coefficients:
 (Intercept)      log(base)   bs(pred, 3)1   bs(pred, 3)2   bs(pred, 3)3
      1.2619         0.7671         0.3914         0.3924         0.7522
```

The coefficients for the spline functions are difficult to interpret directly. But since the B-spline functions are basis functions between the knots, an increasing trend of the coefficient indicates the same trend in the exposure–response relationship. This can also be seen from the previous B-spline examples in (0,1).

In a semiparametric model, it could be that the nonparametric or the parametric part is of the primary interest. For example, to explore the dose-response relationship in argatroban data, one may fit a model with spline functions for the time profile using dose as the covariate. Since this analysis involves repeated measurements and needs a linear mixed model, it will be postponed until the next chapter.

Semiparametric methods can also be used to extend GLMs by using a smoothing method on the linear predictor. For example, one can add spline functions for exposure to a GLM model and obtain

$$g(E(y_i)) = \mathbf{X}_i^T \beta + \sum_{k=1}^{K} B_k(c_i)\eta_k. \tag{2.29}$$

This type of model is known as a generalized additive model, as the spline function part is additive in the linear predictor. We have seen a large difference between the fitted logistic models with and without placebo data, indicating that the data present a more complex ER relationship than a linear model can describe. Fitting a logistic model with natural cubic spline functions using

```
fit=gam(qtp~ns(conc,3),family=binomial,data=moxi[moxi$time==2,])
```

where qtq and conc are the moxifloxacin concentration and response, respectively, at 2 hours, gives a bell-shaped curve for the relationship between the concentration and logOR (Figure 2.1). The results indicate a nonlinear relationship between moxifloxacin exposure and QT prolongation globally. The model fitted to all the data was driven by the strong impact of the risk increase from zero to low concentration, while that fitted to the moxifloxacin data was affected by the right end of the curve and resulted in a negative logOR. The lower and higher ends of the curve are very uncertain, as there was no event between concentration ranges of zero and 1870 ng/mL and higher than 3000 ng/mL. Although there is an initial increase in the risk along with the exposure increase, the left part of the curve cannot be quantified. There is no evidence of an ER relationship in the higher concentration range. In summary, there is not sufficient information to quantify the overall ER relationship based on the parametric and semiparametric models. We will show

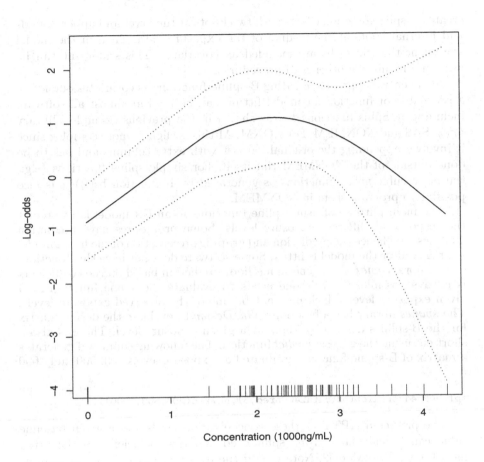

FIGURE 2.1
Nonparametric estimate of the relationship between moxifloxacin concentration and logOR of >20 ms prolongation in QTcF at 2 hours post dosing using natural cubic spline functions. The dotted lines are 95% CI.

other approaches that use more information in the data in later parts of the book.

Using spline functions in SAS is also easy for some procedures such as proc GLIMMIX. The EFFECT statement

```
EFFECT bs = spline(exposure / knotmethod=percentiles(3));
```

creates B-spline function effects with two knots at the lower and upper bounds and internal knots at the median of the exposure data. Then in the model statement the effects bs are included as covariates. This statement can be used in a number of other SAS procedures.

As a general approach, creating B-spline functions as covariates outside of a procedure or function for model fitting can be used in almost all software including R/Splus functions *lm* (as shown in the previous example), library *lme4*, SAS and NONMEM. For NONMEM this is slightly more complex since it involves appending the original dataset with extra columns and has to be done outside of the NONMEM run itself. For simple spline functions (e.g., the basic cubic spline functions as generated by R function bs(.)) it is also possible to program them in NONMEM.

The main purpose of using spline functions in an ER model is to predict the response at different exposure levels. Some procedures have prediction facilities, and hence the prediction and graphic presentation of the fitted model are ready after the model is fitted. Some software does not have this function, e.g., R library *lme4*. As a general method, prediction based on the coefficients is always possible. For this one needs to evaluate the spline function at a given exposure level, which may not be one of the observed exposure levels. The *splines* library has a function *splineDesign* to evaluate the design matrix for the B-splines defined by knots at a given exposure level. There is also a short cut using the generic *predict* function. The following command generates a matrix of B-spline functions evaluated at exposure levels 200, 500 and 1000

```
splinePred=predict(bs(data240$pred,3),newx=c(200,500,1000)).
```

The predicted APTT at these concentration levels and a given baseline value can be calculated using splinePred and the coefficients in the fitted model, e.g., fittedModel2. Note that if the exposure range for prediction, as specified with "newx=", is outside of the data distribution range, a warning " some 'x' values beyond boundary knots may cause ill-conditioned bases" will occur. It warns the user of the fact that spline functions are developed as a tool for interpolation rather than extrapolation.

There are a number of parameters to choose in the spline functions. In general, the higher the order, or the more knots, the more flexible the functions. A recommendation for the number of knots that works well in common situations is to add one knot for every 4 data points, with an upper bound set at 35 (Ruppert et al., 2003). The knots can be distributed at equally spaced percentiles of the exposure data. We will come back to this topic in later chapters where further applications of spline functions can be found.

2.5 Comments and bibliographic notes

There are a number of books about PKPD modeling. As a textbook, Bonate (2006) covers most of the important topics, is easy to read with a large number of worked examples, hence is highly recommended to statisticians and modelers. Senn (2007) gives an excellent summary for statistical issues in drug development. McCullagh and Nelder (1989) and Bates and Watts (1988) are the classical texts for GLMs and nonlinear regression models, respectively. Many examples of using GLMs and thier extensions can be find in Lindsey (1997). Details about count data including over-dispersion and zero-inflated models can be found in Cameron and Trivedi (1998). Details about transformation can be found in Carroll and Ruppert (1988), although quite a few are far beyond our scope. For model fitting using spline functions, Ruppert et al. (2003) provides a good introduction to statisticians and modelers. For readers who want to know mathematical details, de Boor (1978) is the classical text for spline functions.

3

Dose–exposure and exposure–response models for longitudinal data

This chapter introduces the most commonly used models in dose–exposure (DE) and exposure–response (ER) modeling for repeated measures from multiple individuals. For ER models we assume that the response and exposures are measured at the same time. The role of the exposure model in ER modeling may not be obvious at this stage, but in the next chapters its importance will be seen clearly. This chapter will cover several common types of response measures, including longitudinal continuous measures and categorical outcomes such as tumor sizes, biomarker measures and occurrence of AEs.

3.1 Linear mixed models for exposure–response relationships

A linear mixed model (LMM) is a linear model with some factors considered as fixed and some others considered as random, often referred to as fixed and random effects, respectively. It can also be considered an extension of the simple linear model (2.8) to allow the parameters to be random variables. We use the theophylline and PEFR data described in Chapter 1 as an example to show the structure of linear mixed models. In the data both theophylline concentrations and PEFR were measured repeatedly from each patient. Let y_{ij} and c_{ij} be the PEFR and concentration at time $t_j, j = 1, ..., r$ from patient i. Assuming a linear relationship between them, one may write

$$y_{ij} = \beta_0 + \beta c_{ij} + \varepsilon_{ij}, \qquad (3.1)$$

where β_0 is the mean PEFR and β measures the change in mean PEFR for a 1 mg/L theophylline increase. However, one missing factor is the patient's underlining characteristics such as lung function that affect the PEFR. Adding this factor, one may write

$$y_{ij} = \beta_1 + \beta_2 c_{ij} + u_i + \varepsilon_{ij} \qquad (3.2)$$

where u_i is a latent variable representing an individual's variability in PEFR, and ε_{ij} represents within-patient variation in PEFR. If r is sufficiently large, further assumptions on u_i may not be needed, since the u_is can be considered as unknown parameters and can be estimated when fitting the model. However, in many cases r is not large enough to estimate the u_is individually. In these cases, an LMM has the advantage of borrowing information from between patients and can provide better inference for both β and the u_is. For this purpose one needs to assume that u_i is a random variable. Often a normal distribution is assumed, so $u_i \sim N(0, \sigma_u^2)$, and we follow this convention for the moment. The issue of specifying the distribution for the random effects will be discussed later. The distribution of ε_{ij} is generally less important but we also assume that $\varepsilon_{ij} \sim N(0, \sigma_e^2)$. When the ε_{ij}s are independent, model (3.2) is among the simplest linear mixed models, but it does contain key features of LMMs. Based on this model one can calculate the correlation between y_{ij}s in the same patient as $\sqrt{\text{cov}(y_{ij}, y_{ij'})/\text{var}(y_{ij})} = \sqrt{\sigma_u^2/(\sigma_u^2 + \sigma_e^2)}$. The correlation does not depend on the assumed normality. Note that the same correlation occurs between any pair of y_{ij} and $y_{ij'}$. The PEFRs from different patients remain independent. This structure is known as compound symmetry, and is among the simplest and most commonly used correlation structures in LMMs.

The basic model (3.2) can be extended in a number of ways. Intuitively two PEFR measures 5 minutes apart probably have a higher correlation than that between two measures 5 hours apart. Therefore, it is often reasonable to assume that the correlation between ε_{ij} and $\varepsilon_{ij'}$ at t_j and $t_{j'}$, respectively, $\rho(\varepsilon_{ij}, \varepsilon_{ij'})$ is a function of $t_j - t_{j'}$, for example, $\rho(\varepsilon_{ij}, \varepsilon_{ij'}) = a(|t_j - t_{j'}|)^b$ where a and b are parameters to be estimated when fitting the model. This type of correlation is known as serial correlation, with the simplest case being the autoregressive error of order 1 (AR(1)) given by $\varepsilon_{ij+1} = a\varepsilon_{ij} + \varepsilon_{ij}^*$, where the ε_{ij}^*s are independent. Serial correlation and correlation due to patient level random effects may coexist and can be modeled if data are sufficient (e.g., when the number of repeated measures is not small). Models with both correlations can be fitted easily by SAS proc MIXED and the R library nlme (Pinheiro and Bates, 2000).

In model (3.2) we have assumed that a common slope β represents the effect of the concentration on PEFR for **all** patients. This may not be the case, as often treatment effects (e.g., PEFR increase per 1 mg/L concentration increase) may be different among patients. Another extension of model (3.2) is to allow β to be a random variable. This extension leads to a random coefficient model (Longford, 1993)

$$y_{ij} = \beta_0 + \beta_i c_{ij} + u_i + \varepsilon_{ij} \qquad (3.3)$$

where $\beta_i \sim (\beta, \sigma_\beta^2)$ is a random coefficient. A general model form can be written as

$$y_{ij} = \mathbf{X}_{ij}^T \boldsymbol{\beta} + \mathbf{Z}_{ij}^T \mathbf{u}_i + \varepsilon_{ij} \qquad (3.4)$$

where $\mathbf{X}_{ij}$ and $\mathbf{Z}_{ij}$ are vectors that may include c_{ij}, additional covariates and a constant, and $\mathbf{u}_i \sim N(0, \boldsymbol{\Sigma}_u)$ are random subject effects. In the rest of the book we will follow this general model form, and only write the c_{ij} term separately when needed. When $\mathbf{Z}_{ij} = \mathbf{X}_{ij}$, the model is a random coefficient model. $\mathbf{Z}_{ij}$ may only contain a constant, e.g., to form model (3.3). In most mixed model texts, the model is written in vector form

$$\mathbf{y}_i = \mathbf{X}_i\boldsymbol{\beta} + \mathbf{Z}_i\mathbf{u}_i + \boldsymbol{\varepsilon}_i \tag{3.5}$$

where $\mathbf{y}_i = y_{i1}, ..., y_{ir}, \boldsymbol{\varepsilon}_i = \varepsilon_{i1}, ..., \varepsilon_{ir}$ and $\mathbf{X}_i$ and $\mathbf{Z}_i$ are matrices formed by stacking $\mathbf{X}_{ij}^T$s and $\mathbf{Z}_{ij}^T$s. Elements in $\boldsymbol{\Sigma}_u$ are called variance components, which are important parameters in a mixed model. Based on normality and the model structure, the marginal model has a simple form

$$\mathbf{y}_i = \mathbf{X}_i\boldsymbol{\beta} + \boldsymbol{\varepsilon}_i^* \tag{3.6}$$

with $\boldsymbol{\varepsilon}_* \sim N(0, \mathbf{V}_i)$ where $\mathbf{V}_i = \text{var}(\mathbf{y}_i) = \mathbf{Z}_i\boldsymbol{\Sigma}_u\mathbf{Z}_i^T + \sigma_e^2\mathbf{I}$ and $\mathbf{I}$ is the r-dimensional identity matrix.

With the marginal model (3.6), we can estimate parameters $\boldsymbol{\beta}$ using MLE. An alternative estimate for $\boldsymbol{\beta}$ (which is also based on the marginal model and is equivalent to the MLE approach) is the generalized LS estimate as the solution to

$$S(\boldsymbol{\beta}) = \sum_{i=1}^{n} \mathbf{X}_i^T \mathbf{V}_i^{-1}(\mathbf{y}_i - \mathbf{X}_i\boldsymbol{\beta}) = 0. \tag{3.7}$$

The variance components can be estimated in a number of ways. The MLE may underestimate them, since it does not take the impact of estimating $\boldsymbol{\beta}$ into account. A correction for this impact is to use the residual (or restricted) maximum likelihood (REML), particularly when the sample size is small. The REML approach uses a penalty term for the estimation of $\boldsymbol{\beta}$ to correct the bias. For more details about fitting mixed effect models the reader is referred to Chapter 10. Both software packages we use in this book can fit LMMs with a complex structure, using MLE or REML and more options for the estimation of the variance components. R and SAS example codes for fitting LMMs of different correlation structures can be found in Appendix B.

Since in LMMs u_i is an unknown random variable rather than a parameter, one key role of using the model for ER modeling is to predict u_i. To understand the role of distributional assumptions on u_i, let us consider a simple model

$$y_{ij} = u_i + \varepsilon_{ij}. \tag{3.8}$$

If u_i is considered an unknown parameter, then its estimate is the individual mean $\bar{y}_{i.} = \sum_{j=1}^{r} y_{ij}/r$, in which the model and data from other patients are ignored. However, with the assumption $u_i \sim N(0, \sigma_u)$, a better estimation for u_i is

$$\hat{u}_i = \bar{y}_i\sigma_u^2/(\sigma_u^2 + \sigma_e^2). \tag{3.9}$$

Therefore, the assumption of random u_i centered at zero makes the estimate (or prediction) shrink toward zero. Based on the general model with normally distributed $\mathbf{u}_i$ and ε_{ij}, the conditional mean of $\mathbf{u}_i$ is

$$E(\mathbf{u}_i|\mathbf{y}_i) = \boldsymbol{\Sigma}_u \mathbf{Z}_i^T \mathbf{V}_i^{-1}(\mathbf{y}_i - \mathbf{X}_i\boldsymbol{\beta}). \qquad (3.10)$$

Therefore, the right-hand side with unknown parameters replaced by their estimates is used for the prediction of $\mathbf{u}_i$. With estimated parameters, (3.10) is no longer an unbiased conditional mean, but is often a good estimation for it, if the overall sample size is large. Mixed model software normally gives a $\mathbf{u}_i$ prediction based on this.

The LMM can accommodate complex structure in ER models, e.g., when the ER relationship depends on covariates, by properly constructing matrices $\mathbf{X}_i$ and $\mathbf{Z}_i$. For example, the response to an increase in exposure may be different among males and females, hence the model becomes

$$y_{ij} = \beta_0 + \beta_{1i}c_{ij} + Male_i\beta_{2i}c_{ij} + u_i + \varepsilon_{ij}, \qquad (3.11)$$

where $Male_i = 1$ for male patients and $Male_i = 0$ otherwise, and $Male_i\beta_{2i}c_{ij}$ is often called an interaction term. In this model the β_{1i}s are the exposure coefficients for females and β_{2i}s are the coefficients for male/female difference, respectively, and $\beta_{ki} \sim N(\beta_k, \sigma_b^2)$. This model can be written in the form of (3.4) by adding interaction terms between covariates and c_{ij} in $\mathbf{X}_i$. In the previous model two variables in $\mathbf{X}_i$ are the concentration in males and the other the concentration in females. Both R and SAS can form interactions between variables automatically with similar syntaxes such as Male*conc in the model statement, but in NONMEM interaction terms have to be formed explicitly. Special attention needs to be paid to the interpretation of these models. For example, to estimate the effect of a 1 unit increase in c_{ij} in a population, the male/female proportion is needed and the effect is a weighted average of β_1 and β_2 based on this proportion. In general, including the interaction term reduces the random variation in $\mathbf{u}_i$ since the variation between genders is now explained by β_2.

Apart from parameter estimation, we also need statistical inference tools for model comparison, hypothesis testing and estimating CIs for the parameters in the model, and among them, hypothesis testing is a key one since it can be used for the other purposes, such as constructing a CI. Hypothesis testing for LMMs in principle is not much different from that for other models, with one exception: the testing of variance components. Since ML or REML is the main tool for model fitting, it is natural to base statistical inference on one of them. A hypothesis can be tested by using one of the three tests: Wald test, likelihood ratio (LR) test and score test. The Wald test for a parameter is based on the assumption that $(\hat{\beta} - \beta)/SE(\hat{\beta}) \sim N(0,1)$ approximately. The LR test needs models fitted under the null and the alternative hypotheses. The 2 log-LR of the model under the alternative to that under the null follows approximately a chi-square distribution with degrees of freedom equal to the difference in the numbers of parameters included in the two models.

The score test needs the calculation of the score function based on the model under the null only, which is often a big advantage when the model under the alternative is much more complex. Both the Wald and LR tests can be implemented easily using the three software, but interestingly none of them offers an option for the score test.

Since for LMMs there are more alternative models than for a simple linear model, model selection is also a key task. Some may prefer to use a hypothesis test for model selection following the principle that a complex model with an extra parameter or component should show a statistically significant difference, e.g., the LR test suggests a better fit at the 5% level than a simpler one. Some other model selection indices such as the Akaike information criterion (AIC, Akaike (1974)) or the Bayesian information criterion (BIC, Schwarz (1978)) where for both the lower the value the better the model, can also be used. All three software packages used for this book provide these indices. Note that the LR test cannot be based on REML since the penalty term may change with different models.

Applying the LMM approach to the theophylline dataset, we use the R library lme4. See the appendix for a brief description of model implementation in the software. The results from fitting the basic model are, where THEO and ID are the theophylline concentration and patient's ID, respectively,

```
Linear mixed model fit by REML
Formula: PEFR ~ THEO + (1 | ID)
   Data: theo
 AIC  BIC logLik deviance REMLdev
6895 6913  -3444     6893    6887
Random effects:
 Groups   Name        Variance Std.Dev.
 ID       (Intercept) 4055.6   63.683
 Residual             7267.6   85.250
Number of obs: 574, groups: ID, 153

Fixed effects:
            Estimate Std. Error t value
(Intercept) 204.0624    9.5036  21.472
THEO          2.9877    0.5255   5.686
```

where we find that the effect of the concentration on PEFR is 2.99 L/(1mg/L) with SE = 0.526. The residual variance is 7268 and the between-patient variance is 4056. Adding random effects to β seems to have only a small impact, with an estimate of σ_β^2 equal to 2.82, and the LR test results in the test statistic $\chi^2 = 3.33$ with 1 degree of freedom (d.f.).

Adding all the covariates to the model leads to the conclusion that all but race are significant at the 5% level. The between-patient variance is reduced to 2007.

```
Linear mixed model fit by REML
Formula: PEFR ~ THEO+AGE+WT+factor(SEX)+factor(RACE)+factor(DIAG)+(1|ID)
```

```
   Data: theo
   AIC  BIC logLik deviance REMLdev
   6788 6836  -3383    6808    6766
Random effects:
 Groups    Name         Variance Std.Dev.
 ID        (Intercept)  1541.5   39.262
 Residual               7252.2   85.160
Number of obs: 574, groups: ID, 153


Fixed effects:
               Estimate Std. Error t value
(Intercept)    192.7639   27.0525   7.126
THEO             3.1088    0.4957    6.271
AGE             -1.5694    0.3543   -4.429
WT               0.7961    0.3310    2.405
factor(SEX)1    35.7653   11.1587    3.205
factor(RACE)1   29.3233   17.4045    1.685
factor(RACE)2   -8.0572   19.3195   -0.417
factor(DIAG)2  -73.0507   17.7482   -4.116
factor(DIAG)3   19.2605   38.7723    0.497
```

The covariates are coded as Sex: Male = 1 Female = 0; Race: Caucasian = 1, Polynesian = 2, Other = 3 and Diag(diagnosis): Asthma = 1, COPD = 2, Asthma + COPD = 3.

The scatter plot (Figure 1.3) suggests that a log-normal distribution may fit the PEFRs better; so does the nature of PEFR being positive. Fitting log-PEFR to a model with the concentration and all the covariates gives AIC=586. Note that this model cannot be compared with the model fitting PEFR using the LR test or the AIC or BIC indices. One has to judge which model fits the data better empirically. We can refine this model by applying transformations to the concentration and covariates. We used the TBS technique to fit the following multiplicative model and it shows a significantly better fit to the data with a 29.2 increase in 2 log-likelihood, and with no additional parameter.

```
Linear mixed model fit by REML
Formula: log(PEFR) ~ log(THEO) + log(AGE) + log(WT) + factor(SEX)
   + factor(RACE) + factor(DIAG) + (1 | ID)
   Data: theo
   AIC BIC logLik deviance REMLdev
  579.3 627 -278.6   520.8   557.3
Random effects:
 Groups   Name         Variance  Std.Dev.
 ID       (Intercept)  0.032747  0.18096
 Residual              0.124930  0.35345
Number of obs: 569, groups: ID, 152


Fixed effects:
             Estimate Std. Error t value
(Intercept)  4.41107    0.45412   9.714
log(THEO)    0.15016    0.01601   9.380
```

```
log(AGE)        -0.24731    0.05427   -4.557
log(WT)          0.35052    0.11029    3.178
factor(SEX)1     0.13419    0.04891    2.744
factor(RACE)1    0.11214    0.07507    1.494
factor(RACE)2   -0.04028    0.08420   -0.478
factor(DIAG)2   -0.36011    0.07507   -4.797
factor(DIAG)3    0.11500    0.16880    0.681
```

To use the fitted model, one has to transform it back to the original scale. Doing so, we have

$$
\begin{aligned}
\text{PEFR}_{ij} \;=\; & \exp(4.41)\text{THEO}_{ij}^{0.150}\text{AGE}_i^{-0.247}\text{WT}_i^{0.397}\exp(0.121)^{\text{Male}} \\
& \exp(0.112)^{\text{Caucasian}}\exp(0.040)^{\text{Polynesian}}\exp(-0.360)^{\text{COPD}} \\
& \exp(0.115)^{\text{Athma+COPD}}\exp(u_i + \varepsilon_{ij})
\end{aligned}
\tag{3.12}
$$

where the reference category is a female with asthma only and the categorical variables take value 1 if the ith patient belongs to that category, e.g., Male $= 1$ and COPD $= 1$ if patient i is a male with COPD but no asthma. Note that the random effect u_i makes the mean PEFR $\exp(\sigma_u^2/2) - 1$ higher than the geometric mean of PEFRs. In our case it is only $\exp(0.033/2) - 1$ (1.7%) higher, hence has almost no impact.

3.2 Modeling exposures with linear mixed models

In many situations, longitudinal exposure measures can be modeled by an LMM. The exposure measure may be an index of air pollution in occupational epidemiology or drug prescriptions in pharmacoepidemiology, and typically drug concentrations. We will use modeling drug concentrations as an example to develop models and approaches, but they also apply to other areas.

Sometimes PK samples may be taken repeatedly under the same condition, e.g., 24 hours after dosing. These data are not sufficient to fit a popPK model. These situations typically occur in a Phase III study, in which often only trough PK samples (taken just before the next dose in a repeated dosing scheme, also known as C_{min}) are practically feasible. Although one may combine them with external data to fit a popPK model, the value of this approach is often limited, since with trough samples only, the estimated full PK profile is mostly driven by the external data. In this case, the Cmin is widely used as a surrogate exposure measure for PKPD modeling. Since the samples are taken under the same conditions, an LMM is sufficient to model them. A commonly used model is the power model, i.e., an LMM for log concentration and logdose and covariates. Letting c_{ij} be the PK sample with corresponding d_{ij} at visit j from subject i, the power model can be written as

$$
\log(c_{ij}) = \mathbf{X}_{ij}^T\boldsymbol{\theta} + \mathbf{Z}_{ij}^T\mathbf{v}_i + e_{ij}
\tag{3.13}
$$

where $\mathbf{X}_{ij}$ and $\mathbf{Z}_{ij}$ are sets of factors including a constant as well as log-dose $\log(d_{ij})$, and $\boldsymbol{\theta}$ are parameters for the covariates and dose and $\mathbf{v}_i \sim N(0, \boldsymbol{\Sigma}_v)$ are random subject effects. This model is exactly the same as model (3.4) and interpretation of its components is also the same, although in the PK context. It can also be written in the vector form (3.5).

To understand how this general model (3.13) is used, consider the following example:

$$\log(c_{ij}) = \theta_1 + \theta_2 Male_i + \theta_3 \log(Age_i) + \theta_4 \log(d_{ij}) + v_{1i} + e_{ij}, \qquad (3.14)$$

in which we have $\mathbf{X}_{ij} = (1, Male_i, Age_i, \log(d_{ij}))^T$, $\mathbf{Z}_{ij} = \mathbf{X}_{ij}$ and $\mathbf{v}_i = (v_{1i}, 0, 0, 0)$, or equivalently $\boldsymbol{\Sigma}_v$ only has one nonzero parameter at the position of $(1,1)$. Of course the term $\mathbf{Z}_{ij}\mathbf{v}_i$ can also be written directly as v_{1i}. The model can be used to predict the concentration for subject i at a given dose d,

$$\hat{c}_i(d) = \exp(\hat{\theta}_1) \exp(\hat{\theta}_2)^{Male_i} Age_i^{\hat{\theta}_3} d^{\hat{\theta}_4} \exp(\hat{v}_{1i}), \qquad (3.15)$$

where $\hat{\theta}_k, k = 1, ..., 4$, are the estimated parameters and $\hat{v}_{1i}$ is the predicted subject effect v_{1i}.

A special case of these models is drugs with dose-proportional exposure, i.e., $\theta_4 = 1$. In this case, often one chooses to directly model the dose normalized exposure $c_{ij}/Dose_{ij}$. Even when $\theta_4 \neq 1$, the relationship between c_i and dose is simple since $c_i(Dose) \propto Dose^{\theta_4}$. However, this simple relationship is based on the assumption of constant $\mathrm{var}(v_{1i})$. Recall that when modeling c_{ij} directly one almost always needs to consider the variance as a function of the mean. Using a log-transformation helps to stablize the variance, hence it is often assumed that $\mathrm{var}(v_{1i})$ is a constant. Nevertheless, it is worthwhile to check if this assumption holds in an individual dataset. The impact of variance heterogeneity on the estimation of model parameters and prediction of $\log c_i(Dose)$ is often small and can be ignored. However, it may have a large impact on prediction of c_i, since $E(c_i(Dose)) \propto \exp(\mathrm{var}(v_{1i})/2)$. Furthermore, if $\mathrm{var}(v_{1i})$ depends on, e.g., the dose, the simple relationship of $E(c_i(Dose)) \propto Dose^{\theta_4}$ does not hold any more, and the exposure is no longer dose proportional even if $\hat{\theta}_4 = 1$. Its impact will be clear when we consider PKPD modeling using predicted concentrations as the exposure.

3.3   Nonlinear mixed ER models

Nonlinear mixed models (NLMM) describe nonlinear relationships in data with repeated exposures and responses from multiple subjects. Let $y_{ij}, i = 1, ..., n, j = 1, ..., r$, be the response (e.g., a biomarker measure) and c_{ij} be the exposure measured (e.g., the trough concentration) at time t_{ij} (e.g., at day 1 of treatment cycle j) from subject i. A general additive NLMM for y_{ij} can be written as

$$y_{ij} = g(c_{ij}, \boldsymbol{\beta}_i) + \varepsilon_{ij} \qquad (3.16)$$

where $g(c_{ij}, \boldsymbol{\beta}_i)$ is a nonlinear function either based on a pharmacological mechanism or as an empirical model, $\boldsymbol{\beta}_i$ is a set of parameters for subject i and $\varepsilon_{ij} \sim N(0, \sigma_\varepsilon^2)$ is a within-subject measurement error. This model can be adapted to fit positive variables such as a percentage by using a log-transformation, and the model then becomes multiplicative, since the model and error components are multiplicative in the original scale. $\boldsymbol{\beta}_i$ may also depend on covariates. A standard model is

$$\boldsymbol{\beta}_i = \mathbf{X}_i \boldsymbol{\beta} + \mathbf{Z}_i \mathbf{u}_i \qquad (3.17)$$

where $\mathbf{X}_i$ and $\mathbf{Z}_i$ are matrices of categorical and/or continuous covariates, and $\boldsymbol{\beta}$ is a vector of population parameters for $\boldsymbol{\beta}_i$. Often we assume a joint normal (or log-normal) distribution for $\mathbf{u}_i$: $\mathbf{u}_i \sim N(0, \Sigma_b)$. In practice, the most commonly used one is

$$\boldsymbol{\beta}_i = \boldsymbol{\beta} + \mathbf{u}_i \qquad (3.18)$$

where $\boldsymbol{\beta}$ is the population mean of $\boldsymbol{\beta}_i$ for the whole population.

One typical NLMM for PKPD modeling is the Emax model

$$y_{ij} = E_{maxi}/(1 + (EC_{50i}/c_{ij})^{\rho_i}) + \varepsilon_{ij} \qquad (3.19)$$

where E_{maxi}, EC_{50i} and ρ_i are, respectively, the maximum effect, exposure to reach 50% maximum effect and the Hill parameter determining the shape of the curve for subject i. To write the Emax model in the form of (3.16), we define $\boldsymbol{\beta}_i = (E_{maxi}, EC_{50i}, \rho_i)^T$, which may follow the simple model (3.18) or the complex one (3.17). If, for example, EC_{50i} depends on sex, then we need (3.17) and a binary variable (e.g., $Male_i = 1$ for males) in $\mathbf{X}_i$. Some parameters such as EC_{50i} should be positive; hence a log-normal distribution is more appropriate for all or a part of the components of $\boldsymbol{\beta}_i$. With this example we may write $\boldsymbol{\beta}_i = (E_{maxi}, \log(EC_{50i}), \rho_i, \log(EC_{50iMale}))^T$, where the last one is the difference between male and female in $\log(EC_{50})$, and

$$\mathbf{X}_i = \begin{pmatrix} 1 & 0 & 0 & 0 \\ 0 & 1 & 0 & Male_i \\ 0 & 0 & 1 & 0 \end{pmatrix}. \qquad (3.20)$$

The model can be written in the form of (3.17) with $\mathbf{Z}_i = \mathbf{X}_i$ if all parameters are random, or in the form of $\mathbf{X}_i \boldsymbol{\beta} + \mathbf{u}_i$ if the gender effect is fixed. Note that the model for EC_{50i} is a multiplicative model with females as baseline:

$$EC_{50i} = \exp(\beta_{2i}) \exp(Male_i \beta_{4i}). \qquad (3.21)$$

The model is similar to that for trough concentrations.

Fitting an NLMM is more difficult than fitting an LMM, as it generally does not have a simple marginal model such as model (3.6) for LMMs. To use MLE, in principle we can write the likelihood function as

$$L(\boldsymbol{\beta}, \boldsymbol{\Omega}) = \prod_{i=1}^{n} \int_{u} f_y(\mathbf{y}_i|\mathbf{u}_i) f_u(\mathbf{u}_i|\boldsymbol{\Omega}) d\mathbf{u}_i \qquad (3.22)$$

where $f_y(\mathbf{y}_i|\mathbf{u}_i)$ and $f_u(\mathbf{u}_i|\boldsymbol{\Omega})$ are the densities of $\mathbf{y}_i$ and $\mathbf{u}_i$, respectively. To calculate it, one needs a numerical integration method, such as Gaussian quadratures.

As an alternative to the MLE, one can use approximate marginal models such as

$$\mathbf{y}_i \approx g(c_{ij}, \mathbf{X}_i\boldsymbol{\beta}) + \frac{\partial g(c_{ij}, \mathbf{X}_i\boldsymbol{\beta} + \mathbf{Z}_i\mathbf{u}_i)}{\partial \mathbf{u}_i}\Big|_{u_i=0}\mathbf{u}_i + \varepsilon_{ij}, \qquad (3.23)$$

where $f(u)|_{u=u_0}$ means $f(u)$ evaluated at $u = u_0$. Then the MLE or generalized LS approaches can be based on this model. As it is based on a first order Taylor expansion of the mean function around $\mathbf{u}_i = 0$, this model is often called the first order approximation (Beal and Sheiner, 1982). The approximation is poor when $\text{var}(\mathbf{u}_i)$ is large. A more commonly used approximation is proposed by Lindstrom and Bates (1990):

$$\mathbf{y}_i \approx g(c_{ij}, \mathbf{X}_i\boldsymbol{\beta} + \mathbf{Z}_i\hat{\mathbf{u}}_i) + \frac{\partial g(c_{ij}, \mathbf{X}_i\boldsymbol{\beta} + \mathbf{Z}_i\mathbf{u}_i)}{\partial \mathbf{u}_i}\Big|_{u_i=\hat{u}_i}(\mathbf{u}_i - \hat{\mathbf{u}}_i) + \varepsilon_{ij} \qquad (3.24)$$

where $\hat{\mathbf{u}}_i$ is a prediction of $\mathbf{u}_i$. Since the approximation needs $\hat{\mathbf{u}}_i$, the algorithm involves alternating estimate $\mathbf{u}_i$ and the model parameters. Both approximations have been implemented in NONMEM and SAS proc NLMIXED, but R library nlme only has the second one. SAS proc NLMIXED and NONMEM also have facility to use quadrature to calculate the likelihood directly.

3.4   Modeling exposure with a population PK model

Population PK (popPK) modeling has been widely used to determine the relationship between drug exposure and dose and other factors in a population with repeated measures from multiple subjects using NLMMs. Technically they are similar to using NLMMs for ER modeling, but have a number of specific characteristics. Predicted exposures based on exposure models are commonly used in PKPD modeling. One reason is that population PK modeling is often needed for the purpose of understanding the population pharmacokinetics of a drug. Therefore, a popPK model may already be well developed at the stage when PKPD analyses are needed. Typical popPK models are compartmental models such as (2.5), but as in the NLMM for ER modeling, we need to introduce random effects into the parameters, since these parameters vary between subjects. For example, the model (2.5) with random variations in the parameters can be written as

$$c_i(t) = \frac{DK_{ai}F}{V_i(K_{ai} - Cl_i/V_i)}(\exp(-Cl_i/V_i t) - \exp(-K_{ai}t)). \tag{3.25}$$

In general, we can write a model for $c_i(t)$ as

$$c_i(t) = h(t, \boldsymbol{\theta}_i) \tag{3.26}$$

where $\boldsymbol{\theta}_i$ is a set of parameters specifically for this subject. The parameters may depend on factors such as age, body weight etc., hence one may write

$$\boldsymbol{\theta}_i = \mathbf{X}_i \boldsymbol{\theta} + \mathbf{Z}_i \mathbf{v}_i \tag{3.27}$$

where $\mathbf{X}_i$ and $\mathbf{Z}_i$ are factors including the intercept and $\boldsymbol{\theta}$ are population parameters. $\mathbf{v}_i$ represents the remaining variability in $\boldsymbol{\theta}_i$, after accounting for those explained by factors in $\mathbf{X}_i$, and it is often assumed that $\mathbf{v}_i \sim N(0, \boldsymbol{\Sigma}_v)$. This structure is the same as that in model (3.13), except here it applies to parameters in an NLMM. This model may be extended to include time varying covariates in $\mathbf{X}_i$ and $\mathbf{Z}_i$, but they are not common in popPK models.

A transformation may be needed to keep $\boldsymbol{\theta}_i$ in a valid value range. For example, all three parameters K_{ai}, V_i and Cl_i in model (3.25) should be positive for any subject, hence a log-transformation is often used to enforce the correct value range. Taking Cl_i as an example, one may write

$$\log(Cl_i) = \theta_1 + \theta_2 Weight_i + \theta_3 \log(Age_i) + v_{cli} \tag{3.28}$$

where $Weight_i$ and Age_i are the weight and age of subject i, often centered at reference values, e.g., 75 kg and 40 years old for adults. This is equivalent to

$$Cl_i = \exp(\theta_1)\exp(\theta_2)^{Weight_i} Age_i^{\theta_3} \exp(v_{cli})) \tag{3.29}$$

where $\exp(\theta_1)$ is the clearance of the reference subject of, e.g, 75 kg in weight and 40 years old. Categorical factors can be included in the model too. If the clearance depends on sex, with $Male_i = 1$ for males and $Male_i = 0$ otherwise, an additional term $\theta_4 Male_i$ in model (3.28) represents the impact of sex on the clearance. In model (3.29) $\exp(\theta_4)$ is multiplied to the right hand side if subject i is a male. We can write a model for $\boldsymbol{\theta}_i^T = \log(K_{ai}, V_i, Cl_i)$ in the general form (3.27), using the same approach as writing the parameters in the Emax model in the form of (3.17).

The concentration can only be measured by bioassay using a PK sample. Therefore, we only observe $c_i(t)$ with an error. Nevertheless we can add an error in the model to accommodate it. Let c_{ij} be the concentration measure under dose d_i at time t_j from subject i, with the model for $\boldsymbol{\theta}_i$ that

$$c_{ij} = h(t_j, d_i, \boldsymbol{\theta}_i) + e_{ij} \tag{3.30}$$

where e_{ij} contains assay errors but may also include within-subject variations in $c_i(t)$ not described in the model. At this stage we have reached a model of the same type as model (3.16). Indeed, model fitting and statistical inference including prediction based on this model, model selection and goodness of fit are the same as for (3.16). However, it is worthwhile to emphasize that the purpose of exposure modeling here is mainly exposure prediction for ER models, hence a good prediction rather than, e.g., correct estimation of PK parameters, is the primary objective. Therefore, it may not be necessary to use a compartmental model for exposure modeling. A semiparametric or nonparametric approach sometimes is a better option or a more robust alternative, for PKPD modeling, even when a popPK model has been developed.

3.4.1　The moxifloxacin example

Now we consider modeling the moxifloxacin concentration data using a popPK model. Figure 3.4.1 shows the individual PK profiles in this dataset. Florian (2011) has used a one-compartment model with first-order absorption to model the moxifloxacin concentrations. Here we follow the same approach using model (3.25), but parameterize the model in terms of K_a, K_e and Cl.

$$c_i(t) = \frac{DK_{ai}K_{ei}}{Cl_i(K_{ai} - K_{ei})}(\exp(-K_{ei}t) - \exp(-K_{ai}t)). \qquad (3.31)$$

We do not fit covariates in the model parameters since subjects in this trial were healthy volunteers and were expected to have similar PK characters. Although all three parameters may vary among subjects, it may not be necessary or even possible to fit a model with random effects on all parameters. Since moxifloxacin has a quick absorption phase, the first sampling point could not capture the absorption (ascending) phase of the profile for most subjects. Therefore, it is difficult to fit a model with a random effect on K_a. Models with random effects on log-K_e and log-Cl have been fitted, but the final model only needs a random effect on log-Cl. The other two parameters were also modeled on the log-scale. One advantage is that the parameter estimates on the log-scale in this situation often have a distribution closer to normal, and confidence intervals based on a normal approximation are more accurate. The fitted model is

```
Random effects:
 Formula: lcl ~ 1 | Subject
            lcl Residual
StdDev: 0.1850413 472.6988


Fixed effects: lka + lke + lcl ~ 1
          Value   Std.Error  DF    t-value
lka    1.332993  0.08766193 474    15.20607
lke   -2.537591  0.04551668 474   -55.75079
lcl  -10.489785  0.04369915 474  -240.04550.
```

The variation in the clearance is better measured by the coefficient of variation

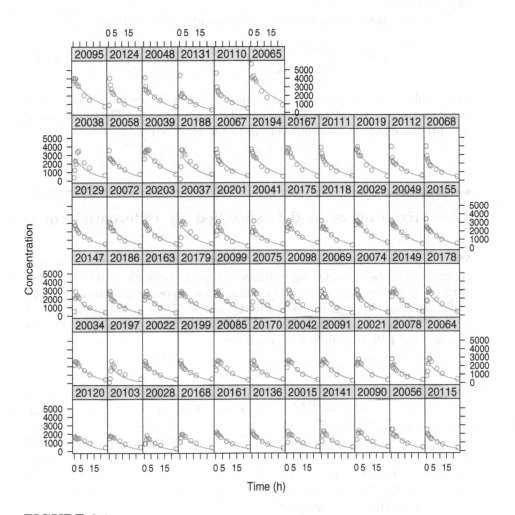

FIGURE 3.1
Observed (circles) and predicted (lines) individual moxifloxacin concentration
profiles using a one-compartment model with first order absorption.

(CV). Using the fact that if $\log(\theta_i) \sim N(\mu, \sigma^2)$ then $CV = \sqrt{\exp(\sigma^2) - 1}$, we can estimate that for clearance. The CV is $\sqrt{\exp(0.185^2) - 1} = 18.7\%$, which is quite moderate by pharmacological standards. It also reflects the fact that subjects in this trial are healthy volunteers of similar ages and had similar PK characteristics.

With the fitted model, we can predict the individual curves as

$$\hat{c}_i(t) = \frac{D\hat{K}_a\hat{K}_e F}{\hat{Cl}_i(\hat{K}_a - \hat{K}_e)}(\exp(-\hat{K}_e t) - \exp(-\hat{K}_a t)) \qquad (3.32)$$

where only $\hat{Cl}_i$ is an individual prediction. Individual time profile prediction based on the fitted model is also presented in Figure (3.4.1). The plots showed well fitted individual PK profile curves to the observed data.

3.5   Mixed effect models specified by differential equations

Many exposure and exposure–response models, particularly mechanistic models are, or can be, specified by ordinary differential equations (ODE). Compartmental PK models are typical examples for exposure models, and model (2.1) is an example for mechanistic PKPD models. The open one-compartment model with first order absorption has a very simple analytical form, but its source is the following differential equation:

$$\begin{aligned}
\frac{dA_1(t)}{dt} &= -k_a A_1(t) \\
\frac{dA_2(t)}{dt} &= k_a A_1(t) - k_e A_2(t)
\end{aligned} \qquad (3.33)$$

where $A_1(t)$ and $A_2(t)$ are the amount of drug at the absorption site and in the central circulation, respectively. A solution to this system depends on the state at $t = 0$ (initial values). For a bolus injection, the amount to be absorbed is $F \times Dose$ where F is bioavaiability, which we set at 1 without loss of generality. Therefore, the initial values are

$$\begin{aligned}
A_1(0) &= Dose \\
A_2(0) &= 0.
\end{aligned}$$

Solving the ODEs (3.33) gives the analytical form (2.5) for the concentration $c(t) = A_2(t)/V$, or $c(t) = A_2(t)K_e/Cl$, as we parameterized for the example in the last section.

For repeated measurement data with multiple subjects, we need to extend the ODE in a similar way as the extension from nonlinear regression models to NLME. To this end, we assume that, in model (3.33), the three parameters k_a, k_e and Cl are random variables with their values for the ith subject as k_{ai}, k_{ei} and Cl_i. Let $A_{2i}(t)$ be the the amount of drug in the central compartment of subject i. One can write the sample observed from subject i at time t_j as $c_{ij} = c_i(t_j) + e_{ij}$. In summary, the open one-compartment model specified by the ODEs for repeated measurement data is

$$\frac{dA_{1i}(t)}{dt} = -k_{ai}A_{1i}(t)$$
$$\frac{dA_{2i}(t)}{dt} = k_{ai}A_{1i}(t) - k_{ei}A_{2i}(t)$$
$$c_{ij} = A_{2i}(t_j)k_{ei}/Cl_i + e_{ij} \qquad (3.34)$$

These equations are among the simplest form of a general class of models known as stochastic differential equations (SDEs), or particularly ODEs with random coefficients. The theory of SDEs can be technically very involved. However, in practice, we can use numerical approaches described below to avoid dealing with the theory of SDEs. Nevertheless, we note that more complex SDEs may also find application in exposure and exposure–response modeling. See, e.g., Overgaard et al. (2005).

For drugs with linear kinetics, their models are linear ODEs with constant coefficients, and solutions to these ODEs have an analytical form. Therefore, fitting these models is not different from fitting an ordinary NLMM as described in this chapter. However, ODEs without analytical solutions are far more common than those with an analytical solution. In this case, a straightforward solution is to solve the ODEs numerically and to feed the solution into software for fitting NLMMs. NONMEM has a built-in ODE solver and one can specify the ODE as a part of the model. There is also an R-library nlmeODE that provides an interface between an ODE solver and the NLMM fitting software. Note that both NONMEM and nlmeODE use the same Fortran routine 'lsoda' in ODEPACK (Hindmarsh, 1983). The efficiency of using ODE solvers normally is much lower than using an analytical function, due to a far heavier computational burden. Another problem is the numerical accuracy and stability when ODEs form a stiff system. It is not possible to give details about stiff systems, but as an easy concept for compartmental models, a stiff system consists of very quick and very slow systems, which makes solving such a system numerically difficult since the two systems need different step sizes. Some ODE solvers such as lsoda have adaptive selection of the step size to control the accuracy, but there is no guarantee that it works under all situations. In particular, the ODE solver is called for in fitting models to individual curves for each subject, and some may have extreme parameter values. It is important to examine individual fits to check if there is any numerical problem in the finally fitted model. However, problems during the search for the MLE are difficult to detect, since it is not feasible to track the accuracy individually because of the huge number of calls on the ODE solver.

The following program uses the R function nlmeODE() to fit the moxifloxacin PK data. The function nlmeODE(.) is an interface from the NLMM function nlme(.) to the ODE solver, taking the ODE and initial value specifications from Onecomp.

```
bdata=bdata[order(bdata$Subject,bdata$Time),]
gdata=groupedData(conc~Time|Subject,bdata)

OneComp <- list(DiffEq=list(
     dy1dt = ~ -ka*y1 ,     #Absoption dynamic
     dy2dt = ~ ka*y1-ke*y2), #Central comp. with elimination rate ke
  ObsEq=list(
     c1 = ~ 0,              #No measure for absorption
     c2 = ~ y2/Cl*ke),   #Observe concentration c2=y2/volume
                          #volume=ke/Cl, y2=amount
  Parms=c("ka","ke","Cl"),
  States=c("y1","y2"),
  Init=list(0,0))  #Initial states for the ODE

moximodel=nlmeODE(OneComp,gdata) #generate function for nlme() call

jk=nlme(conc~moximodel(ka,ke,Cl,Time,Subject)   ,fixed=ka+ke+CL~1,
   random=pdDiag(Cl~1), start=c(ka=2,ke=-1,Cl=-3.5), data=gdata,
   control=list(returnObject=TRUE,msVerbose=TRUE),
   verbose=TRUE)
```

It produces a fitted model:

```
Nonlinear mixed-effects model fit by maximum likelihood
  Model: conc ~ moximodel(ka, ke, Cl, Time, Subject)
 Data: gdata
        AIC      BIC    logLik
 9094.531 9116.483 -4542.266


Random effects:
 Formula: Cl ~ 1 | Subject
               Cl Residual
StdDev: 0.186819 445.4036


Fixed effects: ka + ke + Cl ~ 1
         Value Std.Error  DF    t-value p-value
ka    1.333360 0.08211898 534   16.23692       0
ke   -2.536335 0.05378410 534  -47.15772       0
Cl  -10.487757 0.04901029 534 -213.99091       0
 Correlation:
    ka      ke
ke -0.609
Cl -0.471   0.846
```

Note that in nlmeODE, designed for the purpose of PK modeling, parameters are log-transformed as a default. Hence the three parameters labeled ka, ke and Cl are on the log-scale. The parameter estimates are very similar to those using the analytical form, as given previously. But when the starting values are far away from the estimated values, the nlmeODE fitted model may be different. These scenarios show that the ODE solver based approach might have numerical issues and may go wrong during model fitting.

Although using the ODE format in a model is very convenient, it is far less efficient than using the analytical format. For example, the CPU time was 29.92 seconds for the nlmeODE run, while it was only 0.17 seconds when using the analytical format.

3.6   Generalized linear mixed model and generalized estimating equation

A class of models not included in the NLMM framework is the generalized linear mixed models (GLMM), which are developed from generalized linear models (GLM). Based on a GLM model, a GLMM can be constructed to fit longitudinal data by introducing random components in the linear predictor. Taking the Poisson regression model as an example, letting $y_{ij} \sim \text{Poisson}(\lambda_{ij})$ be the count of events on patient i at visit j, the linear predictor for a mixed Poisson model is

$$\log(\lambda_{ij}) = \mathbf{X}_{ij}^T \boldsymbol{\beta} + \mathbf{Z}_{ij}^T \mathbf{u}_i \qquad (3.35)$$

where $\mathbf{u}_i$ is a set of random variables representing variations in the baseline risk among subjects. In this model $\mathbf{X}_{ij}$ will include c_{ij} with or without a transformation, as well as covariates. A general form of GLMM can be written as

$$g^{-1}(y_{ij}|\mathbf{u}_i) = \mathbf{X}_{ij}^T \boldsymbol{\beta} + \mathbf{Z}_{ij}^T \mathbf{u}_i \qquad (3.36)$$

where y_{ij} follows a distribution belonging to the exponential family, conditional on $\mathbf{u}_i$. In other words, a GLMM is a GLM for any given value of $\mathbf{u}_i$. Apart from a few exceptions, a GLMM does not have an analytical marginal likelihood function (i.e., the likelihood after integrating on $\mathbf{u}_i$). Therefore, approximations have been widely used since the early development of GLMMs and have been implemented in, for example, SAS proc GLIMMIX. Alternatively, numerical calculation of the marginal likelihood has become a feasible method, thanks to the multifold increase in computer power in recent years.

Analysis of longitudinal data may not require a full model specification. The generalized estimating equations (GEE) method is a convenient way to analyze longitudinal data based on a model for the mean of the response only. Taking the Poisson regression as an example and with $\mathbf{y}_i$ and $\mathbf{X}_i$ as for the LMM (3.5), an GEE estimate $\hat{\boldsymbol{\beta}}_{GEE}$ is the solution to the following estimating equation

$$S(\boldsymbol{\beta}) = \sum_{i=1}^{n} \mathbf{X}_i^T (\mathbf{y}_i - \exp(\mathbf{X}_i\boldsymbol{\beta})) = 0. \tag{3.37}$$

A general form of GEE is

$$S(\boldsymbol{\beta}) = \sum_{i=1}^{n} \mathbf{X}_i^T \dot{\boldsymbol{\mu}}_i \mathbf{W}_i (\mathbf{y}_i - \mu(\mathbf{X}_i\boldsymbol{\beta})) = 0, \tag{3.38}$$

where $\mu(\mathbf{X}_i\boldsymbol{\beta}) = E(\mathbf{y}_i|\mathbf{X}_i\boldsymbol{\beta})$ and $\dot{\boldsymbol{\mu}}_i$ are its derivatives with respect to $\mathbf{X}_i\boldsymbol{\beta}$. $\mathbf{W}_i = \mathbf{V}_i^{1/2}\mathbf{R}\mathbf{V}_i^{1/2}$ is a weight matrix in which $\mathbf{V}_i = \mathrm{diag}(\mathrm{var}(\mathbf{y}_i))$ and $\mathbf{R}$ is known as the working correlation matrix, which should be chosen as close to the real correlation matrix of $\mathbf{y}_i$ as possible, but is not necessary for consistent estimation of $\boldsymbol{\beta}$. For consistent estimation of $\boldsymbol{\beta}$, only $\mu(\mathbf{X}_i\boldsymbol{\beta}) = E(\mathbf{y}_i)$ has to be correctly specified, in addition to some technical conditions. See the appendix for more details.

Without random effects, the GEE approach is often equivalent to the ML approach. In the above example, (3.37) is also the score function of the Poisson model ($\mathbf{u}_i = 0$ in model (3.35)), thus the GEE estimate is the MLE. However, when the mixed Poisson model is assumed, $E(\mathbf{y}_i) = \exp(\mathbf{X}_i\boldsymbol{\beta})E(\exp(\mathbf{Z}_i\mathbf{u}_i))$. Therefore, when fitting a model with $E(\mathbf{y}_i) = \exp(\mathbf{X}_i\boldsymbol{\beta})$ with GEE, the $E(\exp(\mathbf{Z}_i\mathbf{b}_i))$ part causes an overestimation of the baseline risk if it is included in $\mathbf{X}_i^T\boldsymbol{\beta}$ as a constant. In fact, the GEE baseline estimate is the population average baseline. In general, the GEE method uses the marginal mean model $\mu(\mathbf{X}_i\boldsymbol{\beta})$ while GLMMs are subject specific models. It is important to distinguish the two types of model for longitudinal data. GLMMs are subject specific models since the linear predictor is explicitly a function of subject specific effects for a specific $\mathbf{u}_i$, while the GEE uses a marginal or population average model to describe the population average ER relationship. For the Poisson model, the difference is in the baseline risk estimate while the treatment effect can be estimated consistently. This property also holds for multiplicative models which are linear after a log-transformation. The difference is more problematic for some other models as, e.g., if a logit link function is specified for the GLMM. The marginal model does not have the same link function since $E_{\mathbf{b}_i}(1/(1 + \exp(-\mathbf{X}_i\boldsymbol{\beta} - \mathbf{Z}_i\mathbf{b}_i)))$ is not a logistic function with respect to $\mathbf{X}_{ij}^T\boldsymbol{\beta}$.

The difference between the population average and subject specific models should not stop us from using any of them. However, it is important to interpret the results correctly. For example, in the mixed logistic model with $\mathbf{X}_i^T\boldsymbol{\beta}$ containing an exposure measure term $\beta_c c_{ij}$, β_c is the subject specific log-odds ratio. The population average odds ratio will be lower than the subject specific one $\exp(\beta_c)$, as the variation in $\mathbf{u}_i$ dilutes the individual effects. As a consequence, with the same dataset, the estimate of the population odds ratio is higher than the subject specific one. Both models have their own merits and could provide more relevant results than the other for specific situations.

For example, the subject specific odds ratio is more relevant when considering treatment decisions as it represents the consequence of changing individual exposure. The population average odds ratio is more relevant when evaluating the treatment effects for a population.

These GLMM approaches can also be used to analyze longitudinal zero-inflated models. Let y_{ij}, λ_{ij} and ρ_{ij} be those in (2.15) and (2.16) at time t_j. Since model (2.16) has two parts, each of which may depend on a random effect, we may construct

$$
\begin{aligned}
\text{logit}(\rho_{ij}) &= \mathbf{X}_{ij}^T\boldsymbol{\beta}_a + c_{ij}\gamma_a + a_i \\
\log(\lambda_i) &= \mathbf{X}_{ij}^T\boldsymbol{\beta}_b + c_{ij}\gamma_b + b_i
\end{aligned}
\tag{3.39}
$$

where $(a_i, b_i) \sim F(\boldsymbol{\Sigma})$ are random subject effects. The most common choice is a multivariate normal distribution $F(\boldsymbol{\Sigma}) = N(0, \boldsymbol{\Sigma})$. The likelihood function for subject i, conditional on a_i and b_i, can be written according to the distribution (2.15) and the overall likelihood function in principle can be obtained by averaging it over a_i and b_i. SAS proc NLMIXED can be used to fit these models.

3.7   Generalized nonlinear mixed models

We have presented models for a continuous response with a linear or nonlinear relationship and for other types of responses that follow distributions belonging to the exponential family, but with the mean connected to a linear function of the exposure and covariates. One situation we have not covered is when the linear function is not sufficient or is inappropriate. For example, let the outcome y_{ij} be the number of asthma attacks and λ_{ij} be the risk within the jth period on patient i, $y_{ij} \sim Poisson(\lambda_{ij})$. Suppose the exposure effect on log-risk follows the Emax model:

$$
\log(\lambda_{ij}) = \mathbf{X}_{ij}^T\boldsymbol{\beta}_{xi} + \frac{E_{maxi}}{1 + EC_{50i}/c_{ij}}
\tag{3.40}
$$

which is a simple combination of a GLMM and an Emax model. There are two sources of random effects, one in the linear part $\mathbf{X}_{ij}^T\boldsymbol{\beta}_{xi}$ and another in the Emax part and parameters in that model. $\boldsymbol{\beta}_i = (\boldsymbol{\beta}_{xi}, E_{maxi}, EC_{50i})^T$ or its transformation can be written in the general form as $\boldsymbol{\beta}_i = \mathbf{X}_i\boldsymbol{\beta} + \mathbf{u}_i$. However, the parameters in the Emax model are not linear with respect to $\log(\lambda_{ij})$. The major technical consequence is that algorithms for fitting GLMMs cannot be used anymore. Therefore, the fitting of (3.40) needs general algorithms for fitting an NLMM specified by the likelihood function conditional on $\mathbf{u}_i$. For a number of y_{ij} distributions, SAS proc NLMIXED can be used without specifying the likelihood function explicitly. For example, to fit model (3.40), the following dummy codes can be used:

```
proc nlmixed;
parms b0=3, b1=1, Emax=0, EC50=0, sigu=1,siguv=0,sigv=1;
Emaxi=exp(Emax+ui);
EC50i=exp(EC50+vi);
linp = b0 + b1*x +conc*Emaxi/(conc+EC50i);
mu = exp(linp);
model y ~ poisson(mu);
random (ui, vi) ~normal([0,0],[sigu,siguv,sigv]) subject=i; run;
```

where for simplicity we assume that the random effects are only in the Emax model part.

3.8   Testing variance components in mixed models

Variance components (VC) are key elements in all types of mixed effects models. Due to the complexity of these models, in particular GLMMs and NLMMs, using a model with fewer random effects is technically beneficial. Therefore, testing variance components is an important tool for model selection. With regard to software for fitting mixed models, hypothesis testing tools commonly used are the Wald and likelihood ratio tests (LRT). Although score tests are often more efficient to calculate, they have not been implemented in common software for mixed models.

The most common test for VCs is on its existence, i.e., we can test $H_0 : \sigma_{uk}^2 = 0$ vs. $\sigma_{uk}^2 > 0$ to decide if the kth VC in $\mathbf{u}_i$ should be included in the model. The σ_{uk}^2 value under H_0 is at the boundary of the parameter space since σ_{uk}^2 cannot be negative. This constraint may be enforced during model fitting. Testing a parameter at the boundary does not cause problems if the distribution of the test statistic is not affected by the constraint. But the LRT statistic is affected by the boundary constraint since forcing $\hat{\sigma}_{uk}^2 \geq 0$ changes the distribution. Under some simple situations the distribution of the LRT statistic under the boundary constraint can be derived. In general the LRT statistic for testing if K VCs or $K + 1$ VCs are needed follows a mixture distribution of one χ_K^2 and one χ_{K+1}^2 with probability $1/2$ each (Stram and Lee, 1994). A special case is the test for no random effect vs one random effect, for which the LRT statistic has a mixture of χ_0^2 and χ_1^2 distributions with $1/2$ probability each. Since χ_0^2 has a unit mass at 0, the type I error in the situation of $\hat{\sigma}_1^2 > 0$ is $1/2$ of the tail probability of χ_1^2. Therefore, in this situation the p-value given by the χ_1^2 distribution for LRTs should be halved if the variance estimate is positive. Otherwise, the p-value is 0.5. SAS proc GLIMMIX has adapted these simple rules for VC tests. However, it should be noted these simple rules may not be accurate for complex models, particularly models with nonlinearity. For other situations, often simulation is needed to find the distribution or to calculate accurate p-values. Recent research has proposed score tests for VCs with boundary constraints, but some require simulations. Practical applications are limited due to difficulties in implementation in commonly used software.

Now we come back to the test in Section 3.1 where the LR test resulted in the test statistic $\chi^2 = 3.33$ with 1 d.f. Without considering the boundary constraint, it is equivalent to a p-value 0.188. However, with the approximate formula under the constraint, the p-value becomes 0.188/2=0.094.

3.9   Nonparametric and semiparametric models with random effects

The regression spline approach introduced in Chapter 2 can be easily extended to model repeated measure data. To this end, we need to write model (2.27) in a mixed model format and allow the coefficients to be random. A simple extension of model (2.27) is

$$y_{ij} = \sum_{k=1}^{K} B_k(c_{ij})\eta_{ik} + \varepsilon_i \qquad (3.41)$$

where the $B_k(c)$s are the base spline functions as defined for model (2.28) and $\beta_i = (\eta_{i1}, ..., \eta_{iK})^T \sim N(\eta, \Sigma)$ are the random coefficients for subject i. In fact, it is not necessary to use the same bases $B_k(c)$s for the fixed and random parts. A general model is

$$y_{ij} = \sum_{k=1}^{K} B_k(c_{ij})\eta_k + \sum_{k=1}^{K_2} C_k(c_{ij})u_{ik} + \varepsilon_i \qquad (3.42)$$

where $\mathbf{u}_i = (u_{i1}, ..., u_{iK_2})^T \sim N(0, \Sigma)$. This model can be written in the same form as model (3.5) by stacking the $\sum_{k=1}^{K} B_k(c_{ij})$ part to form $\mathbf{X}_i$ and the $\sum_{k=1}^{K_2} C_k(c_{ij})$ part to form $\mathbf{Z}_i$. Most statistical software allows a different design matrix for the fixed and random parts of the model, hence the general model can be fitted as easily as the simple one (3.41).

The semiparametric models using the spline functions discussed in Chapter 2 can also be extended to include mixed effects. For example, model (2.27) can be extended to model longitudinal data

$$y_{ij} = \mathbf{X}_{ij}^T \beta_i + \mathbf{B}(c_{ij})^T \eta_i + \varepsilon_{ij}, \qquad (3.43)$$

where $\eta_i \sim N(0, \Sigma_\eta)$ are random coefficients and $\mathbf{B}(c_{ij})$ is a vector of spline functions, as in model (2.27), but evaluated at the exposure level c_{ij}. Spline functions used here can be constructed in the same way as in (2.27). For longitudinal models the spline function part may represent an exposure response relationship, but it may also represent a common time profile in a model such as

$$y_{ij} = \mathbf{X}_{ij}^T \boldsymbol{\beta}_i + \mathbf{B}(t_{ij})^T \boldsymbol{\eta}_i + \varepsilon_{ij}. \tag{3.44}$$

where $\mathbf{X}_{ij}$ may include the exposure variable.

For the argatroban dataset, one may assume a common time profile for APTT but the size depends on the dose. This assumption leads to the power model

$$\log(y_{ij}) = \beta_1 \log(y_{i0}) + \beta_2 \log(d_i) + \mathbf{B}(t_{ij})^T \boldsymbol{\eta} + u_i \tag{3.45}$$

and the fitted model is

```
Formula: log(resp)~log(base)+bs(time,3,knots=c(240))
               +log(dose)+(1|subject)
   AIC    BIC logLik deviance REMLdev
 -200.1 -169.6  109.1  -245.2  -218.1
Random effects:
 Groups   Name        Variance Std.Dev.
 subject  (Intercept) 0.013317 0.1154
 Residual             0.014424 0.1201
Number of obs: 219, groups: subject, 37

Fixed effects:
                                  Estimate Std. Error t value
(Intercept)                        0.15322    0.65377   0.234
log(base)                          0.89956    0.19748   4.555
bs(time, 3, knots = c(240))1       0.91317    0.05241  17.423
bs(time, 3, knots = c(240))2       0.77683    0.07071  10.986
bs(time, 3, knots = c(240))3      -0.25797    0.07300  -3.534
bs(time, 3, knots = c(240))4       0.11119    0.11316   0.983
log(dose)                          0.17640    0.04099   4.304
```

where β_1, β_2 and $\boldsymbol{\eta}$ are labeled as log(base), log(dose) and the coefficients for the spline functions, respectively. Note that this model shows a low between-patient variation (a variance = 0.0133 is equivalent to 12% CV), and a similar size of within-patient variation. Figure 3.2 gives predicted response profiles by dose levels.

To allow more flexibility for the dose level, we can fit a model:

```
lmer(log(resp)~ log(base)+bs(time,3,knots=c(240))+bs(log(dose),3)
            +(1|subject),data=gpd2)
```

in which a smoothed exposure effect is fitted as bs(log(dose),3), and is multiplicative to the other components. One can decide if the extra flexibility is necessary by comparing the two models, which turns out to be unnecessary here ($\chi^2 = 0.18$ for d.f. = 2).

As said in Section 2.4, smoothness of the curves (now there are curves of the fixed and the random parts) is controlled by the number and location of the knots of the spline functions. Compared with those used for model (2.27), they may be more difficult to choose, as the observed points are not well designed.

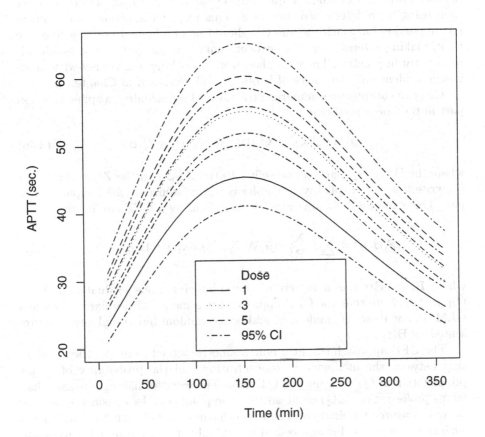

FIGURE 3.2
Predicted APTT for patients with baseline APTT = 25 sec and argatroban dose at 1, 3, and 5 μg/kg/min.

In general, for fitting PK profiles there is no additional complication, since the sampling points are, in most cases, designed to be the same or similar for all subjects. However, when c_{ij} are observed exposure levels, it is very likely their distribution varies. In general, the knots should cover the whole range of observed c_{ij}. However, the equal percentile locations may not work well, particularly when the distribution of c_{ij} varies significantly among subjects. It is advisable to examine the distribution carefully for knot selection rather than using a predetermined approach. This may cause problems in clinical trial reporting for which the analysis should be preplanned and timelines are tight, making extensive data dependent changes in analysis methods difficult to perform in practice. There are, however, smoothing approaches with much less dependence on the choice of knots, as will be shown in Chapter 4.

One can construct a semiparametric GLMM by including a spline function part in the linear predictor

$$g^{-1}(y_{ij}|\mathbf{b}_i) = \mathbf{X}_{ij}^T\boldsymbol{\beta} + \mathbf{B}(t_{ij})^T\boldsymbol{\eta}_i + \mathbf{Z}_{ij}^T\mathbf{b}_i \qquad (3.46)$$

where the $\mathbf{B}(t_{ij})^T\boldsymbol{\eta}_i$ part can be split into the $\mathbf{X}_{ij}^T\boldsymbol{\beta}$ and the $\mathbf{Z}_{ij}^T\mathbf{b}_i$ parts. But we write this part explicitly to emphasis the existence of the nonparametric part. One can also fit semiparametric models using a GEE similar to

$$S(\boldsymbol{\beta},\boldsymbol{\eta}) = \sum_{i=1}^{n}\begin{pmatrix} \mathbf{X}_i^T \\ \mathbf{B}_i^T \end{pmatrix}\dot{\boldsymbol{\mu}}_i\mathbf{W}_i(\mathbf{y}_i - \mu(\mathbf{X}_i\boldsymbol{\beta} + \mathbf{B}_i\boldsymbol{\eta})) = 0, \qquad (3.47)$$

where $\mathbf{B}_i = \mathbf{B}(\mathbf{t}_i)$ is a matrix of the spline functions evaluated at $\mathbf{t}_i = (t_{i1}, ..., t_{ir})$. Note that the GEE approach fits a model with a fixed $\boldsymbol{\eta}$, while a GLMM may allow for random $\boldsymbol{\eta}_i$ and hence random individual curves represented by $\mathbf{B}(t)\boldsymbol{\beta}_i$.

The GEE approach can fit a nonparametric model to model the relationship between the moxifloxacin concentration and the probability of 20 ms prolongation in QTcF using the QT data. The event of having a higher than 20 ms prolongation in QTcF at all the time points can be considered as a repeated measure of a binary outcome, which can be fitted with a logistic model with spline functions for unspecified individual exposure–response curves and correlations between the repeated measures taken care of with the robust variance estimate. We can use the gee(.) function in R in combination with spline function generators:

```
fit3=gee(qtp~bs(conc,3),family=binomial,id=SID1A,data=moxi).
```

The model was fitted separately to all data and data from the moxifloxacin period only. The fitted models give the estimated ER curve (Figure 3.3) with 95% CI based on the robust variance estimate for β. The two fitted models show not only uncertainty at the high exposure end, but also on the low exposure end. But both models suggest that the maximum risk seems to be reached when the concentration is around 3000 ng/mL but the data could

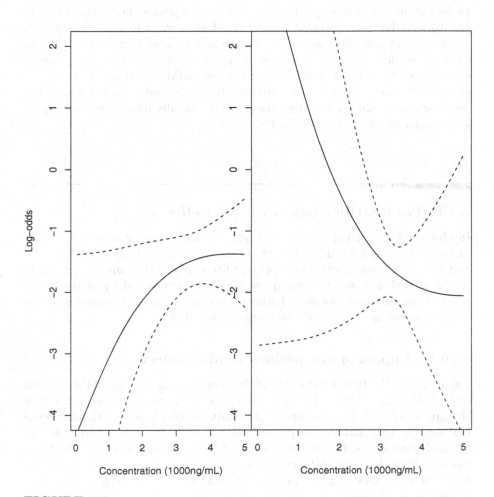

FIGURE 3.3
Nonparametric estimate of the relationship between moxifloxacin concentration and logOR of the probability of 20 ms prolongation in QTcF using natural cubic spline functions. Left panel: all data including placebo. Right panel: moxifloxacin period data only. The dotted lines are 95% CI.

not determine if a further exposure increase will further increase the risk, nor quantify the risk at a very low exposure level. Therefore, the repeated measure data did not provide much more information than the analysis of the events and exposure at 2 hours in Chapter 2. One reason is that extreme QT prolongation occurs mostly around the peak of exposure, hence the majority of subjects who had no event at 2 hours also had no event at other time points.

As a final note for non- and semiparametric models, we would like to draw the reader's attention to advanced approaches to this topic, in particular the relationship between spline smoothing and LMM, functional ANOVA and smoothing based approaches for fitting ODE specified models without the need for solving the ODEs. Since these are technically more advanced topics, we leave them for a separate chapter.

3.10   On distributions of random effects

In almost all the models we have introduced, univariate or multivariate normal distributions for random effects are assumed. Indeed, these models were first developed under this assumption. Often in practice one cannot verify this assumption, hence recently questions have been asked by methodology researchers, and modelers as well, about the consequences of misspecified distributions and alternatives to the assumed distributions.

3.10.1   Impact of misspecified random effects

Suppose that the true distribution of the random effects in an LMM, NLMM or GLMM is $u_i \sim F(u)$, where $F(u)$ is a general distribution function, rather than $u_i \sim N(0, \Sigma)$, as we assumed. Our main interest is in how the difference between the true and assumed distributions affects the fitted model and estimated parameters. In order to know that, we need to know how the fitted model behaves if a part of the model (in our case it is $F(u)$) is misspecified. In fact, results for general model misspecification are available (White, 1982) and are readily applicable to our situation. Putting his results in an intuitive form and omitting technical conditions, we can say that the estimated parameters in the misspecified model converge to the values that minimize the difference between the misspecified and true models. Intuitively, when fitting a linear model to a quadratic curve, the model parameters would make the line as close to the curve as possible, e.g., in terms of minimum LS. The results seem to be not practically useful since to assess the impact of the misspecification we have to know the true model. However, some general results can be used as guidance in practice, and we summarize them as follows.

For LMMs, the $\hat{\beta}$ by either MLE or REML is consistent regardless of the true distribution $F(u)$ (Verbeke and Lesaffre, 1997). In the context of White

(1982), for LMMs, $\hat{\beta}$ converges to the true value when minimizing the distance between the likelihood functions of the misspecified and true models. However, the statistical inference based on the likelihood function, in particular the CI constructed using the LRT or score test, needs correction.

For NLMMs, with a small r, the number of repeated measures, $\hat{\beta}$ is not consistent due to nonlinearity in the model. An evaluation of the impact based on the conditional first order approximation is quite complex, and hence has limited practical use. However, with either the exact MLE or the conditional first order approximation, $\hat{\beta} \rightarrow \beta$ when $r \rightarrow \infty$, regardless of $F(\mathbf{u})$, as long as the model for β_i is correctly specified. In practice $\hat{\beta}$ is robust to the misspecification when r is large.

The same conclusion applies to most GLMMs. For example, to obtain a consistent $\hat{\beta}$, r needs to be very large for a logistic model, if the random effects are misspecified. However, models with a multiplicative structure, such as the Poisson mixed model, are an exception. All the elements in $\hat{\beta}$, except the baseline risk estimate, are consistent for even small r.

3.10.2 Working with misspecified models

In the situation that a misspecified model may still give a valid estimate for β, one still needs correct estimates of SE and CI for statistical inference. A useful result given by White (1982) is the asymptotic variance-covariance matrix for the estimated parameters based on a misspecified model. Therefore, if the misspecification does not affect consistency of the estimates, the asymptotic variance-covariance matrix can be used in inference of the parameters. In fact, for simple models the variance-covariance matrix can be estimated by a sandwich estimate very similar to that of the GEE estimates. The GEE approach can be considered as using a possibly misspecified model to fit the data yet still yields consistent estimates for the parameters of interest. These estimates are typically provided by GEE software such as SAS proc GENMOD and the R function gee(). For LMMs, proc GENMOD provides a robust alternative to proc MIXED, which bases statistical inference on MLE or REML. For NLMMs, and most GLMMs, a robust estimate for var($\hat{\beta}$) is complex and is not available in common software. There is not much use of this robust estimate since $\hat{\beta}$ is inconsistent. However, in the case of a Poisson mixed model and other models of similar multiplicative structure, the estimated risk ratio in $\hat{\beta}$ is consistent, and the sandwich estimate for var($\hat{\beta}$) from GEE software (e.g., SAS proc GENMOD) is valid and can be used together with $\hat{\beta}$.

3.10.3 Using models with nonnormal or flexible distributions

There are many alternatives to the normal distribution for random effects. Although common software for mixed models only allows normally distributed random effects, there are approaches to using this software to fit models with alternative distributions. An approach that can be implemented in, e.g., SAS proc NLMIXED, is the integration transformation. First we consider the situation of only one random effect u_i. Suppose $u_i \sim F(u, \gamma)$, then the marginal likelihood function for subject i can be written as

$$
\begin{aligned}
L_i(\mathbf{y}_i) &= \int_{u_i} L(\mathbf{y}_i | \mathbf{X}_i, u_i, \boldsymbol{\beta}) dF(u, \gamma) \\
&= \int_{s_i} L(\mathbf{y}_i | \mathbf{X}_i, F^{-1}(\Phi(s_i), \gamma), \boldsymbol{\beta}) d\Phi(s_i)
\end{aligned}
\tag{3.48}
$$

where $\Phi(.)$ and $\phi(.)$ are the distribution and density functions of the standard normal distribution. This approach uses the fact that if $s_i \sim N(0,1)$ then $\Phi(s_i) \sim U(0,1)$ and u_i can be generated from a $U(0,1)$ variable by a simple transformation with its inverse distribution function $F^{-1}(.)$, if it has an analytical form. This approach can be easily implemented by the following statements to fit a model with u_i following a gamma distribution: for SAS proc NLMIXED (adapted from Nelson et al., 2006):

```
pi = CDF(NORMAL,si) ;
ui = quantile(GAMMA,pi,1/theta1) ;
...
random si  normal(0,1) subject=id;
```

where theta1 and theta2 are the two parameters in the gamma distribution to be estimated.

For multivariate random effects, we can use a similar approach and specify the correlation between the effects within a multivariate normal distribution. Let $\mathbf{u}_i \sim \mathbf{F}(\mathbf{u}, \gamma)$ where $\mathbf{F}(.,.)$ is a vector of marginal distribution functions for each element in $\mathbf{u}_i$. $\mathbf{F}(.,.)$ may consist of different marginal distributions. Suppose that u_{1i} and u_{2i} are the random variations in $\log(Cl_i)$ and $\log(V_i)$, respectively, and we may assume that u_{1i} follows a gamma distribution and u_{2i} follows a Weibull distribution. In this case, the likelihood function can be written as

$$
\begin{aligned}
L_i(\mathbf{y}_i) &= \int_{u_i} L(\mathbf{y}_i | \mathbf{X}_i, u_i, \boldsymbol{\beta}) d\mathbf{F}(u, \gamma) \\
&= \int_{s_i} L(\mathbf{y}_i | \mathbf{X}_i, \mathbf{F}^{-1}(\Phi(\mathbf{s}_i), \gamma)) d\Phi(\mathbf{s}_i, \boldsymbol{\Sigma})
\end{aligned}
\tag{3.49}
$$

where $\Phi(\mathbf{s}, \boldsymbol{\Sigma})$ is the distribution function of a multivariate normal distribution with unity variance and correlation $\boldsymbol{\Sigma}$. It is almost the same as (3.48) but with a multivariate $\mathbf{u}_i$. In fact, $\mathbf{u}_i$ can be considered as constructed by the Gaussian copula (Nelsen, 1999) defined as

$$
C(\mathbf{w}) = \boldsymbol{\Phi}^{-1}(\phi(w_1), ..., \phi(w_k), \boldsymbol{\Sigma})
\tag{3.50}
$$

where **w** is a vector of variables distributed on (0,1). The $\mathbf{u}_i$ is generated by applying $\mathbf{F}^{-1}(.)$ to **w**. A model with bivariate $\mathbf{u}_i$, all with a gamma marginal distribution, can be fitted with

```
p1 = CDF(NORMAL,s1) ;
p2 = CDF(NORMAL,s2) ;
u1 = quantile(GAMMA,p1,1/theta1) ;
u2 = quantile(GAMMA,p2,1/theta2) ;
...
random s1, s2  normal([0,0],[1,rho,1]) subject=id;
```

Copula theory provides a wide range of nonnormal multivariate distributions. A more general class is known as the Archimedean copula, generated in the same way as the Gaussian ones, but using other functions to replace the normal distribution functions. However, its implementation in mixed models is not as easy as the Gaussian one.

3.11   Bibliographic Notes

There is a vast volume of literature on modeling longitudinal data, hence a complete list of books on this topic cannot fit in this book. The most relevant ones for exposure–response modeling and the software we are using are Davidian and Giltinan (1995) for NLMMs, Pinheiro and Bates (2000) for LMMs and NLMMs using R library nlme, and McCulloch et al. (2008) for an excellent introduction to GLMMs. Bonate (2006) has chapters on mixed models focusing on PKPD modeling using NONMEM. For recent developments, Fitzmaurice et al. (2006) cover almost all the important areas in longitudinal data modeling within one book, which is worthwhile to read through.

4

Sequential and simultaneous exposure–response modeling

In the last chapter we assumed that exposures are available wherever needed in the ER model and can be used in the model directly. Here we focus on other situations when exposures in the ER model are not directly available and have to be estimated. Two main approaches under these situations are known as sequential and simultaneous modeling in pharmacometrics, and two-stage and joint modeling in statistics, although in most cases there is no exact correspondence between these methods in the two fields.

4.1 Joint models for exposure and response

In the previous chapter we introduced the following two models separately:

$$\begin{aligned}
y_{ij} &= g(c_{ij}, \beta_i) + \varepsilon_{ij} \\
c_{ij} &= h(t_{ij}, d_{ij}, \theta_i) + e_{ij}.
\end{aligned} \tag{4.1}$$

As in the previous chapters, we assume that e_{ij} and ε_{ij} follow normal distributions with zero mean and variances σ_e^2 and σ_ε^2, respectively, but may be correlated. The assumption of normality is mainly for simplicity, although it may be utilized in maximum likelihood (ML) approaches. In the following we will refer to the first model in (4.1) as the y-part and the second one as the c-part. One may think that the second model is not necessary, since if c_{ij} is observed, as we assumed in previous chapters, one may simply fit the first model. However, this approach results in a biased parameter estimate when θ_i and β_i or ε_{ij} and e_{ij} are correlated. We will discuss this issue in detail in Chapter 8.

The y-part model in (4.1) is based on a key assumption that it is the observed exposure c_{ij} which directly causes the response. This assumption is often questionable. An alternative to model (4.1) is

$$\begin{aligned}
y_{ij} &= g(c_{ij}^*, \beta_i) + \varepsilon_{ij} \\
c_{ij}^* &= h(t_{ij}, d_{ij}, \theta_i),
\end{aligned} \tag{4.2}$$

but we can only observe $c_{ij} = c_{ij}^* + e_{ij}$. This model is often considered more reasonable, especially in pharmacometrics, since in this field one may consider that it is the latent exposure c_{ij} based on a popPK model, rather than the observed exposure, that directly drives the response. Therefore, fitting both models may be necessary or may bring a significant advantage over fitting the y-part only.

In practice, a careful consideration of the selection of the two models is often worthwhile. Consider the argatroban example. The following are the PK (concentration) and PD (APTT) data from patient 1 in the dataset (PK=1 for PK measures and PK=0 for PD measures). Although the times of the PK and PD measures are generally different, apart from 240 when PK and PD are both measured, the times PK measures at 275 and 295 were only 5 apart from the PD measurement time 270 and 300, respectively. Therefore, one may question whether using a model for prediction of the exposure is a better option than using the observed ones. Later on we will show that sometimes one may be able to find a compromise that fits the ER model better.

Argatroban data from patient 1

Patient	Time	PD	PK
1	30	95.700	1
1	60	122.000	1
1	90	133.000	1
1	160	162.000	1
1	200	200.000	1
1	240	172.000	1
1	245	122.000	1
1	250	120.000	1
1	260	60.600	1
1	275	70.000	1
1	295	47.300	1
1	0	26.100	0
1	120	42.700	0
1	240	43.000	0
1	270	35.400	0
1	300	33.900	0
1	360	26.300	0

To further discuss the selection between the two models, consider the situation of a clinical trial with a fixed dose for each patient where repeated measures of the exposure (e.g., the trough concentration) and response (e.g., the blood pressure change) are measured. Using the second model, we assume that any exposure variation within a patient is irrelevant to the response, unless it can be predicted by the model for c_{ij}^*. This requires the inclusion of time varying factors in the model in a correct manner. However, whether all such factors have been included is not testable.

There is a technical advantage in using the models in (4.1), as model (4.2) cannot be fitted separately since $h(t_{ij}, d_{ij}, \boldsymbol{\theta}_i)$ has unknown parameters $\boldsymbol{\theta}_i$ to be predicted by the c-part. Model (4.2) normally requires simultaneous modeling, with a few exceptions. A large part of this chapter is devoted to discussing these exceptions, as they lead to much easier and more robust model fitting.

The y-part of the joint models may also be a GLMM, as discussed in Section 3.6, so the joint model can be written as

$$
\begin{aligned}
g(E(y_{ij})) &= \mathbf{X}_{ij}^T \boldsymbol{\beta}_{xi} + \beta_{ci} h(t_{ij}, d_{ij}, \boldsymbol{\theta}_i) \\
c_{ij} &= h(t_{ij}, d_{ij}, \boldsymbol{\theta}_i) + e_{ij}
\end{aligned} \tag{4.3}
$$

where $\boldsymbol{\beta}_i = (\boldsymbol{\beta}_{xi}, \beta_{ci})$ are random parameters and $g(E(y))$ is a link function. This model is similar to model (4.2). It is straightforward to construct a GLMM similar to model (4.2) by replacing $h(t_{ij}, d_{ij}, \boldsymbol{\theta}_i)$ with c_{ij}. The two types of model need quite different model fitting methods, as we will see later. Furthermore, one can also expand the model into a generalized nonlinear mixed model by replacing the $\beta_{ci} h(t_{ij}, \boldsymbol{\theta}_i)$ part with a nonlinear function such as the Emax model: $E_{maxi}/(1 + EC_{50i}/h(t_{ij}, d_{ij}, \boldsymbol{\theta}_i))$.

4.2 Simultaneous modeling of exposure and response models

4.2.1 Maximizing the joint likelihood function

The most common approach to fitting both models (4.1) and (4.2) simultaneously is to maximize the joint likelihood function for both exposure and response. Using model (4.1) or (4.2) leads to different likelihood functions. For the former, the conditional density for $\mathbf{y}_i$ given $\mathbf{c}_i$ and $\boldsymbol{\beta}_i$, may be written as $f_y(\mathbf{y}_i|\mathbf{c}_i, \boldsymbol{\beta}_i, \sigma_\varepsilon^2)$, while for the latter, since we do not observe $\mathbf{c}_i^*$, we have to use the density conditional on $\boldsymbol{\theta}_i$ through $c_i^*(t) = h(t_{ij}, \boldsymbol{\theta}_i)$, and hence we need to write $f_y(\mathbf{y}_i|\boldsymbol{\theta}_i, \boldsymbol{\beta}_i, \sigma_\varepsilon^2)$. We will consider the latter case in the general framework, as model fitting for the former is simpler. One can write formally the joint likelihood as

$$
L(\boldsymbol{\theta}, \boldsymbol{\beta}) = \prod_{i=1}^{n} \int_{\theta_i} \int_{\beta_i} f_y(\mathbf{y}_i|\boldsymbol{\beta}_i, \boldsymbol{\theta}_i, \sigma_\varepsilon^2) f_c(\mathbf{c}_i|\boldsymbol{\beta}_i, \sigma_e^2) dF(\boldsymbol{\theta}_i, \boldsymbol{\beta}_i|\boldsymbol{\theta}, \boldsymbol{\beta}, \boldsymbol{\Omega}) \tag{4.4}
$$

where $f_c(\mathbf{c}_i|\boldsymbol{\beta}_i, \sigma_e^2)$ is the density function of $\mathbf{c}_i$, $F(\boldsymbol{\theta}_i, \boldsymbol{\beta}_i|\boldsymbol{\theta}, \boldsymbol{\beta}, \boldsymbol{\Omega})$ is the joint distribution of $\boldsymbol{\theta}_i$ and $\boldsymbol{\beta}_i$ with parameters $\boldsymbol{\theta}$ and $\boldsymbol{\beta}$ for the means and $\boldsymbol{\Omega}$ for the variances. Note that this model does not cover the scenario of correlated ε_{ij} and e_{ij}, as in this case, the density function for $\mathbf{y}_i$ and $\mathbf{c}_i$ cannot be separated even conditionally on $\boldsymbol{\theta}_i$ and $\boldsymbol{\beta}_i$. In pharmacometrics the main interest is often on models with independent PK and PKPD parts, i.e., $\boldsymbol{\beta}_i$ and $\boldsymbol{\theta}_i$, and ε_{ij} and e_{ij} are all independent. For example, a frequently cited work in pharmacometrics (Zhang et al., 2003), which compares the performance of simultaneous and sequential modeling, only considered this situation.

Fitting the joint model is computationally demanding but there is software, e.g., NONMEM and SAS proc NLMIXED, designed for this purpose. In the next subsection, one can find details about programming and discussion on other technical issues. In the case of independent distributions of $\boldsymbol{\theta}_i$ and $\boldsymbol{\beta}_i$, i.e., when $F(\boldsymbol{\theta}_i, \boldsymbol{\beta}_i | \boldsymbol{\theta}, \boldsymbol{\beta}, \boldsymbol{\Omega}) = F(\boldsymbol{\theta}_i | \boldsymbol{\theta}, \boldsymbol{\Omega}_\theta) F(\boldsymbol{\beta}_i | \boldsymbol{\beta}, \boldsymbol{\Omega}_\beta)$, the likelihood function (4.4) can be separated into PK and PKPD parts, and the PK part can be maximized separately. However, the PKPD part still depends on the $\boldsymbol{\theta}_i$s. The sequential modeling (SM) method follows this idea to fit the PK part, then plugs in predicted $\boldsymbol{\theta}_i$s in the PKPD part to estimate the parameters in this part.

One important advantage of joint modeling is that the information matrix gives variance estimates for the MLE $\hat{\boldsymbol{\theta}}$ and $\hat{\boldsymbol{\beta}}$, taking the uncertainties in fitting the other model into account. In contrast, in the SM approach one can only obtain the variance estimates for $\boldsymbol{\theta}$, conditional on the predicted c_{ij} and the uncertainty in the prediction of c_{ij} is often ignored.

The joint modeling approach estimates both $\boldsymbol{\theta}$ and $\boldsymbol{\beta}$. Theoretically, joint model fitting is more efficient in terms of the precision of estimated parameters than sequential model fittings, even for the parameters in the PK model, if both models are correctly specified. However, the estimates of parameters in the PK model also depend on PD outcomes, a feature that is not often desirable. As the PKPD model is often empirical while the PK model is often mechanism based, it is prudent not to base the estimation of the PK parameters on assumptions that are difficult to verify. Consider the following hypothetical situation. Suppose with a popPK model, patient i's clearance is predicted to be 10 L/h, whereas when fitting a joint popPK and PKPD model the estimated clearance is 15 L/h. Do we believe that the response data help us to predict the clearance better? This depends on how certain we are with the PKPD model. This example suggests that when the estimates in the PK part of the joint model are quite different from those when the PK model is fitted separately, the models and assumptions should be examined carefully.

Another advantage of the likelihood approach is that the theory and model fitting algorithms apply to a wide range of models for which the likelihood function can be specified. These models include LMM, NLMM, GLMM and GNLMM, and software such as SAS proc NLMIXED and GLIMMIX can be used for normally or log-normally distributed random effects. However, numerical problems often occur due to the relatively large number of parameters to be estimated in the joint model.

4.2.2 Implementation of simultaneous modeling

General purpose software is not designed to fit multiple models to multiple responses. However, SAS, R and NONMEM have facilities to maximize joint likelihood functions. As described in Chapter 10, the user needs to specify the conditional density function

$$f_y(\mathbf{y}_i|\boldsymbol{\beta}_i, \boldsymbol{\theta}_i)f_c(\mathbf{c}_i|\boldsymbol{\beta}_i) \tag{4.5}$$

in (4.4), then the software calculates the marginal likelihood function based on $F(\boldsymbol{\theta}_i, \boldsymbol{\beta}_i|\boldsymbol{\theta}, \boldsymbol{\beta}, \boldsymbol{\Omega})$. The likelihood function may be specified implicitly via the mean function or specified explicitly by the conditional density functions. For example, in a program for jointly modeling argatroban concentration and APTT, the following part specifies the models for the PK (ind=1) and PKPD (ind=0) parts in SAS proc NLMIXED. Both PK and PD data are in variable resp and flagged by the variable ind.

```
...
pk=(dose/cl)*(1-exp(-cl*t2/v))*exp(-cl*(1-t1)*(time-tinf)/v);
var=(sig**2)*(pred**(2*0.22));
if ind=1 then pred=pk;
else if ind=0 then do;
    emaxi=exp(emax+u1);
    ec50i=exp(ec50);
    pred=E0+(emaxi-E0)*pk/(pk+ec50i);
    var=sig2;
end;
model resp ~ normal(pred,var);
...
```

Note that both the mean pred and variance var are needed for each of the models.

The following SAS program is an example of an explicit specification for the same model. Letting sige and sigv be the variances of the PK and PD responses and ci and yi be the PK and PD measures, we can specify

```
...sige=(sig**2)*(pred**(2*0.22));
    resc=ci-(dose/cl)*(1-exp(-cl*t2/v))*exp(-cl*(1-t1)*(time-tinf)/v);
    resy=yi-E0+(emaxi-E0)*pk/(pk+ec50i);
    if ind=0 then do;
        if (abs(resy) > 1E50) or (sige < 1e-12) then lly = -1e20;
        else   lly = -0.5*(resy**2 / sige  + log(sige)); llw = -1e20;
    end;
    if ind=1 thendo;
        if (abs(resc) > 1E50) or (sigv < 1e-12) then llw = -1e20;
        else  llw = -0.5*(resw**2 / sigv  + log(sigv)); lly = -1e20;
    end;
    model yi ~general(lly+llw);   *yi here is symbolic;
.....
```

where, in order to prevent numerical problems when the residuals are too large or the residual variances are too small, the contribution of these data to the likelihood is set to a very small value (10^{-20}).

4.2.3   The argatroban example

We use the argatroban dataset to illustrate applications of joint modeling of longitudinal data. SAS proc NLMIXED is used for the analysis but the approach of implementation can be applied to similar packages such as NON-MEM and R. Detailed example codes can be found in the appendices. In this example, both exposure and response are continuous variables and can be assumed to have a normal distribution. Therefore, we can use the mean and variance in the model switching between the exposure and response. The following codes run the joint model:

$$c_{ij}^* = \begin{cases} d_i(1 - \exp(-t_{ij}Cl_i/V_i)) & t_{ij} < t_f \\ d_i(1 - \exp(-t_fCl_i/V_i))\exp(-(t_{ij} - t_f)Cl_i/V_i) & t_{ij} \geq t_f \end{cases}$$

$$c_{ij} = c_{ij} + e_{ij}$$

$$y_{ik} = E_0 + (E_{maxi} - E_0)/(EC_{50}/c_{ij} + 1) + \varepsilon_{ij} \qquad (4.6)$$

SAS proc NLMIXED was used to fit the model. After a number of trial-and-error attempts, the following program and parameter setting leads to a successful fit.

```
proc nlmixed data=arg qpoints=2 ebsubsteps=70 ebsteps=100 ebssfrac=0.8
      ebopt ebtol=1e-4 ebsstol=1e-3 ABSXTOL=1e-5 abstol=1e-5 ftol=1e-5
      gtol=1e-3;
parms beta1=-5.4 beta2=-2.0 E0=30 emax=4.5 ec50=5  logb1=-1 logb2=-3
      sig=17 sig2=23 logu=-2;
bounds  sig sig2>0;
cl=exp(beta1+b1);
v=exp(beta2+b2);
sigu1=exp(logu);
s2b1=exp(logb1);
s2b2=exp(logb2);
pk=(dose/cl)*(1-exp(-cl*t2/v))*exp(-cl*(1-t1)*(time-tinf)/v);
var=(sig**2)*(pred**(2*0.22));
if ind=1 then pred=pk;
else if ind=0 then do;
   emaxi=exp(emax+u1);
   ec50i=exp(ec50);
   pred=E0+(emaxi-E0)*pk/(pk+ec50i);
   var=sig2;
end;
model resp ~ normal(pred,var);
random b1 b2 u1 ~ normal([0,0,0],[s2b1,0,s2b2,0,0,sigu1])
                                                 subject=subject;
run;
```

which generates the following outputs:

Parameter	Estimate	Standard Error	DF	t Value
beta1	-5.4329	0.08878	34	-61.19
beta2	-1.9516	0.02707	34	-72.09
E0	29.6918	0.5647	34	52.58
emax	4.4003	0.06318	34	69.65
ec50	6.3800	0.1902	34	33.55
logb1	-1.9051	0.2490	34	-7.65
logb2	-5.0309	1.2982	34	-3.88
sig	23.3031	1.0813	34	21.55
sig2	22.3237	3.6708	34	6.08
logu	-3.7331	0.3140	34	-11.89

with 2 log-likelihood $= 7044$. This fit uses a two-point quadrature (qpoints$=2$) to evaluate the likelihood function numerically. Note that two key parameters for the ER relationship are emax and ec50, which are on the log-scale in the model.

To use the first order approximation we replace "qpoints $= 2$" with "method $=$ firo" and obtain similar parameter estimates:

Parameter	Estimate	Standard Error	DF	t Value
beta1	-5.6498	0.03501	34	-161.40
beta2	-1.7763	0.03066	34	-57.94
E0	30.1260	0.5531	34	54.47
emax	4.5060	0.07850	34	57.40
ec50	6.7661	0.1959	34	34.53
logb1	-1.9536	0.2383	34	-8.20
logb2	-4.2040	0.4737	34	-8.88
sig	24.3275	0.9007	34	27.01
sig2	20.7338	2.2675	34	9.14
logu	-3.5179	0.3002	34	-11.72

The estimates are rather similar to those from the previous fit, but some SEs are considerably different.

4.2.4 Alternatives to the likelihood approach

Although fitting the joint model by the ML approach seems most natural, other approaches may be more convenient to fit some types of model jointly. Recall that the GEE approach as an alternative to the ML approach for mixed models has been used widely for longitudinal data analysis. It is also possible to use a similar approach for joint modeling the exposure and response data. First we write the model (4.1) or (4.2) in a compact form. Let $\mathbf{S}_i = (\mathbf{y}_i, \mathbf{c}_i)$ be the vector containing all the response and exposure measures, $\boldsymbol{\Phi}_i = (\boldsymbol{\beta}_i, \boldsymbol{\theta}_i)$, $\boldsymbol{\Phi} = (\boldsymbol{\beta}, \boldsymbol{\theta})$, and $E(\mathbf{S}_i|\boldsymbol{\Phi}_i) = \mathbf{H}^*(\mathbf{c}_i, d_i, \boldsymbol{\Phi}_i)$ be the conditional means of $\mathbf{S}_i$.

We assume that the marginal mean $\mathbf{H}(\mathbf{c}_i, \mathbf{\Phi}) = E(\mathbf{H}^*(\mathbf{c}_i, \mathbf{\Phi}_i))$, where the expectation is taken over the random components in $\mathbf{\Phi}_i$, can be calculated either analytically or numerically, so we can write the residuals as

$$\mathbf{R}_i = \mathbf{S}_i - \mathbf{H}(\mathbf{c}_i, \mathbf{\Phi}). \tag{4.7}$$

A GEE approach to estimate β and θ is to solve the estimating equation (EE)

$$\sum_{i=1}^n \frac{\partial \mathbf{H}(\mathbf{c}_i, \mathbf{\Phi})}{\partial \mathbf{\Phi}} \mathbf{W}_i \mathbf{R}_i = 0, \tag{4.8}$$

where $\mathbf{W}_i$ is a working weight matrix.

The generalized methods of moments (GMM) approach is more general, and a SAS proc MODEL implements this approach with a number of useful options for, e.g., model fitting approaches and calculation of $\mathbf{H}(\mathbf{c}_i, \mathbf{\Phi})$ from $\mathbf{H}^*(\mathbf{c}_i, \mathbf{\Phi}_i)$ if it does not have a closed form. The GMM approach estimates $\mathbf{\Phi}$ by finding $\hat{\mathbf{\Phi}}_{GMM}$ that minimizes

$$Q(\mathbf{\Phi}) = \mathbf{R}^T(\mathbf{\Phi})\mathbf{A}\mathbf{R}(\mathbf{\Phi}), \tag{4.9}$$

where $\mathbf{R}(\mathbf{\Phi})$ in our case can be $\mathbf{R}(\mathbf{\Phi}) = \sum_{i=1}^n \mathbf{W}_i \mathbf{R}_i / n$, or, in general, any set of statistics that have zero means, and $\mathbf{A}$ is a positive matrix of the user's choice. But the optimal one is $\mathrm{var}(\mathbf{R}(\mathbf{\Phi}))^{-1}$. It can be shown that the GMM estimate that minimizes $Q(\mathbf{\Phi})$ is a consistent estimate for $\mathbf{\Phi}$, if there is only a unique $\mathbf{\Phi}$ such that $E(\mathbf{R}(\mathbf{\Phi})) = 0$. Although a formal proof of the consistency is out of our scope, we note that it follows the fact that for large samples $\mathbf{R}(\mathbf{\Phi}) \to E(\mathbf{R}(\mathbf{\Phi})) = 0$, hence $Q(\mathbf{\Phi}) \to 0$, its minimum. Therefore, $\hat{\mathbf{\Phi}}_{GMM} \to \mathbf{\Phi}$. The GMM approach has similar advantages as the GEE approach, in particular, robust variance-covariance matrices for $\hat{\mathbf{\Phi}}_{GMM}$. See the appendix for details.

As GMM is a very big topic, readers interested in this area are referred to Cameron and Trivedi (2005). Here we concentrate on implementation of this approach. The following is a SAS program that estimates the relative risk for a Poisson model with overdispersion due to a normally distributed subject effect. We assume that three repeated measures for the exposure and the event count are taken, together with the dose levels at each time. The mean functions $\mathbf{H}(\mathbf{\Phi})$ consist of

$$E(y_{ij}) = \exp(\beta_0 + \beta c_{ij})$$
$$\text{and}$$
$$E(\log(c_{ij})) = \theta_0 + \theta \log(d_{ij}). \tag{4.10}$$

The following program uses the residuals to form $Q(\mathbf{\Phi})$, where $\mathbf{A}$ is an inverse empirical variance-covariance of $\mathbf{R}$ estimated based on a preliminary least squares fit. The individual residuals of y_{ij} are weighted by $1/\exp(\beta_0 + \beta c_{ij})$ to take variance heterogeneity into account.

```
proc model data=simu;
  EXOGENOUS dose1 dose2 dose3;
  parms base=-3 beta=0.2 a=0 b=1;
  eq.c1=a+b*dose1-lc1;
  eq.c2=a+b*dose2-lc2;
  eq.c3=a+b*dose3-lc3;
  means=exp(base+beta*exp(a+b*dose1));
  eq.y1=(means-resp1)/means;
  means=exp(base+beta*exp(a+b*dose2));
  eq.y2=(means-resp2)/means;
  means=exp(base+beta*exp(a+b*dose3));
  eq.y3=(means-resp3)/means;
  fit c1 c2 c3 y1  y2 y3/ gmm;
run;
```

Applying this program to a dataset generated by this code, with correlation between the *c*- and *y*-parts,

```
%let nsimu=1; %let nsub=100; %let cdose=log(1.3); %let bound=log(10);
%let base=-2; %let dose1=log(5); %let sigv=0.5; %let sigu=0.3;
%let sige=0.2; %let nsub=100; %let nrep=3; %let beta=0.25;

data simu;
  array conc{&nrep} lc1-lc&nrep;
  array ddose{&nrep} dose1-dose&nrep;
  array  resp{&nrep} resp1-resp&nrep;
 do simu=1 to &nsimu;
  do group=0 to 1;
  do sub=1 to &nsub;
    ui=rannor(1)*&sigu;
    vi=rannor(1)*&sigv;
    nn=0;
    dose=&dose1;
    do i=1 to &nrep;
    lci=dose+ui;
    lce=lci+rannor(12)*&sige;
*    trend=0.1*i-0.4;
    trend=0;
    if group=1 then lmd=exp(&base+&beta*exp(lci)+vi+trend);
    else lmd=exp(&base+vi+trend);
    yi=ranpoi(1,lmd);
    ddose{i}=dose; conc{i}=lce;  resp{i}=yi;
    if lce<&bound then dose=dose+&cdose;
      end;
      output;
    end;
  end;
end; run;
```

we obtain the following estimates, which are reasonably close to the parameters used for the data generation. Here the approximate SEs are from the robust variance-covariance matrix of $\hat{\Phi}_{GMM}$.

Nonlinear GMM Parameter Estimates

Parameter	Estimate	Approx Std Err	t Value	Approx Pr > |t|
base	-2.07026	0.2992	-6.92	<.0001
beta	0.229641	0.0463	4.96	<.0001
a	0.052874	0.0789	0.67	0.5035
b	0.96653	0.0412	23.46	<.0001

The analysis does not need to specify the form of correlations between the repeated measures. Although the generated data may be considered as from an ideal situation, the robustness of GMM approaches has been well documented and verified in practical applications. Minimization of $Q(\Phi)$ can also be performed using either SAS or R using a general purpose optimization software, or SAS proc NLMIXED treating $Q(\Phi)$ as a quasi-likelihood function. But robust variance-covariance matrices have to be calculated separately.

4.2.5   Simultaneously fitting simplified joint models

Since β_i and θ_i are multidimensional, a full joint distribution is hard to specify and hard to fit. However, the general models (4.1) and (4.2) may have simple forms that allow easier model fitting approaches. For example, the correlation between the exposure and response may be in the baseline only. Accordingly, models (4.1) may be simplified to become

$$
\begin{aligned}
y_{ij} &= g(c_{ij}, \beta_i) + u_i + \varepsilon_{ij} \\
c_{ij} &= h(t_{ij}, d_{ij}, \theta_i) + v_i + e_{ij}
\end{aligned}
\tag{4.11}
$$

where β_i and θ_i are independent, u_i and v_i are random subject baselines and may be correlated. To estimate β it seems there is no need to fit the second model at all. But this is true only when u_i and v_i are independent. When u_i and v_i are correlated, as the exposure c_{ij} depends on v_i, u_i in the first model becomes a confounding factor. The implication of confounding will be discussed separately.

When the two individual models in (4.11) can be fitted easily, for example, if they are linear or generalized linear mixed models, the joint model may also be fitted with some standard software, such as SAS proc MIXED or proc GLIMMIX without extra programming. For example, to fit a joint linear mixed model

$$
\begin{aligned}
y_{ij} &= \beta_0 + \beta c_{ij} + u_i + \varepsilon_{ij} \\
c_{ij} &= \theta_0 + \theta d_{ij} + v_i + e_{ij},
\end{aligned}
\tag{4.12}
$$

where θ_j is the mean exposure at time point j, the following code shows how to organize the data and use the proc MIXED options.

```
data pkpd;
  set pkdata(in=pk) pddata(in=pd);
  if pk then do;
      yv=conc;
  ind="pk";
    conc=0; end;
  else if pd then do;
      yv=QTcF;
ind="pd";
    dose=0;   end; run;

proc mixed data=pkpd;
  class sid1a ind;
  model yv=ind dose conc/solution;
  random ind /type=un subject=sid1a;
  repeated /group=ind; run;
```

where pkdata has variables conc and dose and pddata has QTcF and conc. In the data step, a flag "ind" is created and used in proc MIXED to specify if the intercept is θ_0 or β_0 and the corresponding variances to PK and PD (with "group = ind"). For a PD record the dose is set 0 so it has no impact on the PKPD model. The same is true for conc in the PK record. This approach also applies to GLMM models.

This linear mixed model approach cannot be applied to the y-part in (4.2) even when both PK and PD models are linear, e.g., when

$$y_{ij} = (\theta_i d_{ij})\beta_i + u_i + e_{ij}$$
$$c_{ij} = \theta_i d_{ij} + v_i + \varepsilon_{ij}, \tag{4.13}$$

and $\theta_i = \theta + v_{\theta i}$ and $\beta_i = \beta + u_{\beta i}$ are independent. Note that the first equation can be written as

$$y_{ij} = \theta\beta d_{ij} + u_i^* + e_{ij} \tag{4.14}$$

where $u_i^* = d_{ij}(\beta u_{\theta i} + \theta v_{\beta i} + u_{\theta i} v_{\beta i}) + u_i$ has $E(u_i^*) = 0$. However, both θ and β appear in the d_{ij} term, so they cannot be estimated in a linear mixed model. One needs to use, e.g., proc NLMIXED to fit it. Alternatively, one may estimate $\theta\beta$ as whole in this model, then divide it by $\hat{\theta}$ from the c-part fitted separately. But this approach often has bad properties, particularly when $\hat{\theta}$ has a large variation.

4.3   Sequential exposure–response modeling

Sequential modeling (SM) is a routine approach in pharmacometrics to analyze PKPD relationships when a PK model is available. The logic of the SM approach is obvious: to replace the exposure in the ER model by a model based prediction. Therefore, this approach is based on the assumption that the y-part model is (4.2), i.e., we assume

$$
\begin{aligned}
y_{ij} &= g(h(t_{ij}, d_{ij}, \boldsymbol{\theta}_i), \boldsymbol{\beta}_i) + \varepsilon_{ij} \\
c_{ij} &= h(t_{ij}, d_{ij}, \boldsymbol{\theta}_i) + e_{ij}.
\end{aligned} \tag{4.15}
$$

The SM approach first fits the PK model (the c-part), then uses predicted exposures to replace those in the PKPD model (y-part) to fit the response data. Specifically, in the first stage, the c-part in (4.15) is fitted and θ_i are predicted, then c_{ij}^* at time t_{ij} under dose d_{ij} can be predicted as $\hat{c}_{ij}^* = h(t_{ij}, d_{ij}, \hat{\boldsymbol{\theta}}_i)$. In the second stage, $h(t_{ij}, \boldsymbol{\theta}_i)$ in the y-part model is replaced with $\hat{c}_{ij}^*$:

$$
y_{ij} = g(h(t_{ij}, d_{ij}, \hat{\boldsymbol{\theta}}_i), \boldsymbol{\beta}_i) + \varepsilon_{ij} \tag{4.16}
$$

where $\hat{\boldsymbol{\theta}}_i = \mathbf{X}_i \hat{\boldsymbol{\theta}} + \mathbf{Z}_i \hat{\mathbf{v}}_i$. The model can be fitted in the same way as described in Chapter 3.

The major advantage of this approach is its simplicity, since it can easily be implemented with standard software. Another advantage is that the PK model parameter estimates are consistent as long as the PK model is correct and fitted properly. Often an intensive popPK modeling may have been carried out and information is available to assess the PK model and model fitting. Therefore, a good prediction of the PK data is likely to be available.

Now we apply the SM approach to the QT dataset based on the PK model fitted in Section 3.4.1. Following the SM approach we fit a linear mixed model for $y_{ij} = \Delta\Delta QTcF_i(t_j)$:

$$
y_{ij} = \beta_0 + \beta \hat{c}_{ij} + u_i + \varepsilon_{ij} \tag{4.17}
$$

where $\hat{c}_{ij}$ is the the predicted $c_i(t_j)$ based on model (2.5) in Chapter 2 with random effects on the clearance only. The fitted model is summarized by the following output:

```
Random effects:
 Formula: ~1 | Subject
         (Intercept) Residual
StdDev:     5.11129 7.442359

Fixed effects: dd_tavg ~ pred
              Value Std.Error  DF  t-value p-value
(Intercept) 7.252284 0.9351954 535 7.754832       0
pred        1.726746 0.3298987 535 5.234171       0
```

However, since concentrations were also measured at all time points, they might fit the the QT data better than the predicted ones. In fact, using the observed moxifloxacin concentration to fit model (3.16) we get $-$ log-likelihood $= 2087$, while using the predicted ones we get 2097, showing a better fit as well as a stronger exposure–response relationship. The following is the fitted model:

```
Random effects:
 Formula: ~1 | Subject
          (Intercept) Residual
StdDev:    5.113733 7.339934

Fixed effects: dd_tavg ~ concmu
               Value Std.Error  DF   t-value p-value
(Intercept) 6.868198 0.8962787 535 7.663016       0
concmu      1.948925 0.2958786 535 6.586909       0
```

The true exposure in the PKPD model (denoted as c_{ij}^*) is unknown and both c_{ij} and $h(t_{ij}, d_{ij}, \hat{\boldsymbol{\theta}}_i)$ are surrogates for c_{ij}^* but subject to random errors, which we will refer to as measurement error from now on. The impact of measurement errors is the main problem in SM. For a full understanding, we will need knowledge of the ME models, to be introduced in Section 4.4.

A number of approaches have been developed by pharmacometricians to improve the performance of the SM procedures. One approach, referred to as PPP&D, (Lacroix et al., 2012) is to fix the population parameters in the model but allow $\boldsymbol{\theta}_i$ to be reestimated when fitting the ER model. This approach is to reduce the complexity of parameter estimation in the joint model, yet still allow for individual random effects in the parameters to be estimated jointly. A number of variants of this approach have also been developed. One will be introduced in Section 4.4.

4.4 Measurement error models and regression calibration

Measurement error (ME) models describe the relationship between observed measures with error and the unobserved true value. In ER analysis, although both exposures and responses may be measured with error, we will concentrate on the former. To see why ME in exposures is an important issue, before going to technical details, consider the following hypothetical example. Suppose a drug which can increase the response from 0 to 1 units is compared with placebo in a trial with 20 hypothetical patients. However, two patients, one from each group, have their treatment group recorded wrongly, which can be considered as a measurement error in the exposure. Then the mean treatment difference between the two groups as recorded will be $9/10 - 1/10 = 0.8$ units. Therefore, measurement error in the exposure may reduce the estimated effect. This example also indicates that, if we know the error rate, we may be able to adjust the bias caused by ME.

4.4.1   Measurement error models

We introduce ME models starting from a simple linear model where the co-variate is measured with errors. Details about ME models can be found in standard texts such as Carroll et al. (2006). Let $c_i^*, i = 1, ..., n$ be a covariate in a model for response y_i with coefficient β_0 and β, and error term ε_i

$$y_i = \beta_0 + \beta c_i^* + \varepsilon_i. \tag{4.18}$$

Sometimes c_i^* cannot be measured directly. Instead,

$$c_i = c_i^* + e_i \tag{4.19}$$

is measured, where e_i is an error term. This model is known as the classical ME model. Another type of ME model is the Berkson model:

$$c_i^* = c_i + e_i, \tag{4.20}$$

in which we also only observe c_i, but in contrast to (4.19), here c_i^* is a random variable conditional on c_i.

If c_i, instead of c_i^*, is used to fit model (4.18) using the least squares (LS) method, in the situation of the classical model, the LS estimate for β is attenuated to $\beta\lambda$ where $\lambda = \sigma_{c*}^2/(\sigma_{c*}^2 + \sigma_e^2)$, and σ_{c*}^2 and σ_e^2 are the variances of c_i^* and e_i, respectively. To verify this in the situation of $\beta_0 = 0$, we calculate

$$\begin{aligned}
\hat{\beta} &= \sum_{i=1}^{n} y_i c_i / \sum_{i=1}^{n} c_i^2 \\
&= \beta(1 - \frac{\sum_{i=1}^{n} c_i(e_i + \varepsilon_i)}{\sum_{i=1}^{n} c_i^2})
\end{aligned} \tag{4.21}$$

in which $\sum_{i=1}^{n} c_i(e_i + \varepsilon_i)/\sum_{i=1}^{n} c_i^2 \to \text{cov}(c_i, e_i)/\text{var}(c_i) = \sigma_e^2/(\sigma_e^2 + \sigma_{c*}^2) = 1 - \lambda$, by Slusky theorem (A.4). There is no attenuation in the same LS estimate if the exposure follows the Berkson model, since it leads to $y_i = \beta_0 + \beta c_i + \beta e_i + \varepsilon_i$, and $\beta e_i + \varepsilon_i$ is independent of c_i. However, in both situations the measurement error leads to an increase in residual variance, hence in the variance of the estimate for β. Consequently, the accuracy of estimation and power for testing, e.g., $\beta = 0$, will be reduced.

A third type of ME model is a combination of the first two types, i.e., a mixture of the classical and Berkson models. For this type of ME model we assume that

$$c_i = s_i + u_{bi}, \quad c_i^* = s_i + u_{ci} \tag{4.22}$$

are all random variables, conditional on an unknown variable s_i, and u_{bi} and u_{ci} are independent. It is c_i^* that directly acts in model (4.18); but we can only observe c_i. This model includes both the classical model ($u_{ci} = 0$) and Berkson model ($u_{bi} = 0$) as special cases. If c_i is used for an LS estimate for β, the estimate is biased due to u_{bi}, but u_{ci} does not contribute to the bias.

A numerical example is useful to illustrate the difference between the different ME models. In the following we simulated 1000 samples from a linear regression of unit slope assuming the classical, Berkson and the mixed ME models separately. The classical ME model leads to a shrinkage to 0.53 in the estimated slope, the Berkson model results in no shrinkage and the mixed ME model gives almost the same estimated slope as the classical model, since only the classical part of ME causes bias.

```
> nsimu=1000
%Classical ME
> ci=rnorm(nsimu)
> xi=ci+rnorm(nsimu)
> yi=ci+rnorm(nsimu)
> lm(yi~xi)
(Intercept)            xi
  -0.03009        0.53276
%Berkson
> ci=xi+rnorm(nsimu)
> yi=ci+rnorm(nsimu)
> lm(yi~xi)
(Intercept)            xi
   0.03645        0.96008
%Mixed
> ui=rnorm(nsimu)
> ci=ui+rnorm(nsimu)
> xi=ui+rnorm(nsimu)
> yi=ci+rnorm(nsimu)
> lm(yi~xi)
(Intercept)            xi
   0.07416        0.53546
```

We have not specified a dose–exposure model for how c_i^* changes with, e.g., dose, in the ME model. The model can be used to adjust the bias in sequential modeling using predicted concentrations based on a fitted dose–exposure model. For example, when the PKPD model is linear, the bias can be corrected as $\hat{\beta}_c = \hat{\beta}_{LS}/\hat{\lambda}$ where $\hat{\lambda}$ is the estimate of λ using the variance component estimates from the fitted model. For other types of PKPD model, the correction may be more complex. Compared with the direct adjustment, regression calibration (RC) is a more general and powerful approach than the direct correction, as described below.

4.4.2   Regression calibration

RC is a method to fit the y-part model using the expected exposure, conditional on observed data. Taking the linear model example

$$y_i = \beta c_i^* + \varepsilon_i, \tag{4.23}$$

we examine the idea behind the RC approach. Since only c_i is observed, we would like to fit y_i with its conditional mean $E(y_i|c_i) = \beta E(c_i^*|c_i)$, i.e., we fit

$$y_i = \beta E(c_i^*|c_i) + \varepsilon_i^*, \tag{4.24}$$

where $\varepsilon_i^* = \varepsilon_i + \beta(c_i^* - E(c_i^*|c_i))$. It can be verified that as long as $E(c_i^*|c_i)$ is specified correctly, fitting this model gives a consistent estimate for β, although ε_i^* depends on β. To this end, note that the RC estimate $\hat{\beta}_{RC}$ is the solution to

$$S_{RC}(\beta) = \sum_{i=1}^{n} E(c_i^*|c_i)(y_i - \beta E(c_i^*|c_i))/n = 0, \tag{4.25}$$

we check $E(S_{RC}(\beta)) = 0$, which is the key condition for consistency of $\hat{\beta}_{RC}$ (A.6). $E(S_{RC}(\beta))$ can be calculated by first conditioning on c_i:

$$E(S_{RC}(\beta)) = E_c(E(c_i^*|c_i)E(y_i - \beta E(c_i^*|c_i)|c_i)) \tag{4.26}$$

in which $E(y_i - \beta E(c_i^*|c_i)|c_i) = \beta E(c_i^* - E(c_i^*|c_i)|c_i) = 0$. Therefore, $E(S_{RC}(\beta)) = 0$ as we expected. For a Berkson model $c_i^* = c_i + e_i$, $E(c_i^*|c_i) = c_i$, so the RC approach is just to fit the model with c_i. With the classical model, when c_i and c_i^* are jointly normally distributed, it is also easy to find $E(c_i^*|c_i) = c_i/\lambda$, following their conditional distribution properties.

If the latent exposure measure c_i^* or s_i in (4.22) has a structure and can be modeled, e.g., as

$$s_i = h(d_i, \boldsymbol{\theta}), \tag{4.27}$$

the RC approach can be based on $E(c_i^*|d_i, c_i) = E(s_i|d_i, c_i)$. Since given $\boldsymbol{\theta}$, $E(s_i|d_i, c_i) = h(d_i, \boldsymbol{\theta})$, we can estimate $E(s_i|d_i, c_i)$ by $h(d_i, \hat{\boldsymbol{\theta}})$, where $\hat{\boldsymbol{\theta}}$ is estimated by the exposure data. Again the RC approach leads to replacing c_i^* with $h(d_i, \hat{\boldsymbol{\theta}})$, which is the SM approach.

The following example shows that when there are repeated measures for c_i^*, even without a dose–exposure model for c_i^*, RC still leads to the same approach as SM when the first stage uses a linear mixed model prediction. Suppose that $c_i^* \sim N(0, \sigma_v^2)$ and it is measured repeatedly with error as $c_{ij} = c_i^* + e_{ij}, j = 1, ..., r$, where $e_{ij} \sim N(0, \sigma_e^2)$. Letting $\bar{c}_{i\cdot}$ be the mean of the c_{ij}s, then $\bar{c}_{i\cdot}|c_i^* \sim N(c_i^*, \sigma_e^2/r)$ and unconditionally $\bar{c}_{i\cdot} \sim (0, \sigma_v^2 + \sigma_e^2/r)$. Therefore, a linear mixed model with only subject effect v_i would predict c_i^* as

$$\hat{c}_i \equiv E(c_i^*|\bar{c}_{i\cdot}) = \lambda \bar{c}_{i\cdot}. \tag{4.28}$$

where $\lambda = \sigma_v^2/(\sigma_v^2 + \sigma_e^2/r)$, assuming the variances are known. The λ is just the correction factor for β. The RC approach uses $\hat{\lambda}\bar{c}_{i\cdot}$, which is exactly the SM approach as $\hat{c}_i$ is the linear mixed model prediction from the first stage. From here we can also find that when r is large, $\bar{c}_{i\cdot}$ is a very good estimate for c_i^*, and $\lambda \approx 1$, hence we may simply replace c_i^* with $\bar{c}_{i\cdot}$.

The RC approach is based on the prediction $E(c_i^*|c_i)$ or $E(c_i^*|d_i, c_i)$ for c_i^*, so one may write

$$c_i^* = E(c_i^*|d_i, c_i) + e_i \tag{4.29}$$

where e_i is the prediction error. This model can be considered a Berkson model with measurement error w_i. Hence, using $E(c_i^*|d_i, c_i)$ to replace c_i^* in a linear model, one obtains a consistent estimate for β. However, this is generally not true for nonlinear models. For a nonlinear model $y_i = g(c_i^*, \beta)$, RC is generally more complex since one needs to calculate

$$E(y_i|c_i) = E_{c_i^*}(g(c_i^*, \beta)|c_i) \qquad (4.30)$$

during the model fitting for different β values. One may use a simulation approach to calculate it, but since c_i^* is univariate, it is often more efficient to use Gaussian quadrature

$$E_{c_i^*}(g(c_i^*, \beta)|c_i) \approx \sum_{k=1}^{K} w_k g(c_i + x_k, \beta) \qquad (4.31)$$

where w_k and x_k are the weights and points that can be found by, e.g., the R package gaussquad.

For some special nonlinear models, $E_{c_i^*}(g(c_i^*, \beta)|c_i)$ can be calculated analytically and the RC method is very simple to use. One example is a multiplicative model with conditional mean

$$E(y_i|c_i^*) = \exp(\beta_0 + \beta c_i^*). \qquad (4.32)$$

Adding, for example, covariates or overdispersion does not change the results below. We will assume the Berkson model, since the RC approach also leads to this model. Given $c_i^* = c_i + e_i$,

$$E(y_i|c_i) = E(\exp(\beta_0 + \beta(c_i + e_i))|c_i) = \exp(\beta c_i)K \qquad (4.33)$$

where K may have a simple form. When $e_i \sim N(0, \sigma_e^2)$, e.g., $K = \exp(\beta^2 \sigma_e^2/2)$. But if we are only concerned with β (not β_0) there is no need to find K. For example, if y_i follows Poisson distribution $y_i \sim Poisson(\lambda_i)$ with

$$\lambda_i = \exp(\beta_0 + \beta c_i^*), \qquad (4.34)$$

when the model is fitted with c_i^* replaced by c_i, the term K goes to the estimated β_0. The estimate for β remains consistent. The same conclusion holds for using the RC approach based on repeated exposure measures and/or based on a dose–exposure relationship.

4.4.3 Models for longitudinal data

We consider different mixed models for the analysis of longitudinal data, and will not include covariates for simplicity. Let $\mathbf{y}_i$ and $\mathbf{c}_i^*$ be the repeated response and the true exposure measures, respectively. We first consider the linear mixed model

$$\mathbf{y}_i = \beta_0 + \beta \mathbf{c}_i^* + u_i + \varepsilon_i \qquad (4.35)$$

where ε_i is a vector of identical independent (IID) random errors. We further assume that we observe $\mathbf{c}_i = \mathbf{c}_i^* + \mathbf{e}_i$ with $\mathbf{e}_i \sim N(0, \sigma_e^2 \mathbf{I})$ as measurement errors. As before, the approach of using the RC method is to construct a model based on $E(\mathbf{y}_i|\mathbf{c}_i)$ and take the nature of longitudinal data into account. For the linear model this is simple since, as long as $E(\mathbf{c}_i^*|\mathbf{c}_i)$ can be estimated by an unbiased estimate $\hat{E}(\mathbf{c}_i^*|\mathbf{c}_i)$, we can use the following linear mixed model:

$$\mathbf{y}_i = \beta \hat{E}(\mathbf{c}_i^*|\mathbf{c}_i) + \mathbf{u}_i^* + \mathbf{e}_i^*, \tag{4.36}$$

where $\mathbf{u}_i^*$ and $\mathbf{e}_i^*$ may contain error terms from $c_i^* - \hat{E}(\mathbf{c}_i^*|\mathbf{c}_i)$. This model can be fitted using standard mixed model software such as SAS proc MIXED, or R function lme(), if $\hat{E}(\mathbf{c}_i^*|\mathbf{c}_i)$ is available.

Therefore, the key step in the RC approach is to estimate $E(\mathbf{c}_i^*|\mathbf{c}_i)$. When there are repeated exposure measurements, but the exposure is stationary, one can use a mixed model to predict an individual average exposure and replace $\mathbf{c}_i^*$ with it, or with the sample average $\bar{\mathbf{c}}_{i.}$, if appropriate. When $\mathbf{c}_i^*$ depends on dose through a model $h(\boldsymbol{\theta}_i, \mathbf{d}_i)$, a model-based RC is often a better choice. $E(\mathbf{c}_i^*|\mathbf{c}_i)$ can be estimated by $h(\hat{\boldsymbol{\theta}}_i, \mathbf{d}_i)$ if repeated exposure measures are available, or $h(\hat{\boldsymbol{\theta}}, d_i)$ if the exposure model is not a mixed model. This approach works for both dose changes over time within subjects and a constant dose with multiple dose groups, or a mixture of the two situations.

This approach can be extended to some GLMMs and GEEs. For example, for repeated count data with y_{ij} being the count of events at time interval j, and risk

$$\lambda_i = \exp(\beta_0 + \beta \mathbf{c}_i^*), \tag{4.37}$$

one can simply replace $\mathbf{c}_i^*$ with $\hat{E}(\mathbf{c}_i^*|\mathbf{c}_i)$ in a GEE to obtain a consistent estimate for β. The reason is that the conditional mean of $\mathbf{y}_i$ after the replacement is

$$\begin{aligned} E(\mathbf{y}_i|\mathbf{c}_i) &= E(\exp(\beta_0 + \hat{E}(\mathbf{c}_i^*|\mathbf{c}_i) + \mathbf{e}_i)) \\ &= E(\exp(\beta_0 + K + \hat{E}(\mathbf{c}_i^*|\mathbf{c}_i))) \end{aligned} \tag{4.38}$$

where $K = \log(E(\exp(\mathbf{e}_i)))$. Therefore, replacing c_i^* with $E(c_i^*|c_i)$ leads to a consistent estimate for β, and a robust SE estimate can be obtained by the sandwich robust variance estimate option in, e.g., SAS proc GENMOD or the R library geepack. In addition to the RC based on dose–exposure models, a commonly used model for RC in longitudinal data modeling includes a time profile, e.g., as in Tosteson et al. (1998), where they used a linear time trend for the underlining exposure. Extending the RC approach to fit other models such as the generalized logistic model is more complex, as $E(\mathbf{y}_i|\mathbf{c}_i)$ generally is no longer a logistic function of $E(\mathbf{c}_i^*|\mathbf{c}_i)$, and one has to calculate $E(\mathbf{y}_i|\mathbf{c}_i)$ numerically.

Measurement errors in NLMMs are more difficult to deal with, due to the nonlinearity. Consider the model

$$\mathbf{y}_i = g(\mathbf{c}_i^*, \boldsymbol{\beta}_i) + \varepsilon_i. \tag{4.39}$$

It is much more difficult to construct a model based on $E(c_i^*|c_i)$, since in general $E(g(c_i^*, \beta_i)|c_i) \neq g(E(c_i^*|c_i), \beta_i)$. Higgins et al. (1997) used a time profile model for an RC, assuming $E(g(c_i^*, \beta_i)|c_i) \approx g(E(c_i^*|c_i), \beta_i)$. More complex approaches were investigated by Wang and Davidian (1996) and Ko and Davidian (2000). Both used linear expansions on the NLMM and derived asymptotic results and corrections for ME. But the approach has not been widely used, perhaps due to the complexity of the formulae and lack of software implementation.

In fact, with the availability of software for calculating a marginal likelihood either exactly or approximately, such as SAS proc NLMIXED or NONMEM, one practical approach is to consider e_i as a random effect in the NLMM. This approach is based on the following approximation:

$$\eta_i \equiv \hat{\theta}_i - \theta_i \sim N(0, S_i) \tag{4.40}$$

so that

$$y_i = g(h(t_{ij}, \hat{\theta}_i - \eta_i), \beta_i) + \varepsilon_i. \tag{4.41}$$

To fit this model, one adds η_i as jointly normally distributed variables with fixed variance S_i. This approach demands intensive calculation, but the number of parameters to be estimated is not increased. This approach was named IPPSE by Lacroix et al. (2012) and implemented in NONMEM. They assumed that S_i is a diagonal matrix of squared SEs of each parameter for subject i. One practical reason is that NONMEM only provides SEs for $\hat{\theta}_i$. Nevertheless, the approach seems a useful way to address one of the issues and has been shown to have a better performance than ignoring the estimation error. The approach can be implemented in SAS proc NLMIXED, as shown in the next subsection.

4.4.4 Sequential modeling for argatroban data

We illustrate the use of SM approaches by applying them to the argatroban data. First we add the time points when PD is measured as additional variables ptime=time, p1=t1 and p2=t2 so that the predicted concentrations can be generated from the fitted model.

```
data pkpred;
  merge arg(where=(ind=1) in=a) arg(where=(ind=0) keep=subject t1 t2
    time resp ind rename=(resp=pd time=ptime t1=p1 t2=p2));
  by subject;  if a;
run;
```

The following program fits the c-part in (4.6):

```
proc nlmixed data=pkpred method=gauss tech=trureg qmax=101 qtol=1e-5;
parms beta1=-5.5 beta2=-2.0 s2b1=0.14 cb12=0.006 s2b2=0.006 sig=23.0;
cl=exp(beta1+b1);
v=exp(beta2+b2);
```

```
pred=(dose/cl)*(1-exp(-cl*t2/v))*exp(-cl*(1-t1)*(time-tinf)/v);
var=(sig**2)*(pred**(2*0.22));
id pd;
*The following line generate predicted concentration at PD time points;
predict (dose/cl)*(1-exp(-cl*p2/v))*exp(-cl*(1-p1)*(ptime-tinf)/v)
                                                          out=pred;
model resp ~ normal(pred,var);
random b1 b2 ~ normal([0,0],[s2b1,cb12,s2b2]) subject=subject;
run;
```

and gives the following parameter estimates:

Parameter	Estimate	Standard Error	DF	t Value
beta1	-5.4220	0.06423	35	-84.41
beta2	-1.9210	0.03242	35	-59.26
s2b1	0.1473	0.03556	35	4.14
cb12	0.005757	0.01170	35	0.49
s2b2	0.01112	0.007969	35	1.40
sig	23.0211	0.8178	35	28.15

The following data step and program take the predicted $c_i(t_j)$ to fit the ER model in (4.6):

```
data pred;  set pred;
  if ind=0;
  rename pred=conc;
run;
```

```
proc nlmixed data=pred qpoints=16 qtol=1e-3 ebtol=1e-7 EBOPT itdetails;
parms E0=30 emax=5 ec50=3  sig=23.0  sigu1=0.2;
    bounds emax  sig  sigu1>0;
    emaxi=exp(emax+u1);
    ec50i=exp(ec50);
    pred=E0+(emaxi-E0)*conc/(conc+ec50i);
model pd ~ normal(pred,sig);
random u1 ~ normal([0],[sigu1]) subject=subject;
run;
```

The estimated parameters in the ER model are

Parameter	Estimate	Standard Error	DF	t Value
E0	29.5821	0.4843	36	61.09
emax	4.3813	0.05641	36	77.66
ec50	6.3204	0.1573	36	40.19
sig	16.9057	1.7844	36	9.47
sigu1	0.03403	0.009881	36	3.44

The estimated two key parameters emax and ec50 are very similar to those in the joint model (Section 4.2.3). However, the SEs here are lower: 0.157 (rather than 0.196) for ec50 and 0.056 (rather than 0.079) for emax, as the variation in the fitted dose–exposure model has not been counted for.

Next we implement the IPPSE approach in SAS, in a similar way as in NONMEM. To this end, one needs to predict the individual parameters (here cl and v) for each patient. The SEs of the predicted values are generated in the output dataset. In the proc NLMIXED call for fitting the ER model, the calculation of concentrations is repeated with the predicted parameter values plus random effects b1 and b2. Note that they have a different meaning than in the model for the dose-PK relationship, where they are the random subject effects, and here they are the random error terms.

```
proc nlmixed data=pkpred method=gauss tech=trureg qmax=101 qtol=1e-5;
parms beta1=-5.5 beta2=-2.0 s2b1=0.14 cb12=0.006 s2b2=0.006 sig=23.0;
cl=exp(beta1+b1);
v=exp(beta2+b2);
pred=(dose/cl)*(1-exp(-cl*t2/v))*exp(-cl*(1-t1)*(time-tinf)/v);
var=(sig**2)*(pred**(2*0.22));
id pd;
predict beta1+b1 out=cl;
predict beta2+b2 out=v;
model resp ~ normal(pred,var);
random b1 b2 ~ normal([0,0],[s2b1,cb12,s2b2]) subject=subject;
run;


data para;
  merge cl(rename=(pred=lcl StdErrPred=secl)) v(rename=(pred=lv
                                          StdErrPred=sev));
run;


proc nlmixed data=para qpoints=16 qtol=1e-3 ebtol=1e-7 EBOPT itdetails;
parms E0=30 emax=5 ec50=3  sig=23.0  sigu1=0.2;
bounds sig sigu>0;
cl=exp(lcl+b1);
v=exp(lv+b2);
conc=(dose/cl)*(1-exp(-cl*p2/v))*exp(-cl*(1-p1)*(ptime-tinf)/v);
    bounds emax  sig  sigu1>0;
    emaxi=exp(emax+u1);
    ec50i=exp(ec50);
    pred=E0+(emaxi-E0)*conc/(conc+ec50i);
model pd ~ normal(pred,sig);
random u1 b1 b2 ~ normal([0,0,0],[sigu1,0,secl**2,0,0,sev**2])
                                          subject=subject;
run;
```

The fitted model gives similar parameter estimates. The SEs for the two key parameter estimates are only slightly higher: 0.162 (rather than 0.157) for ec50 and 0.059 (rather than 0.056) for emax, but are still considerably lower than the joint modeling estimate.

Parameter	Estimate	Standard Error	DF	t Value
E0	29.4559	0.4739	34	62.15
emax	4.4016	0.05852	34	75.22
ec50	6.3846	0.1620	34	39.41
sig	15.5709	1.7145	34	9.08
sigu1	0.03208	0.009479	34	3.38

In summary, for the argatroban data, the SM approach gave quite similar parameter estimates, but the variability of the estimates seemed underestimated, even with a procedure taking the SEs in the c-part parameters into account compared with the joint approach.

4.4.5   Simulation extrapolation

Simulation extrapolation (SIMEX) is an approach to estimate the exposure–response parameter with an extrapolation from artificially biased estimates. This approach seems not widely used in ER modeling, apart from an application to a PK-QT relationship analysis in Bonate (2013). Consider a simple ME model $c_i = c_i^* + e_i$ with $e_i \sim N(0, \sigma_e^2)$. The idea of SIMEX is to artificially increase the variance of e_i by adding extra variation to assess the relationship between the increased variation and the estimate of β, then extrapolating back to $\sigma_e = 0$. To implement this, one needs to know σ_e in order to generate e_i', which is independent of e_i with $\mathrm{var}(e_i') = \rho\sigma_e^2$. Then $c_i' = c_i + e_i'$ with variance increased to $\sigma_e^2(1+\rho)$ can be generated to replace c_i in the ER model. $\hat{\beta}$ can be obtained after fitting the model. For each ρ, B sets of e_{bi}', $b = 1, ..., B$ samples may be taken to generate $c_{bi}' = c_i + e_{bi}'$. Then B estimates for β: $\hat{\beta}_b(\rho)$ can be obtained and their mean $\hat{\beta}(\rho)$ calculated. These steps are repeated for a number of ρ values so that one can extrapolate with the $\hat{\beta}(\rho)$ to the corresponding value of $\rho = -1$. Hypothetically, at this value $c_i' = c_i^*$, which has no ME, and the corresponding estimate $\hat{\beta}(-1)$ is the SIMEX estimate for β.

This approach can easily be extended to models with heterogeneous variances, in particular, when $\sigma_{ei}^2(\mu_i)$ is a function of the mean μ_i. To this end, one needs to generate $e_{bi}' \sim N(0, \rho\sigma_{ei}^2(\mu_i))$. Otherwise, the approach applies in the same way as for the constant variance case. Another case is that $\sigma_{ei}^2(\mu_i)$ is unknown but there are repeated measures available to construct the variance inflated exposure. This approach is slightly complex and the reader is referred to Carroll et al. (2006). The SE of $\hat{\beta}$ is difficult to find analytically. A technically more straightforward way is to use bootstrap approach. An introduction to this approach can be found in the appendix. In short, the bootstrap approach repeats the above steps on each bootstrapped dataset constructed by resampling from the original data to obtain a SIMEX estimate for β. This approach forms an empirical distribution for $\hat{\beta}$, from which the SE can be obtained or statistical inference can be made based on the distribution.

4.4.6 A compromise between y-part models

We have faced the question which of the two following models is correct,

$$y_{ij} = g(c_{ij}, \boldsymbol{\beta}_i) + \varepsilon_{ij}, \tag{4.42}$$

or

$$y_{ij} = g(c_{ij}^*, \boldsymbol{\beta}_i) + \varepsilon_{ij}, \tag{4.43}$$

with $c_{ij}^* = h(d_{ij}, \boldsymbol{\theta}_i)$, particularly when available data allow fitting both models. As an empirical approach, we may assume that in the second model c_{ij}^* could be a linear combination of the observed and the model-based exposures:

$$c_{ij}^* = \gamma h(t_{ij}, \boldsymbol{\theta}_i) + (1 - \gamma)c_{ij}, \tag{4.44}$$

where γ is a parameter. Setting γ to 1 or 0, we obtain the first or second model. But another γ value might get a better fit. For linear and generalized linear mixed models we can easily estimate γ and test if the linear combination is better than any of them by fitting both $f(t_{ij}, \boldsymbol{\theta}_i)$ and c_{ij} in the same model. For example, if the true model is

$$y_{ij} = \beta_0 + \beta c_{ij}^* + u_i + e_{ij}, \tag{4.45}$$

where c_{ij}^* is given by (4.44), we can fit

$$y_{ij} = \beta_0 + \beta_1 h(t_{ij}, \hat{\boldsymbol{\theta}}_i) + \beta_2 c_{ij} + u_i + e_{ij}, \tag{4.46}$$

and γ can be estimated as $\hat{\gamma} = 1/(1 + \beta_2/\beta_1)$. The two models can also be compared with, e.g., a likelihood ratio test (LRT) or a model selection criterion such as the AIC. The same approach can also be used for models with random coefficients. The model

$$y_{ij} = \beta_0 + \beta_{1i} h(t_{ij}, \hat{\boldsymbol{\theta}}_i) + \beta_{2i} c_{ij} + u_i + e_{ij} \tag{4.47}$$

allows different variances for β_{1i} and β_{2i}. This approach can be an empirical tool to explore the role of the model-based and observed exposures in the model.

Applying this approach to the moxifloxacin-QT data by fitting both observed concentration c_{ij} and its prediction $h(t_{ij}, \hat{\boldsymbol{\theta}}_i)$, we obtain the fitted model

```
    AIC      BIC    logLik
 4180.384 4202.335 -2085.192

Random effects:
 Formula: ~1 | Subject
         (Intercept) Residual
StdDev:    5.116812 7.321278

Fixed effects: dd_tavg ~ concmu + pred
```

```
             Value Std.Error  DF   t-value  p-value
(Intercept)  7.270901 0.9290171 534   7.826444  0.0000
cij          3.011423 0.7094194 534   4.244912  0.0000
pred        -1.285460 0.7802837 534  -1.647426  0.1001
```

The LRT test gave a p-value of 0.099 for the comparison between this model and that fitting c_{ij} only, but the AIC of this model is 4180, slightly lower than AIC = 4181 for the model with c_{ij} only. From the fitted model we can estimate $\hat{\gamma} = 1/(1 + \beta_2/\beta_1) = 1/(1 + 3.01/(-1.29)) = -0.75$. The value suggests that the exposure directly relating to the QTcF is not a compromise between $h(t_{ij}, \hat{\theta}_i)$ and c_{ij}, but is beyond c_{ij}. Although this is mechanistically unlikely, the results suggest that using c_{ij} directly seem to be a better choice, which is consistent with the comparison based on AIC.

4.4.7 When measurement error models are useful?

Since dealing with ME increases complexity in ER modeling, one may ask how much value this approach adds, given that the true exposure is almost never observed? Indeed, if the question is, e.g., "how much is the blood pressure difference between two patients with observed concentration 10 ng/mL and 20 mg/mL of drug A?" there is no need to use the ME approach. The simple modeling approach using the observed exposures can answer this question. However, the ME approach may answer some very important questions, such as, if the true exposure level is increased from 10 ng/mL to 20 mg/mL, how much blood pressure reduction will it cause? To help understand the difference between the two, we recall the example of a trial with treatment wrongly labeled. The first question asks, in this context, what is the average difference between two patients *labeled* as the treatment and placebo, respectively, while the second one asks what is the average difference between a patient who *took* the treatment and a patient who *took* the placebo. This question is often asked during the dose finding process within which an appropriate dose is selected by comparing the response between different doses. Note that an increase of dose from, e.g., 10 mg to 20 mg will increase the *true* exposure from 10 ng/mL to 20 ng/mL. This question is also valid for exposure adjustment for individual patients, since repeated PK measures can be taken and used to estimate the true concentration with reasonable accuracy.

The RC approach may also provide additional interpretation, e.g., dose-response analysis results. For example, if on average the responses under dose levels 10 mg and 20 mg were 1 and 2 units, and the two doses produced exposure levels 10 ng/mL to 20 ng/mL, assuming a linear ER relationship, a 10 ng/mL increase would produce 1 unit response. If the dose level is randomized, then this difference could serve as a gold standard to check confounding biases in PKPD modeling. Details can be found in Chapter 8.

4.4.8 Biases in parameter estimation by sequential modeling

Although the SM approach is simple to use, due to errors in the prediction of c_{ij}^* in the y-part model, the parameter estimates are generally biased if the y-part is a nonlinear regression model, such as, $y_i = g(c_i^*, \beta) + \varepsilon$. Here a bias assessment method based on the theory of model misspecification (White, 1981, 1982) is described. Although a thorough investigation needs to take the approach of predicting c_i^* into account, if the c-part model is linear or log-linear, e.g., the power model, the problem can be much simplified. If the model for exposure is an additive model, we can assume that $\hat{E}(c_i^*|c_i)$ is an unbiased estimate based on a linear mixed model and we can write

$$c_i^* = \hat{E}(c_i^*|c_i) + s_i, \tag{4.48}$$

where $s_i \sim N(0, \sigma_s^2)$ represents the prediction error. In this case,

$$E(y_i) = E_s(g(\hat{E}(c_i^*|c_i) + s_i, \beta)), \tag{4.49}$$

which, in general does not equal $g(\hat{E}(c_i^*|c_i), \beta)$, due to the impact of s_i and the nonlinearity of $g(\hat{E}(c_i^*|c_i) + s_i, \beta)$.

If the exposure model is log-linear, the prediction has a multiplicative error term. We may assume that

$$c_i^* = \exp(\hat{E}(\log(c_i^*)|c_i) + s_i). \tag{4.50}$$

In this case,

$$E(y_i) = E(g(\exp(\hat{E}(\log(c_i^*)|c_i) + s_i), \beta)) \tag{4.51}$$

and we are interested in the bias when fitting model $g(\exp(\hat{E}(\log(c_i^*)|c_i)), \beta)$.

From the theory of model misspecification (White, 1981, 1982), we can determine the bias in the parameter estimate for β if s_i is ignored in the SM. In general, the LS estimates based on model $g(\hat{E}(c_i^*|c_i), \beta)$, $\hat{\beta}_{LS} \to \beta^*$ when $n \to \infty$. β^* minimizes the squared difference between this model and the true model $E(g(c_i^*, \beta))$ at given c_i^* values:

$$D(\beta^*) = \sum_{i=1}^{n} (E(g(c_i^*, \beta)) - g(\hat{E}(c_i^*|c_i), \beta^*))^2. \tag{4.52}$$

Therefore, the bias $\beta^* - \beta$ depends not only on the model parameters but also on the c_i^*s. With given values of $\hat{E}(c_i^*|c_i)$, (4.52) can be calculated numerically and β^* can be found with optimization software. The bias when using a multiplicative exposure model can be determined in the same way, after replacing $g(\hat{E}(c_i^*|c_i), \beta)$ with $g(\hat{E}(\log(c_i^*)|c_i), \beta)$ in (4.52).

One may think that the bias can be evaluated by a large number of simulations so that the small sample properties of the SM approach can also be assessed. However, for a complex nonlinear regression model, numerical problems almost surely occur during simulations, leading to a considerable number of nonconverged runs. The calculation of (4.52) is numerically very stable and easy to perform. The large sample bias, together with the small sample property assessment, provides useful information for the evaluation of the SM approach.

We use the combination of the power model and the Emax model to illustrate this approach to bias evaluation. Consider the Emax model

$$y_i = E_{max}/(1 + EC_{50}/c_i^*) + \varepsilon_i \tag{4.53}$$

with c_i^* given by an power model. For our purpose, we estimate c_i^* from the predicted conditional mean of $\log(c_i^*)$

$$c_i^* = \exp(\hat{E}(\log(c_i^*)|c_i) + s_i). \tag{4.54}$$

The major computational task is to calculate $E(g(\hat{E}(c_i^*|c_i), \boldsymbol{\beta}))$, for which we use a 10-point Gaussian–Hermite quadrature, assuming that $\log(\hat{c}_i^*)$ has a uniform distribution over $(-1, 1)$. Then $\boldsymbol{\beta}^*$ was found for different $\boldsymbol{\beta}$ and σ_s^2 values using the R-function optim(.). Figure 4.1 shows the relative bias as a function of σ_s^2 when $EC_{50} = 1$ and 2. In general, under this setting the bias is small to moderate, unless σ_s is very large. Since $EC_{50} = 1$ is at the center of the $\hat{c}_i^*$ distribution but $EC_{50} = 2$ is far away from it, one can see from the figure the impact of the distribution range for $\log(\hat{c}_i^*)$ relative to the EC_{50} value.

If the y-part is a nonlinear mixed model, the bias may depend on model fitting approaches. When the number of repeated measures is large, the bias can be assessed in a similar way, since, as described in Chapter 3, the model fitting is close to fitting the nonlinear model to individual subject data. When the number of repeated measures is small, one has to calculate the likelihood function directly. In this case, the SM estimates tend to the $\boldsymbol{\beta}$ value that minimizes the difference between the true likelihood with c_i^* in the model and the one using $E(c_i^*|c_i)$. White (1982) gives more details.

4.5   Instrumental variable methods

The instrumental variable (IV) approach (Cameron and Trivedi, 2005) is commonly used to eliminate the bias due to ME. The same IV approaches have been used for causal effect estimation (Chapter 8). To see how the IV approach can eliminate the biases due to ME, it is useful to have another look at the simple models

$$
\begin{aligned}
y_i &= \beta c_i^* + \varepsilon_i \\
c_i &= c_i^* + u_i.
\end{aligned}
\tag{4.55}
$$

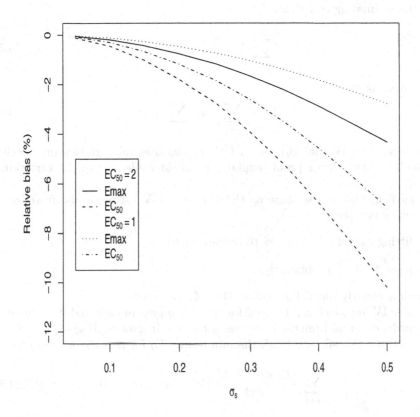

FIGURE 4.1
Relative asymptotic bias in $\hat{\beta}_{LS}$ in the Emax model as a function of σ_s under different parameter settings.

It seems the first one can be written as $y_i = \beta c_i - \beta u_i + \varepsilon_i$, in which $\beta u_i + \varepsilon_i$ has zero mean, so it seems we can consider it an random error. However, the fundamental assumption of independence between c_i and the error terms $\beta u_i + \varepsilon_i$ no longer holds, as c_i depends on u_i, hence the LS estimate for β is inconsistent. An IV is a variable that correlates to the exposure, but not to the response, apart via the exposure, i.e., it is independent of ε_i. When c_i^* is a function of randomized dose d_i, e.g.,

$$c_i^* = \theta d_i \tag{4.56}$$

we can use dose d_i as an IV. It is an IV since it affects c_i^* via the model above, and is independent of ε_i since it is randomized. The IV estimate is the solution to the estimating equation

$$\sum_{i=1}^{n} d_i(y_i - \beta c_i) = 0, \qquad (4.57)$$

which leads to

$$\hat{\beta}_{IV} = \sum_{i=1}^{n} d_i y_i / \sum_{i=1}^{n} d_i c_i. \qquad (4.58)$$

It can be verified that $\hat{\beta}_{IV} \to \beta$. Other instrumental variables may include time if a varying time profile explains a considerable part of the variation in c_i.

In fact, $\hat{\beta}_{IV}$ is the same as the two-stage IV estimate obtained by the following two steps:

- fitting model $c_i = \theta d_i + e_i$ to obtain $\hat{c}_i(d_i)$

- $y_i = \beta \hat{c}_i + \varepsilon_i$ to obtain $\hat{\beta}_{IV}$,

which is exactly the SM as well as the RC approaches.

The IV approach can be used for more complex models and its estimate is generally different from the two-stage estimate. In general, if $y_i = g(c_i^*, \boldsymbol{\beta}) + \varepsilon_i$, the optimal IV estimate (with the minimum SE) for $\boldsymbol{\beta}$ is the solution to

$$\sum_{i=1}^{n} \frac{\partial E(g(c_i^*, \boldsymbol{\beta})|\mathbf{U}_i)}{\partial \boldsymbol{\beta}} (y_i - g(c_i, \boldsymbol{\beta})) = 0 \qquad (4.59)$$

where $\mathbf{U}_i$ are instrumental variables. This approach can be extended to the situation when y_i contains multiple repeated measurements. But in this case, the implementation needs substantial programming effort. Here we consider the linear mixed model

$$\mathbf{y}_i = \beta \mathbf{c}_i^* + u_i + \varepsilon_i \qquad (4.60)$$

where $\mathbf{c}_i^* = h(\mathbf{t}_i, \boldsymbol{\theta}_i)$ and $\mathbf{c}_i = \mathbf{c}_i^* + \mathbf{e}_i$ are measured at time points $\mathbf{t}_i$. In this case, $\mathbf{t}_i$ can be used as IVs if there is sufficient variation in $h(\mathbf{t}_i, \boldsymbol{\theta}_i)$, and the IV estimate for β is the solution to

$$\sum_{i=1}^{n} E(\mathbf{c}_i^*|\mathbf{t}_i)\mathbf{W}_i(\mathbf{y}_i - \beta \mathbf{c}_i) \qquad (4.61)$$

where $\mathbf{W}_i$ is a matrix one can choose based on $\text{var}(\mathbf{y}_i)$ to optimize the IV estimate. $E(\mathbf{c}_i^*|\mathbf{t}_i)$ can be estimated as $h(\mathbf{t}_i, \hat{\boldsymbol{\theta}}_i)$ after fitting the model for $\mathbf{c}_i$ as a function of $\mathbf{t}_i$. Note that, although using $E(\mathbf{c}_i^*|\mathbf{t}_i)$ is optimal, the estimate for β is robust to misspecification of $E(\mathbf{c}_i^*|\mathbf{t}_i)$. This is in contrast to the SM method, which needs correct specification of the c-part model. The following simulation example shows the robustness of the IV approach. In the simulation we generate c_i based on an exponential model $c_i^*(t) = dose \exp(-(k + v_i)t)$ with fixed dose. Then c_i are fitted to a model with a linear (rather than exponential) time trend and the fitted model is used to predict $E(c_i^*|t_i)$ as the IV, hence the IV is generated from a misspecified model $c_i^*(t) = a_i + b_i t$. The following SAS code generates and analyzes the data.

```
%let nsimu=500; %let nsub=50; %let nrep=6; %let sigv=0.2;
%let sigu=0.2; %let ke=-1; %let sig=0.2; %let sigme=0.2;
data simu;
  dose=5;
 do simu=1 to &nsimu;
  do sub=1 to &nsub;
    ui=rannor(12)*&sige;
    vi=rannor(12)*&sigv;
    kei=exp(&ke+ui);
    do rep=1 to &nrep;
      cij=dose*exp(-kei*&nrep);
      csij=cij+rannor(12)*&sigme;
      yij=cij+rannor(12)*&sig;
      output;
    end;
   end;
 end; run;

proc mixed data=simu;
  class sub;
  model csij=rep/outp=pred; *IV in pred;
  random intercept rep/subject=sub;
  by simu;
run;
ods output ParameterEstimates=iv;
proc panel data=pred;
  inst exogenous=(pred);
  id sub rep;
  model yij=csij/ranone gmm;
  by simu;
run;
ods output ParameterEstimates=LSE;
proc panel data=pred;
  id sub rep;
  model yij=csij/ranone;
  by simu;
run;
```

The 500 simulated datasets are analyzed using a linear mixed model with c_i as exposure without adjustment and the IV approach, implemented in SAS proc PANEL in the time series and econometrics (TSE) package. The two estimates for β have means (SD) of 0.354 (0.058) and 0.982 (0.030), respectively. Compared with the true value of 1, and the linear mixed model estimates with the true exposure c_i^* as the gold standard, which have mean (SD) of 1.002 (0.022), the IV estimate is almost unbiased with SD lower than the unadjusted, although $E(c_i^*|t_i)$ is misspecified. The SD is higher than that of the gold standard but the difference is moderate. Therefore, the IV approach provides a robust alternative to the two-stage approach when the specification of the model for c_i is uncertain. Furthermore, unlike the two-stage method for which standard software does not directly provide the SE of the estimate, software for IV estimates gives a valid robust SE. For example, with 50 subjects, the mean of SE estimates from proc PANEL was 0.033, only slightly higher than the SD of the estimates. When the sample size was increased to 100, the mean of the SEs is 0.023, almost the same as the SD of estimates, 0.022.

Proc PANEL is not commonly used by biostatisticians, hence it is worthwhile to give some detailed explanations. The "id" statement specifies individual subjects or cohorts and the time variable. The statement "inst exogenous=(pred)" specifies "pred" as the IV, which is exogenous, as it is independent of the error terms. The option gmm is an important feature in this procedure, as it requests using GMM for parameter estimation. The option oneran specifies a one-way random effect model. The GMM method provides a sandwich robust SE estimate, as the GEE approach does in proc GENMOD. For nonlinear regression models, SAS proc MODEL provides IV estimates. Unfortunately, the procedure does not allow for fitting NLMMs or GLMMs. Special programs are needed to use IV methods for these models.

4.6   Modeling multiple exposures and responses

Sometimes subjects may be exposed to more multiple exposures. A typical example is a clinical trial with two or more drugs used in combination. In clinical practice, it is also common that a patient is prescribed more than one drug. The impact of the exposure of one drug to another is known as a drug-drug interaction (DDI) , and this DDI is particularly referred to as the interaction of exposure. For example, if one drug inhibits the liver enzymes that help the metabolism of the other drug, then the increase of its exposure will also increase the exposure of the other. Let c_{ik} and d_{ik} be the exposure of the kth drug in subject i. If two drugs are administrated together, one may assume that

$$
\begin{aligned}
\log(c_{i1}) &= \theta_{01} + \theta_1 \log(d_{i1}) + \theta_{12} \log(c_{i2}) + v_{i1} \\
\log(c_{i2}) &= \theta_{02} + \theta_2 \log(d_{i2}) + \theta_{21} \log(c_{i1}) + v_{i2},
\end{aligned} \tag{4.62}
$$

where θ_{12} and θ_{21} present DDI effects of the other drug. However, the concentration occurs on both sides of the model. These models are often referred to as simultaneous models. Special modeling procedures have been developed to fit them, e.g., available in SAS proc MODEL. However, since our goal is to predict the concentration and to fit ER models, considering the following marginal model derived from (4.62) is sufficient:

$$\begin{aligned}
\log(c_{i1}) &= \theta_{01} + \theta_1 \log(d_{i1}) + \theta_{12} \log(d_{i2}) + v_{i1} \\
\log(c_{i2}) &= \theta_{02} + \theta_2 \log(d_{i2}) + \theta_{21} \log(d_{i1}) + v_{i2}.
\end{aligned} \tag{4.63}$$

They are almost the same as model (4.62), apart from the interaction term $\theta_{kk'} c_{ik'}, c_{ik}$ being replaced with $d_{ik'}$. The two models are equivalent in terms of the dose–exposure relationship. However, all parameters and v_{ik}s in model (4.63), not only $\theta_{kk'}$, have different meanings from those in model (4.62). But we do not introduce new names for simplicity. Also, even if the v_{i1} and v_{i2} are independent in model (4.62), those in model (4.63) are not. Note that the system described by model (4.62) may not be stable unless the $\theta_{kk'}$s satisfy some conditions. For example, if $\theta_{12} = 1$ and $\theta_{21} = 1$, then the exposure increase of drug 1 stimulates an increase in the exposure of drug 2. The increase in exposure of drug 2, in turn, increases that of drug 1. This positive feedback through the other drug leads to the exposure going, in theory, to infinity, although in reality attenuation happens when the exposures are high enough. This note gives a caution for using the two linear models, as often nonlinearity occurs more often when DDI occurs. The model (4.63) is always stable, as it cannot be derived from model (4.62) if the latter is unstable.

DDIs may also happen in drug effects on the response, that is, when two drugs act together, the joint effect may be higher (or lower) than the additive effects of two single drugs. Letting y_i be the response of subject i under the exposure of two drugs 1 and 2, a simple linear model to describe the joint effect of the two exposures is

$$y_i = \beta_0 + \beta_1 c_{1i} + \beta_2 c_{2i} + \beta_{12} c_{1i} c_{2i} + \varepsilon_i \tag{4.64}$$

where β_1 and β_2 represent the individual drugs, additive effect and β_{12} is a measure of the extra joint effect. As in the previous sections, there is a set of counterpart models to model (4.2), in which we assume

$$\begin{aligned}
\log(c_{ik}^*) &= \theta_{0k} + \theta_k \log(d_{ik}) + \theta_{kk'} \log(d_{ik'}), k \neq k' \\
y_i &= \beta_0 + \beta_1 c_{1i}^* + \beta_2 c_{2i}^* + \beta_{12} c_{1i}^* c_{2i}^* + \varepsilon_i
\end{aligned} \tag{4.65}$$

and we only observe $\log(c_{ik}) = \log(c_{ik}^*) + e_{ik}$. All the models can be extended to include repeated measures by introducing random effects into the βs and θs. In this case, there are two levels of correlation structure in the exposure model, one between the exposures of different drugs at the same time point, and one at the subject level over repeated time points. Other types of ER models can also be introduced to replace (4.64). Possible nonlinearity in the exposure–response relationship may also need flexible models using, e.g., two-dimensional spline functions to replace the linear structure. An application of these models to dose-escalation trials for drug combinations can be found in Cotterill et al. (2015).

4.7   Internal validation data and partially observed and surrogate exposure measures

4.7.1   Exposure measured in a subpopulation

It is common in a study that the exposure measure, e.g., drug concentration, is measured only in a subgroup of the study population. For example, in a trial with 500 patients randomized into two dose levels, PK samples may be taken only from 100 patients. In the ER modeling should we use the whole population, and if so, how? We consider two scenarios: when there is only one exposure and one response measure on each subject, and when there are repeated measures for both exposure and response. The issue here relates to approaches dealing with ME models with internal validation data (Carroll et al., 2006).

We start from the simple scenario of one exposure and one response measure and a linear ER relationship. Let y_i and c_i be the response and exposure of subject $i, i = 1, ..., n$,

$$
\begin{aligned}
y_i &= \beta_0 + \beta c_i + u_i \\
c_i &= h(d_i, \boldsymbol{\theta}) + v_i.
\end{aligned}
\tag{4.66}
$$

We are interested in estimating β, but only $c_i, i = 1, ..., n_1 < n$ are observed. In this case, the complete data analysis using the first n_1 subject gives an unbiased estimate for β. However, one can also use the data from the other subjects in the RC approach, with different mean functions for the first n_1 subjects and the rest:

$$
\begin{aligned}
E(y_i|c_i) &= \beta_0 + \beta c_i \quad i \leq n_1 \\
E(y_i|d_i) &= \beta_0 + \beta E(c_i|d_i) \quad i > n_1
\end{aligned}
\tag{4.67}
$$

where $E(c_i|d_i) = h(d_i, \boldsymbol{\theta})$. Therefore, we can impute $\hat{\boldsymbol{\theta}}$ to form $\hat{c}_i(d_i) = h(d_i, \hat{\boldsymbol{\theta}})$ to predict $E(c_i|d_i)$ for those with missing c_i. Since

$$
\begin{aligned}
\text{var}(y_i|c_i) &= \sigma_u^2 \quad i \leq n_1 \\
\text{var}(y_i|d_i) &= \sigma_u^2 + \beta^2 \sigma_v^2 \quad i > n_1,
\end{aligned}
\tag{4.68}
$$

a weighted LS estimate for β is more efficient than the LS one. It can be obtained by setting weights 1 for those with c_i data, and $1/(1 + \beta^2 \sigma_v^2/\sigma_u^2)$ for those with d_i only. Here β is unknown, but can be replaced with an LS estimate based on the first n_1 subjects. We have also ignored the estimation error in the parameters in the weight, as it is normally much smaller than v_i if n_1 is not very small. Note that σ_v^2 can be estimated from the complete pairs.

Now consider the classical ME model

$$
\begin{aligned}
y_i &= \beta_0 + \beta c_i^* + u_i \\
c_i^* &= h(d_i, \boldsymbol{\theta})
\end{aligned}
\tag{4.69}
$$

in which we observe $c_i = c_i^* + v_i, i = 1, ..., n_1$. With this model, the complete data estimate of β obtained by replacing c_i^* with c_i for $i = 1, ..., n_1$ is biased, even when the missing c_is are completely at random (MCAR). To eliminate the bias, we can use the RC approach to replace c_i^* with $\hat{c}_i = h(d_i, \hat{\boldsymbol{\theta}})$, where $\hat{\boldsymbol{\theta}}$ is estimated based on $c_i, d_i, i = 1, ..., n_1$, since

$$E(y_i | d_i) = \beta_0 + \beta E(h(d_i, \hat{\boldsymbol{\theta}})). \tag{4.70}$$

However, (4.70) holds for $i = n_1 + 1, ..., n$ too. Therefore, we can replace c_i with $\hat{\theta} d_i$ for all subjects. Note that $\mathrm{var}(y_i | d_i)$ is the same for all subjects, so there is no need to weight subjects $i = n_1 + 1, ..., n$ differently from the first n_1 subjects.

If n_1 is sufficiently large, models (4.66) and (4.69) can be compared to assess which one fits the complete data better, using a proper goodness of fit criterion, as described in previous sections. It is likely that none of the assumed models (4.66 and 4.69) is correct, and the mixed ME model is more appropriate. In this case, as shown in the previous section, one may select the best combination of predicted and observed exposure measures.

This RC approach does not work for single dose situations, whichever model is assumed, because if $d_i = d$ for all i, everyone would have the same $E(\hat{c}_i | d_i)$. If there are other factors that have a significant impact on $c^*(t)$, in theory, these factors can be used to predict c_i, but the information gain is often much smaller than using dose in multiple dose studies. If these factors are also predictors of the response, using this approach also has the risk of introducing confounding bias.

It is useful to see how much gain can be made by including data from subjects $i = n_1 + 1, ..., n$. For simplicity, we assume that all subjects come from the same population with randomized dose, and data y_i have been centered to eliminate the intercept. Under this assumption, we denote $\sigma_h^2 = \sum_{i=1}^{n} (h(d_i, \boldsymbol{\theta}) - \bar{h})^2 / n$, where $\bar{h}$ is the average of $h(d_i, \boldsymbol{\theta})$s, for the variation of $h(d_i, \boldsymbol{\theta})$, and assume that it is the same among the first n_1 subjects as among the rest. For model (4.69), if we use the first n_1 subjects only, we obtain an estimate $\hat{\beta}_{n1}$ with $\mathrm{var}(\hat{\beta}_{n1}) = \sigma_u^2 / (n_1 \sigma_h^2)$ while when all data are used, the weighted LS estimate $\hat{\beta}_n$ has variance $\mathrm{var}(\hat{\beta}_n) = (\sigma_u^{-2} n_1 + (n - n_1)/(\sigma_u^2 + \beta^2 \sigma_v^2))^{-1} / \sigma_h^2$, therefore, the information gain can be measured by the variance ratio

$$\frac{\mathrm{var}(\hat{\beta}_{n1})}{\mathrm{var}(\hat{\beta}_n)} = \frac{\sigma_u^{-2} n_1 + (n - n_1)/(\sigma_u^2 + \beta^2 \sigma_v^2)}{n_1 \sigma_u^{-2}}$$

$$= 1 + \frac{(n - n_1)\sigma_u^2}{n_1(\sigma_u^2 + \beta^2 \sigma_v^2)}. \tag{4.71}$$

Therefore, the information increment is proportional to the sample size ratio $(n - n_1)/n_1$ and the variance ratio $\sigma_u^2/(\sigma_u^2 + \beta^2 \sigma_v^2)$. For the classical ME model case, the information gain is simply the sample size ratio, as even for the first n_1 subjects, the predicted $h(d_i, \boldsymbol{\theta})$ has to be used, just as for the rest. Therefore, there is no difference in the variances. This means we simply gain the information of additional $n - n_1$ subjects by including them in the modeling and the information increment is the sample size ratio.

For a multiplicative model such as a Poisson model with a log-link function, the same approach can be applied. Suppose that $y_i \sim Poisson(\lambda_i)$ and

$$
\begin{aligned}
\lambda_i &= \exp(\beta c_i) \\
c_i &= h(d_i, \boldsymbol{\theta}) + v_i
\end{aligned}
\tag{4.72}
$$

and $c_i, i = 1, ..., n_1$, are observed; we can also use $h(d_i, \hat{\boldsymbol{\theta}})$ to replace the missing c_is to fit the Poisson model. To decide how to weight the data with predicted c_i, we need to calculate the conditional variances of y_i. This can be done in the standard way using the well known formula

$$
\mathrm{var}(y_i) = E(\mathrm{var}(y_i|v_i)) + \mathrm{var}(E(y_i|v_i)).
\tag{4.73}
$$

Note that for those with c_i observed, the Poisson model is fitted conditional on c_i and for those without, the model is fitted conditional on $h(d_i, \hat{\boldsymbol{\theta}})$. Ignoring the variability in $\hat{\boldsymbol{\theta}}$, which is negligible compared with that of v_i, the calculation leads to

$$
\mathrm{var}(y_i|d_i) = \bar{\lambda}_i S(1 + \bar{\lambda}_i(S - 1)),
\tag{4.74}
$$

where $S = \exp(\beta^2 \sigma_v^2)$ and $\bar{\lambda}_i = \exp(\beta h(d_i, \boldsymbol{\theta}))$. As the variance depends on an unknown parameter in $\bar{\lambda}_i$, one may need to estimate it with, e.g., the first n_1 subject, then precalculate the weight in the same way as for linear models. Some software such as SAS proc GENMOD allows specifying the variance as a function of $\bar{\lambda}_i$. Hence, if σ_u^2 is known, it is possible to fit the Poisson model with the variance heterogeneity correctly taken into account. Also note that using the option of calculating the overdispersion parameter in proc GENMOD gives a robust SE estimate for β, even when the variances are wrongly specified. For a general GLM such as the logistic model, the variance cannot be easily calculated, hence the weighted estimate may not be possible. Nevertheless, the information gained by using the weight is generally small.

Under the assumption of the classical ME model, that is, when $\lambda_i = \exp(\beta h(d_i, \boldsymbol{\theta}))$ and we only observe $c_i = g(d_i, \boldsymbol{\theta}) + e_i, i = 1, ..., n_1$, the same approach as for the linear model can be used, i.e., $h(d_i, \boldsymbol{\theta})$ can be replaced by $h(d_i, \hat{\boldsymbol{\theta}})$ when fitting the model. Again there is no need to weight the two sets of subjects differently.

When repeated measures $c_{ij} = c_i + e_{ij}$ for $c_i = c_i^* + v_i, i = 1, ..., n_1$, are available, one my fit the dose-PK model in (4.69) to obtain not only $\hat{\theta}$ but also $\hat{v}_i$. Therefore, one can replace c_i with

$$
\hat{E}(c_i|d_i) = \begin{cases} \hat{\theta}d_i + \hat{v}_i & i \le n_1 \\ \hat{\theta}d_i & i > n_1 \end{cases}.
$$

For the linear model

$$\text{var}(y_i|d_i) = \begin{cases} \sigma_u^2 + \beta^2 \text{var}(\hat{v}_i - v_i) & i \le n_1 \\ \beta\sigma_u^2 + \beta^2\sigma_v^2 & i > n_1 \end{cases},$$

and one can weight the data accordingly to increase the efficiency. The same or similar approaches may be used for other types of ER models. But we will postpone the extension to the other model types to the next section, since, as one will find there, the approaches are similar.

Two more complex situations, for which we are not going to give the details, are when the ER model is an NLMM. Fitting these models to all data is less straightforward. However, it is worthwhile to note here that, if it is a classical ME model, fitting all data with the RC approach does not require more effort than fitting the complete data, although the calibration will be based on the complete data only. For a model, e.g., $y_i = h(c_i, \beta) + \varepsilon_i$, the complete data have exposure measured without error, while for the others we use the predicted c_i, which can be considered as from the Berkson model. It is possible to apply the approaches in the previous sections to this situation.

4.7.2 Exposure measured in a subpopulation but a surrogate is measured in the whole population

Another commonly encountered situation is that the exposure (e.g., C_{max}) may only be measured in a subpopulation, while in the whole population, a surrogate (e.g., the concentration at a fixed time point close to T_{max}) is measured. We consider using the RC approaches to utilize the surrogate data in order to fit all the data. Let c_i be the surrogate and c_i^* be the true exposure; again we have two choices for the relationship between them:

$$c_i^* = h(c_i, \boldsymbol{\theta}) + e_i$$

or

$$c_i = h(c_i^*, \boldsymbol{\theta}) + e_i, \tag{4.75}$$

where $\boldsymbol{\theta}$ are parameters that may be estimated from subjects with both c_i and c_i^*. Since our goal is to estimate $E(c_i^*|c_i)$, the first one simply gives $E(c_i^*|c_i) = h(c_i, \boldsymbol{\theta})$. If the second model is assumed, $E(c_i^*|c_i)$ is generally difficult to calculate, even when simple distributions for c_i or c_i^*, and e_i are assumed. One exception is when they follow a joint normal distribution and the model is linear. In this case, the two models are equivalent, i.e., given all parameters in one model, one can derive the parameters in the other one. A significant advantage of having the subgroup (internal validation data) is that we can simply fit model (4.75) without the derivation.

In the following we assume that

$$\begin{aligned} y_i &= g(c_i^*, \boldsymbol{\beta}) + u_i \\ c_i^* &= h(c_i, \boldsymbol{\theta}) + e_i \end{aligned} \tag{4.76}$$

and note that the assumption leads to a simple solution to the problem, but may not be reasonable for some practical scenarios. These models are exactly the same as models (4.66) if $g(c_i^*, \boldsymbol{\beta})$ is linear, although the notation is different, with c_i and d_i replaced with c_i^* and c_i, respectively. Therefore, the same RC approaches for models (4.66), such as the weighted LS approaches described above, apply here. For example, for a linear y-part model, the RC approach first fits model $c_i^* = h(c_i, \boldsymbol{\theta}) + e_i$ and estimate $E(c^*|c_i)$ by $h(c_i, \hat{\boldsymbol{\theta}})$. Then a weighted LS can be used to estimate β.

4.7.3   Using a surrogate for C_{max}: An example

Here we consider a practical example of using a surrogate for C_{max} to illustrate the approaches described above, and to develop further some relevant approaches. First we assume that y_i is a biomarker relating to C_{max} and the exposure follows the open one-compartment model with first order absorption (2.5). Suppose we can use samples c_{i1} and c_{i2} taken at 0.5 and 1 hours, respectively, to obtain the C_{max} as the maximum of the two, e.g., $C_{maxi} = \max(c_{i1}, c_{i2})$. Note that this C_{maxi} is only an estimate based on concentrations at the two time points. We assume that only half of the patients have both c_{1i} and c_{2i} (identified by wei=1 in the data), hence C_{maxi} can be calculated, while the other half only have c_{1i} available. We are interested in using, e.g., c_{i1} as a surrogate for C_{maxi} for those with c_{1i} only. Again we assume that y_is are centered and the relationship between C_{maxi} and y_i is linear:

$$y_i = \beta C_{maxi} + \varepsilon_i. \tag{4.77}$$

The following program simulates a dataset based on a set of parameter values, in particular with $\beta = 1$.

```
nsimu=10000
ka=exp(1+0.5*rnorm(nsimu))
ke=exp(-1+0.5*rnorm(nsimu))
v=exp(-1+rnorm(nsimu)*0.03)
pk=function(t) (exp(-ke*t)-exp(-ka*t))/v*ka/(ka-ke)
c1=pk(0.5)
c2=pk(1)
cmax=pmax(c1,c2)
yi=cmax+rnorm(nsimu)*0.1
wei=rep(c(0,1),rep(nsimu/2,2))
```

To examine the performance of using the surrogate to estimate β, we first use the subgroup with c_{1i} only, and fit the model with c_{1i} rather than C_{maxi}, which results in

```
> summary(lm(yi~c1,weights=1-wei))

lm(formula = yi ~ c1, weights = 1 - wei)
        Estimate Std. Error t value Pr(>|t|)
c1      0.729696   0.006243  116.87   <2e-16 ***,
```

showing a significant shrinkage in $\hat{\beta}$. To use the RC approach, we first need to find $E(C_{maxi}|c_{1i})$. For this we use an empirical model

$$C_{maxi} = a + bc_{1i} + e_i \tag{4.78}$$

instead of a population PK model, since it cannot be fitted with c_{1i} and c_{2i} only. The empirical model can be fitted to those with both c_{1i} and C_{maxi}, then used to predict C_{maxi} for those with c_{1i} only. To implement this approach we use the following R-code:

```
> pcmax=predict(lm(cmax~c1,weights=wei))
> summary(lm(yi~pcmax,weights=1-wei))
```

which uses the complete pairs (wei=1) to fit the model between the c_{1i} and C_{maxi} model, then uses the predicted C_{maxi} in the ER model to fit the data with no observed C_{max}. The results show that this approach gives an almost unbiased estimate for β.

```
       Estimate Std. Error t value Pr(>|t|)
pcmax  1.023243   0.008755 116.874   <2e-16 ***
```

The performance of this approach depends on how good the linear model is. Other models may be used too.

It might be tempting to predict c_{2i} then reconstruct C_{maxi} from c_{1i} and a predicted c_{2i} based on a model $c_{i2} = a + b\,c_{1i} + e_i$. This approach yields

```
> pc2=predict(lm(c2~c1,weights=wei))
> pcmax=pmax(c1,pc2)
> summary(lm(yi~pcmax,weights=1-wei))
```

```
       Estimate Std. Error t value Pr(>|t|)
pcmax  1.073511   0.009438 113.740  < 2e-16 ***
```

There is a considerable upward bias in $\hat{\beta}$ (1.07 rather than 1.00). As we discussed, the performance of this approach depends on the empirical model. But the prediction should be for $E(C_{maxi}|c_{1i})$. Even if model $c_{i2} = a + bc_{1i} + e_i$ is correct, $\max(c_{1i}, \hat{c}_{2i})$ is not an unbiased estimate of $E(C_{maxi}|c_{1i})$, because the maximization and expectation are not exchangeable. Therefore, one should try to predict exactly the term in the y-part model, e.g., if C_{maxi}^2 is in the model then a model between C_{maxi}^2 and c_{1i} should be used to predict $E(C_{maxi}^2|c_{1i})$.

Now we include all data in the model fitting and use the weighted LS for those with predicted C_{max} and 1 for the true C_{max}.

```
> weic=fitm$sigma^2/(fitm$sigma^2+fitm$coef[2,1]^2*fitc$sigma^2)
> weight=ifelse(wei==1,1,weic)
> ncmax=ifelse(wei==1,cmax,pcmax)
> summary(lm(yi~ncmax,weights=weight))
```

```
       Estimate Std. Error t value Pr(>|t|)
ncmax  1.007403   0.004313 233.550   <2e-16 ***
```

Compared with the results of fitting the data with C_{maxi} only, the SE was about 10% smaller.

```
summary(lm(yi~cmax,weights=wei))
```

```
            Estimate Std. Error t value Pr(>|t|)
cmax        1.000413   0.004723 211.795   <2e-16 ***.
```

The key assumption, on which this approach is based, is the empirical model (4.78). One may ask the question "What if the empirical model is

$$c_{i1} = a + b\, C_{maxi} + e_i?" \tag{4.79}$$

Indeed, this is the classical ME model and an adjustment is needed after replacing C_{maxi} with c_{i1} in the model, as we described in the previous sections. Note that the simulation example is based on a real scenario and model (4.78) is used as an empirical model. The example has shown a shrinkage in the estimated β. In fact, both models can be correct, e.g., when c_{i1} and C_{maxi} are jointly normally distributed, and model (4.78) is the prediction for the RC, if model (4.79) is first assumed.

Now suppose that y_i is an indicator of an adverse event, taking value 1 if an event occurs and 0 otherwise. We consider fitting a logistic model linking C_{max} with the risk of event. Let $P(y_i = 1|C_{maxi}) = 1/(1 + \exp(-\beta_0 - \beta C_{maxi}))$. Our goal is to use all the data to estimate β. Using all the data is especially important if the event is rare. The major technical problem in using the RC approach is that $1/(1 + \exp(-\beta_0 - \beta E(C_{maci}|c_{1i}))) \neq E(y_i|c_{1i})$, due to the nonlinearity in the logistic function. If we are willing to assume a normal distribution for C_{maxi}, then a well known approximation (Carroll et al., 2006, Chapter 4),

$$E(1/(1 + \exp(\beta x_i))) \approx 1/(1 + \exp(\beta \mu_x/(1 + \beta^2 \sigma_x^2/1.7^2))), \tag{4.80}$$

where $x_i \sim N(\mu_x, \sigma_x^2)$, can be used. Therefore, let $\hat{\beta}^*$ be the estimate when C_{maxi} is replaced with an unbiased estimate for $E(C_{maxi}|c_{1i})$. One may find a corrected estimate for β by solving the equation

$$\hat{\beta}^* = \beta/(1 + \beta^2 \sigma_c^2/1.7^2)^{1/2} \tag{4.81}$$

where σ_c^2 is the prediction error for C_{maxi}, which can be estimated when fitting the model for C_{maxi} and c_{1i}. Note that the corrected estimate is no longer an unbiased estimate for β. Its SE needs additional calculation, and although a delta approach (appendix) may be useful, it may also be worthwhile to consider a bootstrap approach to apply to both stages. Nevertheless, $(1 + \beta^2 \sigma_c^2/1.7^2)^{1/2}$ is often close to 1, hence $\hat{\beta}^* \approx \beta$. If the AE is rare, i.e., $E(y_i|C_{maxi})$ is small, then $E(y_i|C_{maxi}) \approx \exp(\beta_0 + \beta C_{maxi})$. The variability in C_{maxi} around $E(C_{maxi}|c_{1i})$ goes to the baseline risk estimate if we simply replace C_{maxi} with an estimate for $E(C_{maxi}|c_{1i})$.

The selection of a subgroup in which the exposure is measured may be operational, e.g., taking samples from patients in a subset of study centers, or from patients who give consent to take blood samples. In some cases, one may be able to randomly select a subgroup. In the latter case, using different weights, when applicable, may gain extra efficiency, as shown above. However, in the former case, the subgroup may be remarkably different from the rest. In this case, weighting them differently may introduce biases. Further details about biases due to confounding factors can be found in Chapter 8.

As a final remark, we note that the problems of using partially measured exposure and surrogate measures are similar. The key considerations are the assumption on how to put exposure in the ER model and how to predict it. In the former case the dose–exposure model provides a standard tool, and in the second case, often an empirical model is used. Under some situations, model $c_i = h(c_i^*, \boldsymbol{\theta}) + e_i$ may be more reasonable. Under this model, the calculation of $E(c_i^*|c_i)$ needs a more advanced approach. One possibility is to use a Bayesian approach to get the posterior distribution of c_i^* given c_i, then estimate $E(c_i^*|c_i)$ numerically by, e.g., taking the mean of the posterior samples.

4.8   Comments and bibliographic notes

Joint and sequential modeling approaches in pharamacometrics can be found in Bonate (2006) and Ette and Williams (2007). A systematic comparison was reported in Zhang et al. (2003). For measurement error theory, Carroll et al. (2006) covers a wide range of models yet still start from the very basic concept. Several other books are on the same topic, but mainly linear models. Bennett and Wakefield (2001) applied a Bayesian approach to joint modeling, a topic we have not covered.

The literature in this area comes from both statisticians and pharmaco-metricians, with different focuses. Joint modeling, used by the former, and simultaneous modeling, used by the latter, seem to carry the same meaning, but refer to slightly different approaches. Joint modeling refers to modeling multiple outcomes (e.g., longitudinal and survival data) to explore the relationship between them, including using two-stage approaches. Simultaneous modeling means fitting multiple models jointly in one step, in contrast to the sequential modeling procedure which is the same as the two-stage one. The focus of pharmacometricians is on the direct relationship between the exposure and response, while the statistician's focus is often on the correlation between the outcomes, e.g., time to diagnosed AIDS or death and CD4 counts, tumor size and death.

5

Exposure–risk modeling for time-to-event data

The time to a certain event is often an important outcome in medical or biological research. For exposure–response analysis, our primary interest is on the impact of exposure on the risk of an event, which can be quantified by using a model to fit time-to-event data. In this chapter we consider situations when the ER relationship is instant, i.e., the exposure level at a certain time only affects the risk at the same time. If the exposure has a cumulative effect, some special modeling techniques may be needed and will be described in a later chapter.

5.1 An example

The example arises from a type of study used in drug development. For drugs with highly variable exposures, often therapeutic dose monitoring (TDM) is used to control an individual's exposure in a desired range by adjusting individual doses until the exposure is within this range. For detailed dose escalation approaches, the reader is referred to Chapter 9. The dataset used here is simulated to mimic a real scenario. The dataset contains 100 patients, 50 in the active treatment group and 50 in the placebo group. The dose levels in the active group were adjusted by TDM to reach the target concentration of 10 ng/mL, on a 14 day interval when the drug concentration is measured to decide if a dose escalation is needed. The endpoint is an exposure related AE. The dose, concentration, and event data at each interval, until the the end of study or until the first event from five patients are given below.

	group	sub	dose	conc	start	stop	event
1	1	1	1.61	6.30	0	14.00	0
2	1	1	1.87	9.61	14	28.00	0
3	1	1	2.13	8.42	28	30.04	1
4	1	2	1.61	3.97	0	7.78	1
5	1	3	1.61	5.64	0	14.00	0
6	1	3	1.87	8.33	14	28.00	0
7	1	3	2.13	8.80	28	42.00	0
8	1	3	2.40	12.66	42	56.00	0

9	1	3	2.40	13.75	56	70.00	0
10	1	3	2.40	11.11	70	84.00	0
11	1	3	2.40	9.05	84	98.00	0
12	1	3	2.66	12.60	98	112.00	0
13	1	4	1.61	2.50	0	0.68	1
14	1	5	1.61	2.82	0	4.45	1

The probability of no event until a certain time point, known as the survival curve, for this dataset is plotted in Figure 5.1, along with the distribution of dose level and concentration changes over time. From the dose and concen-

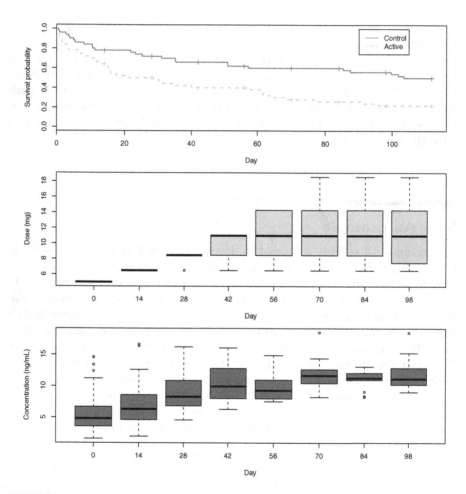

FIGURE 5.1
Survival curve of time to first AE for active and placebo groups in a hypothetical trial with TDM (top panel), and dose and concentration levels (lower panels).

tration plots, the dose–exposure relationship is obvious. But the relationship between the risk of AE and the exposure is not clear, although evidence of the exposure–risk relationship is suggested by the increasing difference in the survival curves between the active and controls groups (top panel). We are interested in how drug exposure is related to the risk of AEs. Depending on the context, we may be interested in 1) the risk of AE for a given dose level; 2) the risk of AE under cumulative exposure following a complex dynamic; 3) how the risk changes with a dose adjustment. The following gives tools to find answers to these questions from analysis of event time aspects. But approaches and tools from other chapters may also be needed. A number of variants of this example, including one with recurrent events and another when the recurrent events are known to occur with time intervals but not the exact occurrence time (interval censored) will also be used.

5.2 Basic concepts and models for time-to-event data

5.2.1 Distributions for time-to-event data

An approach for modeling TTE often starts from specifying its distribution, rather than its mean and variance, as we did in the previous chapters. The distribution of the TTE, T, can be specified by the density and distribution functions $f(t)$ and $F(t)$, respectively. In addition the hazard function

$$h(t) = \lim_{\Delta \to 0} \frac{P(t < T < t + \Delta | T > t)}{\Delta} \qquad (5.1)$$

plays a central role in modeling TTE. The hazard is the risk of having an event just after time t, conditional on having no event up to time t. From the definition, we can derive $h(t) = f(t)/S(t)$ where $S(t) = 1 - F(t)$ is called the survival function, as it is the probability of having no event until time t. From $h(t)$ we can define the cumulative hazard function as

$$H(t) = \int_0^t h(\tau)d\tau, \qquad (5.2)$$

which links to $S(t)$ with $S(t) = \exp(-H(t))$. The relationships between $h(t)$, $H(t)$ and $f(t)$ and $S(t)$ are the key formulas that represent key concepts in TTE modeling.

One might think of the time-to-event as a positively distributed variable, hence its analysis could use, e.g., a log transformation or a generalized linear model. However, one issue makes the analysis of TTE data a very special topic. Suppose that we are interested in the risk of an adverse event under the exposure of drug X in a clinical trial which runs for 6 months. At the end of the trial some patients may never have had an event, so we only know that the

TTE for them is longer than 6 months. In this case, the TTE is considered as (right) *censored* by the end of study. It may also be censored by another event, such as death. Formally, let T_i^* be the true event time, C_i be the censoring time (due to end of study or death), and we only observe $T_i = \min(T_i^*, C_i)$. To identify those who are censored we also need a censoring indicator r_i, with $\delta_i = 0$ if $T_i^* > C_i$ and $\delta_i = 1$ otherwise. In previous chapters, we have almost always linked the exposure and covariates to the mean of the response. For TTE data, we link the exposure and covariates to the $h(t)$, due to reasons that will become clear later.

5.2.2   Parametric models

Some simple models can be constructed by specifying the distribution of the TTE. For example, assume that T follows an exponential distribution: $T \sim Exp(\lambda)$ with $f(t) = \lambda \exp(-\lambda t)$ and $F(t) = 1 - \exp(-\lambda t)$, which leads to $S(t) = 1 - F(t) = \exp(-\lambda t)$. Expressing them as $h(t) = f(t)/S(t)$, we get $h(t) = \lambda$, i.e., the hazard function for the exponential distribution is a constant. To link T to covariates $\mathbf{X}_i$, one can assume $T \sim Exp(\lambda_i)$ with $\lambda_i = \exp(\mathbf{X}_i^T \boldsymbol{\beta})$. If a covariate in $\mathbf{X}_i$ is a group indicator, the corresponding $\boldsymbol{\beta}$ element is the log-risk ratio (RR) of this group to that of the reference. Or, if the covariate is a continuous variable, then the $\boldsymbol{\beta}$ element is the log-RR of one unit increase in the covariate. The use of $\exp(\lambda_i)$ is to ensure that the risk at any $\mathbf{X}_i$ value is nonnegative. This approach simply replaces λ in $f(t) = \lambda \exp(-\lambda t)$ with $\exp(\mathbf{X}_i^T \boldsymbol{\beta})$. However, in the next section, we will show that this approach is not valid if $\mathbf{X}_i$ has time varying components.

The exponential distribution is often too restrictive, since it has only one parameter. In principle, any positive distribution can be used for the TTE, and some other distributions may be transformed to a positive distribution. As a simple example, T_i may follow a log-normal distribution, that is, $\log(T_i) \sim N(\mathbf{X}_i^T \boldsymbol{\beta}, \sigma^2)$. With the additional parameter σ^2, this distribution offers more flexibility. The most commonly used distribution for TTE is the Weibull distribution. Its hazard and survival functions are

$$
\begin{aligned}
h(t) &= \gamma a(at)^{\gamma - 1} \\
S(t) &= \exp(-(at)^\gamma),
\end{aligned}
\tag{5.3}
$$

respectively, where a^{-1} and γ are called the scale and shape parameters. The density function can be found from $f(t) = h(t)S(t)$ easily. To include covariates, one may assume that $\mathbf{X}_i$ has an impact on a, γ or both, but $a_i = \exp(\mathbf{X}_i^T \boldsymbol{\beta})$ is a common choice for a simple model, as a_i is interpreted as an acceleration factor, and $\boldsymbol{\beta}$ are log-TTE ratios of one category to the other if the corresponding $\mathbf{X}_i$ elements are binary.

Based on the observed $T_i, \delta_i, i = 1, ..., n$, we can write the log-likelihood function as

$$
L(\boldsymbol{\beta}) = \prod_{i=1}^{n} f^{\delta_i}(T_i, \boldsymbol{\beta}) S^{1-\delta_i}(T_i, \boldsymbol{\beta}) g(\delta_i),
\tag{5.4}
$$

where $g(\delta_i)$ is the density function of δ_i, and we assume that C_i, and consequently δ_i, are independent of T_i and hence we will ignore the $g(\delta_i)$ part afterwards. This type of censoring is known as noninformative censoring. The MLE maximizes (5.4), which can be done easily under some simple situations, for example, if the exposure is constant and T_i follows the exponential distribution $f(t) = \lambda \exp(-\lambda t)$, and we would like to estimate the risk λ. Taking $f(t)$ and the corresponding survival function $S(t) = \exp(-\lambda t)$ into (5.4) and maximizing it with respect to λ (an easy way is to maximize $\log(L(\lambda))$, we find the MLE for λ is $\hat{\lambda} = n_e / \sum_{i=1}^{n} T_i$, where n_e is the number of events. This estimate is an important measure of exposure–risk relationship, under the assumption of constant hazard, known as event rate per person-year exposure. Later on we will find this measure may not be appropriate for many situations when the exposure or risk are time varying. The ML approach for TTE models is almost the same as the ML approaches elsewhere, except for the extra assumption of independent δ_i. The score equation for MLE can be derived from (5.4) and solved by statistical software. Asymptotic properties, in particular, $\hat{\beta} \sim N(\beta, \Sigma)$, can also be derived, but the large sample condition needs to apply to the data after censoring. For example, asymptotic normality of the log-RR estimate based on the exponential model only holds when the number of events after censoring is sufficiently large.

5.2.3 Cox models

The analysis of TTE data is much more sensitive to the assumption on the distribution than the analyses described in the previous chapters. Therefore, it is desirable to avoid specifying the distribution by using a semiparametric model. The most commonly used semiparametric TTE model is the Cox model, which assumes that

$$h(t, \beta, \mathbf{X}_i) = h_0(t) \exp(\mathbf{X}_i^T \beta), \tag{5.5}$$

where $h_0(t)$ is an unspecified baseline hazard function. This model can be fitted by maximizing a partial likelihood function

$$L(\beta) = \prod_{i=1}^{n} \left(\frac{\exp(\mathbf{X}_i^T \beta)}{\sum_{j \in R_i} \exp(\mathbf{X}_j^T \beta)} \right)^{\delta_i}, \tag{5.6}$$

where $\delta_i = 0$ if t_i is censored and $\delta_i = 1$ if it is an event time, which consists of the likelihood of event occurrence on subject i at a time point t_i as observed, among R_i: those still at risk at t_i, for each i who has had an event. Compared with the ML approach, an advantage is that there is no need to work out $f(t)$ and $S(t)$. Another advantage is that $h_0(t)$ is eliminated, since everyone at risk at time t shares the same $h_0(t)$. The resulting estimating equation by maximizing the partial likelihood is

$$\sum_{i=1}^{n} \delta_i \left(\mathbf{X}_i - \frac{\sum_{j \in R_i} \mathbf{X}_j \exp(\mathbf{X}_j^T \beta)}{\sum_{j \in R_i} \exp(\mathbf{X}_j^T \beta)} \right) = 0. \tag{5.7}$$

Note that the second term in the bracket is a weighted average of $\mathbf{X}_i$. This approach leads to a very efficient algorithm for estimating β and has been implemented in common software such as SAS and R. The most important large sample properties of the estimate $\hat{\beta}$ is $\hat{\beta} \sim N(\beta, \Sigma)$, where Σ is the asymptotic variance of β and is routinely provided by software for fitting the Cox model. This asymptotic distribution is the basis of statistical inference for β. The derivation of the asymptotic properties and calculation of Σ involve complex formulae, hence the reader is referred to a standard textbook such as Chapter 8 of Fleming and Harrington (1991).

In the example of exposure–safety data in the simulated TDM study, one may be interested in comparing the risks between the two treatment groups, although we know that the risk changes during dose escalation. Fitting a Cox model to the data with R using the function call below, we obtain

```
coxph(Surv(start,stop,event)~group,data=Out)
       coef exp(coef) se(coef)    z      p
group 1.65      5.19    0.286 5.76 8.5e-09
```

which shows log-RR= $1.65(SE = 0.286)$ between the active group and the placebo group.

5.2.4 Time-varying covariates

The drug exposure, either as the drug dose or the concentration, is often time-varying. A simple example is the exposure–safety data from the simulated TDM study, in which the dose is piecewise constant. The Cox regression can be easily extended to include time-varying covariates. The corresponding risk function is

$$h(t, \beta, \mathbf{X}_i(t)) = h_0(t) \exp(\mathbf{X}_i^T(t)\beta) \tag{5.8}$$

where $\mathbf{X}_i(t)$ contains time-varying covariates and $h_0(t)$ is the unspecified base-line hazard function. This model can also be fitted by maximizing the partial likelihood function that consists of the likelihood of event occurrence on subject i at a time point t_i as observed, given all those still at risk at this time point. In the partial likelihood, $h_0(t)$ is also eliminated. However, it is worthwhile noticing that for exposure–risk modeling the effect of an exposure time profile shared by all subjects cannot be estimated, since it is confounded with $h_0(t)$. Estimating β for time-varying $\mathbf{X}_i(t)$ follows the approach above. The resulting estimating equation for minimizing the partial likelihood is

$$\sum_{i=1}^{n} \delta_i(\mathbf{X}_i(t_i) - \frac{\sum_{j \in R_i} \mathbf{X}_j(t_i) \exp(\mathbf{X}_j^T(t_i)\beta)}{\sum_{j \in R_i} \exp(\mathbf{X}_j^T(t_i)\beta)}) = 0. \tag{5.9}$$

With time-varying $\mathbf{X}_i(t)$, the same large sample properties of $\hat{\beta}$, in particular $\hat{\beta} \sim N(\beta, \Sigma)$ also hold, under similar assumptions.

The counting process approach has been used increasingly in TTE modeling, because of its theoretical and practical advantages. The counting process can be considered as an indicator of event occurrence. For subject i, its counting process $N_i(t)$ is a step function that is zero at the beginning and jumps to 1 when the event happens. It can be shown, under some technical conditions, that $E(N_i(t)) = h(t)$, if i is at risk. Intuitively, one may consider $N_i(t)$ on discrete times $t = 0, 1, 2, \dots.$ At time t, the probability of $N_i(t) = 1$ is $h(t)$ for those still alive at $t - 1$. This view links the Cox model to a logistic regression model (Efron, 1989). The connection between the two will be further explored later.

The counting process approach also provides a convenient way to implement a model with time-varying covariates. To see how the counting process approach helps to implement a TTE model, we consider the following hypothetical example. To accommodate piecewise exposures, the counting process can be split into multiple pieces according to event time and changes in covariates. For example, if subject i has an event at time 10, and $X_i(t) = 1$ when $t \leq 5$ and $X_i(t) = 2$ when $t > 5$, one can write the data into two records: $s = 0, t = 5, N_i(t) = 0, X_i(t) = 1$ and $s = 5, t = 10, N_i(t) = 1, X_i(t) = 2$, where s and t are the beginning and the end of an interval. A data structure like this is referred to as a counting process structure and is often used for fitting piecewise constant covariate data. Cox model fitting software such as SAS proc PHREG and R-survival library both allow fitting counting process data.

Looking at the example exposure safety dataset in Section 5.1, we find it is already in the counting process form. It is very easy to fit models with time-varying dose or concentration into the model. The R-survival library provides a function Surv(start,stop,event) in which one specifies the start and stop time points (s and t in the example above) of each record and event is an indicator. The following are the function calls for the Cox models fitting dose and concentration, respectively, and the results.

```
> coxph(Surv(start,stop,event)~dose,data=Out)

          coef exp(coef) se(coef)    z       p
dose 0.806      2.24     0.138 5.84 5.3e-09

Likelihood ratio test=37.1 on 1 df, p=1.1e-09 n= 458,
                              number of events= 63
> coxph(Surv(start,stop,event)~conc,data=Out)

          coef exp(coef) se(coef)    z       p
conc 0.156      1.17    0.0209 7.46 8.9e-14
```

in which coef is β and exp(coef) is the risk ratio of one unit dose or concentration change.

Since the exposures (dose and concentration) change over time, one may also be tempted to fit the same models to the active group only to estimate β based on exposure variations within the group. This approach leads to

```
> coxph(Surv(start,stop,event)~dose,data=Out[Out$group==1,])
```

```
      coef exp(coef) se(coef)    z    p
dose 2.09      8.11      1.7 1.23 0.22

Likelihood ratio test=1.77  on 1 df, p=0.183  n= 143,
                                number of events = 45
> coxph(Surv(start,stop,event)~conc,data=Out[Out$group==1,])

      coef exp(coef) se(coef)    z      p
conc 0.133      1.14    0.0344 3.86 0.00011
```

Note that the estimate of β for dose increased more than 100%, indicating possible confounding between the changes in the dose and the risk function.

Including time-varying covariates in a parametric model is often more difficult, unless the covariates are among a few types of simple time functions. For example, for the exponential distribution, it is natural to assume that $\mathbf{X}_i(t)$ has an impact on the hazard $h(t)$. Hence we may assume that the hazard of subject i $\lambda_i(t) = \exp(\boldsymbol{\beta}^T \mathbf{X}_i(t))$. However, to use the ML approach one also needs to find the corresponding $f(t)$ and $S(t)$. For this the basic relationships between $h(t)$, $S(t)$ and $f(t)$ are needed. Since $f(t) = h(t)S(t)$ and $S(t) = \exp(-H(t))$, the key step is to calculate $H_i(t, \mathbf{X}_i) = \int_0^t h_i(\tau)d\tau$ for each i. This is trivial when $\mathbf{X}_i(t)$ is constant, but an analytical form of $H_i(t, \boldsymbol{\beta})$ may not exist for a general $\mathbf{X}_i(t)$, even for the exponential distribution. Therefore, the density function $f(t, \boldsymbol{\beta})$ and survival function $S(t, \boldsymbol{\beta})$ cannot be easily calculated. For some other distributions one can assume that $h(t) = h_0(t) \exp(\boldsymbol{\beta}^T \mathbf{X}_i(t))$. For example, the Weibull distribution has $h_0(t) = t^{\gamma-1}$. In general, the calculation of $H_i(t, \boldsymbol{\beta})$ is more difficult with non-constant $h_0(t)$. Common software for fitting parametric survival models, such as SAS proc LIFEREG and the R function survreg, do not have the option of counting process structures to support fitting time-varying covariates, probably due to the difficulty of deriving the distribution. Therefore, it is not easy to include time-varying covariates even when one is willing to approximate them with a piecewise constant function. In NONMEM and R, a numerical integration algorithm is available, hence one can evaluate $H_i(t)$ numerically. This approach allows using a wide range of exposure models and TTE distributions at the cost of much more intensive computation and potential numerical instability.

5.2.5   Additive models

The Cox model assumes that the effect of each factor is multiplicative to the risk. In contrast, the additive model is based on the assumption that the impact of each factor, including the baseline, is additive:

$$h(t, \mathbf{X}_i) = h_0(t) + \mathbf{X}_i^T(t)\boldsymbol{\beta}, \tag{5.10}$$

where $h_0(t)$ and $\mathbf{X}_i(t)$ have a similar role to play as in the Cox model, except their effects are additive to the risk. Additive models are less commonly used than the Cox model, for a number of reasons. Since the risk should be non-negative, an additive structure does not fit into this scale naturally. With given estimates for $h_0(t)$ and $\boldsymbol{\beta}$, the predicted risk $h(t, \mathbf{X}_i)$ may not be non-negative for all possible $\mathbf{X}_i(t)$, unless appropriate range restrictions are applied on it and $\boldsymbol{\beta}$. Second, the estimation of $h_0(t)$ and $\boldsymbol{\beta}$ cannot be completely separated. The choice of baseline function may have a considerable impact on the estimation of $\boldsymbol{\beta}$. Fitting an additive model needs software such as the R-library timereg (Martinussen and Scheike, 2006), which is less popular than those for fitting Cox models. The function call for using timereg is almost identical to the coxph function call, which also uses the Surv() function.

However, for an additive model, the calculation of $H_i(t, \mathbf{X}_i) = \int_0^t h_i(\tau) d\tau$ is much easier for a given $\mathbf{X}_i(t)$ since it has an analytical solution as long as $\int \mathbf{X}_i(t) dt$ has one. If appropriate, we may consider a mixture of additive and multiplicative models

$$h(t, \mathbf{X}_i) = h_0(t) \exp(\mathbf{X}_i^T(t)\boldsymbol{\beta}) + \boldsymbol{\beta}_c c^*(t) \tag{5.11}$$

with additive exposure and multiplicative covariates effects. It can also be fitted with function timereg. If the exposure is to increase the risk, it is possible to let $\boldsymbol{\beta}_c \geq 0$ to ensure that $h(t) \geq 0$ for all ts. For some typical $c^*(t)$s as a covariate in $\mathbf{X}_i$, $H(t, \mathbf{X}_i)$ can be easily calculated. These include $c^*(t) = \sum_{k=1}^K \theta_k \exp(-C_k t)$ from a linear compartmental model.

5.2.6 Accelerated failure time model

The previous models are mostly specified in terms of the risk function. The accelerated failure time (AFT) model represents another class of TTE models. The most common form for the AFT model can be written as

$$\log(T_i) = \mathbf{X}_i^T \boldsymbol{\beta} + \varepsilon_i, \tag{5.12}$$

where ε_i is a zero mean random variable. An alternative form to (5.12) is

$$T_i = \exp(\mathbf{X}_i^T \boldsymbol{\beta}) T_{i0}, \tag{5.13}$$

where $T_{i0} = \exp(\varepsilon_i)$ is the "baseline" time when $\mathbf{X}_i = 0$. When it is assumed to follow a certain distribution, this AFT model is a parametric model we have already introduced. Suppose that the survival function for $\exp(\varepsilon_i)$ is $S_0(t)$, since $T_i = \exp(\mathbf{X}_i^T \boldsymbol{\beta} + \varepsilon_i)$; one can derive that for T_i as

$$
\begin{aligned}
S(t|\mathbf{X}_i) &= P(\exp(\mathbf{X}_i^T \boldsymbol{\beta} + \varepsilon_i) < t|\mathbf{X}_i) \\
&= P(\exp(\varepsilon_i) < t \exp(-\mathbf{X}_i^T \boldsymbol{\beta})|\mathbf{X}_i) \\
&= S_0(t \exp(-\mathbf{X}_i^T \boldsymbol{\beta}))
\end{aligned}
\tag{5.14}
$$

which is $S_0(t)$ with the time scale changed by the acceleration factor $\exp(-\mathbf{X}_i^T \beta)$. The corresponding $f(t)$ and $h(t)$ can be derived easily if $S_0(t)$ is known. Therefore, the ML approach can be applied to estimate β by constructing a likelihood function based on $f(t)$ and $S(t)$ and maximizing it. In general, an AFT model with an arbitrary distribution for ε_i and acceleration factor $\exp(\mathbf{X}_i^T \beta)$ does not lead to a proportional hazard model. One exception is the Weibull distribution. Details can be find in most textbooks such as Aalen et al. (2008).

For the same reason as that for using Cox models, we often want to leave the distribution of ε_i unspecified. Recall for linear and nonlinear models of continuous response, we sometimes assume the measurement error as a normally distributed variable. However, the asymptotic properties of the fitted model and estimated parameters do not depend on normality, as long as the mean is zero. For the AFT model, the distribution of ε_i plays a far more important role. It is straightforward to introduce a semiparametric model based on (5.12) by leaving the distribution of ε_i unspecified, giving a model without a parametric assumption. However, a similar approach to deriving a model fitting algorithm for the Cox model leads to a rank regression for β. A well known estimate was proposed by Buckley and James (1979). Although software such as the jm library in R is available, the estimate is much less stable than the MLE for a parametric model.

5.3   Dynamic exposure model as a time-varying covariate

Until now we have not explicitly considered the effect of exposure on the risk of TTE. If the exposure is available and its impact on the risk is instantaneous, one may simply include it as a time-varying covariate, as discussed before. But often, even when the exposure is observed, its impact on the risk of events may not be instantaneous. It is often the case that the risk is built by exposure accumulation according to a dynamic system. In this section we assume that the latent cumulative exposure $c_i^*(t)$ has an immediate effect on the risk, e.g.,

$$h_i(t) = h_0(t) \exp(\beta c_i^*(t)), \tag{5.15}$$

where we drop covariates for simplicity. We also assume that the latent exposure $c_i^*(t)$ is related to observed exposures, which may either be drug concentration or dose, via a mechanistic or empirical model. But in this section we concentrate on the situation where no exposure model is needed to be fitted explicitly. The model may have unknown parameters to be estimated when fitting the TTE model. A typical exposure dynamic model is given by the effect compartment model in which

$$c_i^*(t) = \int_0^t g(t - \tau, \boldsymbol{\beta}_i) e_i(\tau) d\tau \qquad (5.16)$$

where we use $e_i(t)$ as general notation for the observed exposure. An example for $g(t, \boldsymbol{\beta})$ is

$$g(t, \boldsymbol{\theta}) = K_{in} \exp(-Kt), \qquad (5.17)$$

in which K_{in} and K are unknown parameters, when the exposure is constant, i.e., $e(t) = E$, $c_i^*(t) = EK_{in}(1 - \exp(-Kt)/K)$. This section explores approaches to fitting TTE models when $c^*(t)$ is known except for a finite number of parameters to be estimated.

5.3.1 Semiparametric models

In some specific cases, one can estimate the exposure dynamic represented by (5.16) with an empirical model approximation while fitting the TTE model. For example, if time t is discrete, taking values $0, 1, 2, 3, ...$, (5.16) can be approximated by a discrete linear combination of past exposures

$$c_i^*(t) = \sum_{j=0}^{t} g_{t-j} e_i(j) \qquad (5.18)$$

where the g_js are unknown coefficients. Including $c_i^*(t)$ in the Cox model leads to

$$h_i(t) = h_i(0) \exp(\beta \sum_{j=0}^{t} g_{t-j} e_i(j)). \qquad (5.19)$$

Note that β cannot be distinguished from the g_js, unless $c^*(t)$ can be measured. However, βg_js are sufficient to represent the dynamic relationship between the observed exposure and its impact on the risk. To fit such a model one simply includes lagged exposures as covariates in the Cox model, hence this is the simplest way to explore the exposure dynamic with minimum assumptions on the model.

A simple Cox model can also be used when $g(t)$ in (5.16) can be approximated by a set of spline functions. Suppose that $g_i(t) \approx \sum_{k=1}^{K} b_k(t) \theta_{ik}$, where the $b_k(t)$s are spline functions (Chapter 3); we can write (5.16) as

$$c_i^*(t) = \sum_{k=1}^{K} B_k(t) \theta_{ik} \qquad (5.20)$$

where $B_k(t) = \int_0^t b_k(t - \tau) e_i(\tau) d\tau$. Since $b_k(t)$ is a piecewise polynomial, for some types of $e_i(t)$, e.g., a sum of exponential functions from a population PK model, $B_k(t)$ can be calculated analytically. This approach has been used in several exposure–risk modelings, e.g., in Sylvestre and Abrahamowicz (2009). Fitting these models using standard Cox model software can be made easy by including precalculated $B_k(t)$ in the input dataset.

Sometimes $c_i^*(t)$ may have parameters that are not linear to it, e.g., $c_i^*(t) = 1 - \exp(-Kt)$ is not a linear function of K. Although the function is very simple, a technical difficulty is that K cannot be estimated by, e.g., software fitting a Cox model. However, these types of exposure models are of particular importance, hence we discuss model fitting for some simple but typical models. Two typical examples are the K-dynamic models in the PKPD modeling literature, and models for the relationship between drug prescription and reported adverse events in pharmacoepidemiology (Abrahamowicz et al, 2006). In both cases, the exposure we observed is the dose. Since often the dose can only be changed at finite time points, one can assume it a piecewise constant function of time. Let e_{ik} be the exposure level for subject i at the kth time interval from t_{ik} to t_{ik+1}:

$$e_i(t) = e_{ik}, \quad t_{ik} \le t < t_{ik+1}. \tag{5.21}$$

We assume that the exposure $c_i^*(t)$ follows the effect compartment model (5.16) with $g(t, \boldsymbol{\theta})$ given in (5.17). In this case, one can calculate

$$
\begin{aligned}
c_i^*(t) &= K_{in} \int_0^t \exp(-K\tau) e_i(t - \tau) d\tau \\
&= \frac{K_{in}}{K} \Big(\sum_{k:t_{ik}<t} e_{ik-1}[1 - \exp(-K(t_{ik} - t_{ik-1}))] \exp(-K(t - t_{ik})) \\
&\quad + e_{ik}[1 - \exp(-K(t - t_{ik}))] \Big).
\end{aligned}
\tag{5.22}
$$

For easy computing, the following recursive formula is more convenient:

$$
\begin{aligned}
c_i^*(t) &= \frac{K_{in}}{K}[c_i^*(t_{ik}) \exp(-K(t - t_{ik})) + e_{ik}(1 - \exp(-K(t - t_{ik})))] \\
& \quad t_{ik} \le t < t_{ik+1}
\end{aligned}
\tag{5.23}
$$

where $c_i^*(t_{ik})$ is the cumulative concentration at time t_{ik}.

A familiar example of this general situation is the cumulative exposure after a constant instant exposure at level e_0 that gives

$$c^*(t) = e_0 K_{in}(1 - \exp(-Kt))/K. \tag{5.24}$$

This formula has been widely used as an approximation to the process of reaching the steady state concentration after repeated dosing. It also helps with interpreting formula (6.33), as the first term represents the contribution of existing exposure $c_i^*(t_{ik})$ decayed to time t with rate K and the second term is the contribution of additional exposure e_{ik}. Figure 5.2 gives an example for the relationship between $c_i^*(t)$ and piecewise constant $e_i(t)$, in which we set

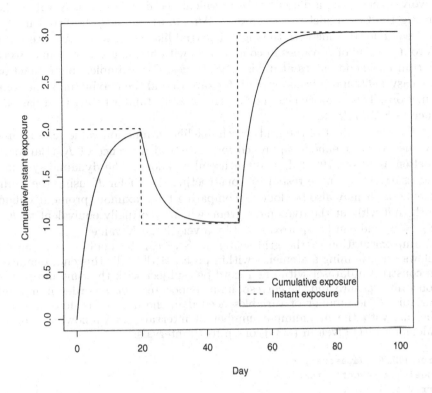

FIGURE 5.2
The relationship between the cumulative exposure and the piecewise constant instant exposure with an effect compartment model ($K = 0.2$ and $K_{in} = 0.2$).

$K = 0.2$ and $K_{in} = 0.2$, so that when reaching steady state, $c_i(t)$ has the same level as $e_i(t)$. Suppose that the TTE follows the Cox model (5.8). Taking $c_i^*(t)$ from model (6.33) or (6.34) into (5.5) we find that K_{in} cannot be distinguished from β, hence (5.8) becomes

$$
\begin{aligned}
h(t|\mathbf{e}_i) \;=\; & h_0(t) \exp[\beta(\sum_{k:t_{ik}<t} \frac{e_{ik-1}}{K}[1 - \exp(-K(t_{ik} - t_{ik-1}))] \\
& \times \exp(-K(t - t_{ik})) + e_{ik}[1 - \exp(-K(t - t_{ik})])])], \quad (5.25)
\end{aligned}
$$

where in the exposure accumulation part only K needs to be estimated.

Fitting model (5.25) directly is not easy with standard Cox regression software, since the log-relative risk is nonlinear with respect to K. However, K can be estimated with a grid search, in which we fix K at a number of different values so that we can calculate $c_i^*(t)$ for each of the values. This step involves generating a dataset with precalculated $c_i^*(t)$ for each K value, then fitting the Cox model to the dataset. Afterwards, among the fitted models, we search for the one with the highest partial likelihood value. An alternative is to use a grid of K values in combination with fitting a smooth curve to the partial likelihood values at different K values. For example, spline functions are easy to fit, and it is also straightforward to find the maximum of the spline functions. This approach can be implemented in standard statistical packages such as SAS, R/Splus.

As a by-product of the grid search, likelihood function values can be used by the profile likelihood approach for statistical inference of K (Barndorff-Nielsen and Cox, 1994). It is often useful to know if the dynamic system is necessary. This can be tested by constructing the CI for K using the profile likelihood. It may also be done by comparing the maximum profile likelihood with that without the dynamic system, which is formally equivalent to $K = K_{in} = \infty$ and can be approximated by a very large K value.

Implementation of the grid search in SAS can be made easier, since it allows programming statements within proc PHREG. To this end, one needs to construct a dataset with one record per subject with the time intervals of constant exposure and the corresponding exposure levels recorded in multiple variables. The number of variables needed in the dataset is determined by the one with the maxmimum number of intervals. The following program calculates $c^*(t)$ given in (6.34) for up to 10 intervals:

```
proc PHREG  data=example;
model Time*event(0)=cstar;
array cc{*} c1-c10;
array tt{*} t1-t10;
  prevc=0; *set start exposure 0;
  do i=1 to nrep;  *nrep is the N of intervals;
  if tt[i] <= Time < tt[i+1] then do;
  cstar= prevc*exp(-&K(Time-tt[i]))+cc[i](1-exp(-&K(Time-tt[i])));
  prevc= exp(-&K(tt[i+1]-tt[i]))+cc[i](1-exp(-&K(tt[i+1]-tt[i])));
end;
run;
```

where the K value is passed on from a macro variable &K. This approach may not be suitable for data with a very large number of intervals, particularly when only a few subjects have them.

5.3.2   Parametric models

The approach of including exposure dynamics in the hazard is generally not easy to apply to a model for a parametric baseline density distribution. To derive the density function $f(t)$ with exposure dynamics, one has to follow the route of calculating $H(t)$, then $S(t)$ and $f(t)$ from the baseline density. One may be able to calculate $H(t)$ analytically for certain types of models and distributions. In an additive hazard model (5.10) the baseline hazard and other covariates can be expressed as $h_i(t) = h_0(t) + \beta c_i(t) + \boldsymbol{\beta}_z^T \mathbf{Z}_i$, but we will omit the covariates for notational simplicity. In this case,

$$H_i(t) = H_0(t) + \beta \int_0^t c_i^*(\tau)d\tau. \tag{5.26}$$

In most cases we can assume that $c^*(t)$ is driven by a linear compartmental model with constant coefficients and can be written as

$$c_i^*(t) = \sum_{l=1}^{L} \theta_{il} \exp(-K_{il}t)), \tag{5.27}$$

where θ_{il} and K_{il} are parameters already estimated from the PK data. In this case, we can derive

$$
\begin{aligned}
H_i(t) &= H_0(t) + \beta \int_0^t \sum_{l=1}^{L} \theta_{il} \exp(-K_{il}\tau))d\tau \\
&= H_0(t) + \beta \sum_{l=1}^{L} \theta_{il}(1 - \exp(-K_{il}t))/K_{il}. \tag{5.28}
\end{aligned}
$$

Hence

$$S_i(t) = \exp(-H(t)) = \exp(-H_0(t) - \beta \sum_{l=1}^{L} \theta_{il}(1 - \exp(-K_{il}t))/K_{il}) \tag{5.29}$$

and $f(t)$ can also be calculated to form the likelihood function (5.4). Note that for the same $c^*(t)$ but with a multiplicative hazard, $H(t)$ cannot be calculated analytically, hence even for $c^*(t)$ as simple as this, a multiplicative model with a parametric baseline density cannot be fitted easily.

For a complex model, e.g., a model with a complex exposure dynamic or TTE distribution, a numerical integration can be used to calculate $H(t)$. Some software such as NONMEM and R has a built-in ODE solver and can be used to calculate $H(t)$ by specifying a dummy ODE, $dH(t)/dt = h(t)$, although this is not an efficient approach from a computational aspect.

Comparing the parametric and semiparametric (Cox model based) modeling approaches, it is clear that the Cox model approach is much simpler. However, this simplicity should not be the reason for using the approach without considering the alternative. In general, the parametric model has higher identifiability. For example, if the exposure effect has a common trend, e.g., if all patients have an exposure increase in a period together with a treatment effect increase, then the effect due to the common exposure increase is eliminated in a Cox model, but may still be identified if a constant baseline risk is assumed.

5.4   Multiple TTE and competing risks

Often multiple events such as adverse events of different types and death can happen to the same subject. To model the exposure–risk relationship for multiple events, a number of issues should be considered. These include the correlation between the risks of different events, termination of other events due to one event (e.g., death), and the change of exposure due to the occurrence of an event (e.g., AE leading to dose reduction). Andersen and Keiding (2002) give a short review of multistate models.

5.4.1   Multistate models

A common situation that involves multiple TTE is when we consider a model with multiple states and are interested in how the exposure affects the transition times between the states. For example, a three-state model consists of stable disease (SD) or a normal health state, disease progression (DP) and death states, as illustrated in Figure 5.3, in which t_{sp}, t_{sd} and t_{pd} are the times from stable disease to disease progression, to death without disease progression, and from disease progression to death, respectively. For each of them, one can specify a model in a similar way as for a single TTE. For example, the hazard functions of t_{sp}, t_{sd} and t_{pd} can be specified as

$$
\begin{aligned}
h_{spi}(t, \mathbf{X}_i) &= h_{sp0}(t) \exp(\mathbf{X}_i^T \boldsymbol{\beta}_{sp}) \\
h_{sdi}(t, \mathbf{X}_i) &= h_{sd0}(t) \exp(\mathbf{X}_i^T \boldsymbol{\beta}_{sd}) \\
h_{pdi}(t, \mathbf{X}_i) &= h_{pd0}(t) \exp(\mathbf{X}_i^T \boldsymbol{\beta}_{pd}),
\end{aligned}
\tag{5.30}
$$

respectively, where $\boldsymbol{\beta}_{sp}$, $\boldsymbol{\beta}_{sd}$ and $\boldsymbol{\beta}_{pd}$ are parameters, with possibly shared components, representing the impacts of $\mathbf{X}_i$, including the exposure effect. This model is based on an important assumption, that t_{sp}, t_{pd} and t_{sd} are independent, which sometimes may not hold. Under this assumption fitting such a joint model for three events with shared parameters is straightforward with common software. The first stage of fitting the model is to prepare the data so that each subject has records for all possible states along the path he reaches the terminal state or until the end of follow up. For example, if patient i had experienced a DP at day 10 then was dead at day 20, with exposure values ci = 1.2 before day 10 and ci = 2.5 afterwards, then the following records are needed:

Id	start	stop	path	event	ci
i	0	10	SP	1	1.2
i	0	10	SD	0	1.2
i	10	20	PD	1	2.5
i	10	20	PS	0	2.5

The last record is needed only if the model allows transitions from DP to SD.

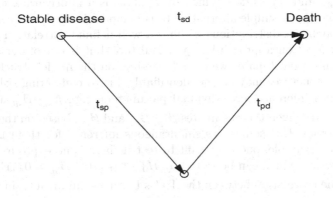

FIGURE 5.3
A three-state model for stable disease, disease progression and death.

The data contribute an event for models SP and PD, and two censored times for the other two models. Then separate models can be fitted for each path.

Under the assumption of independent t_{sp}, t_{sd} and t_{pd}, the multistate models are Markovian with the property of the transition time from one state to another not depending on the history of this subject. The Markovness may not hold due to dependence on the history. However, sometimes one can extend a model to make it a Markov model. For example, if the number of DPs a patient has experienced may influence the risk of death, the model in Figure 5.3 can be extended to include the states of 1st DP, 2nd DP, ..., to allow for different risks of death from these states. Another easy extension of the Markov model is the semi-Markov model in which the transition time from one state to another depends on the time to reach the current state. Fitting this model one only needs to add this time into $\mathbf{X}_i$.

One may also introduce correlation between the transition times by a joint survival function $S(t_{sp}, t_{sd})$ for t_{sp} and t_{sd} and specify a parametric multistate model. The model can include a correlation structure between the transition times. For example, assume $\mathbf{t}_i = (t_{spi}, t_{sdi}, t_{pdi})^T$ follows a log-multinormal distribution

$$\log(\mathbf{t}_i) \sim N(\boldsymbol{\mu}_i, \boldsymbol{\Sigma}) \tag{5.31}$$

where $\boldsymbol{\mu}_i = \mathbf{X}_i^T \boldsymbol{\beta}$ and $\boldsymbol{\beta} = (\boldsymbol{\beta}_{sp}, \boldsymbol{\beta}_{sd}, \boldsymbol{\beta}_{pd})^T$. In this model the correlation between t_{spi} and t_{sdi} is determined by $\boldsymbol{\Sigma}$. This is a multivariate example of the AFT model. A simple alternative to this model is the multivariate exponential model. For distributions without a well defined correlation structure, the copula function approach is a powerful tool, but is out of our scope due to space limitation. Finally, we note depending on the model structure, some correlation structures may not be identifiable due to competing risks.

Statistical inference for individual parameters, e.g., $\boldsymbol{\beta}_{sp}$, $\boldsymbol{\beta}_{sd}$ and $\boldsymbol{\beta}_{pd}$ is straightforward given their estimates $\hat{\boldsymbol{\beta}}_{sp}$, $\hat{\boldsymbol{\beta}}_{sd}$ and $\hat{\boldsymbol{\beta}}_{pd}$, based on their asymptotic properties. But sometimes simultaneous inference for them is also important. For example, one may want to test if there is no exposure effect for overall survival, which can be tested by $H_0 : \boldsymbol{\beta}_{sd} = 0 \cap \boldsymbol{\beta}_{sp} = 0 \cap \boldsymbol{\beta}_{pd} = 0$. In this case the correlation between the TTEs becomes important, since $\hat{\boldsymbol{\beta}}_{sp}, \hat{\boldsymbol{\beta}}_{sd}$ and $\hat{\boldsymbol{\beta}}_{pd}$ are correlated. One may take the correlation into account by explicitly modeling the correlation as above. Alternatively, an estimating equation procedure was developed in Wei, Lin and Weissfeld (1989) that fits the individual models to obtain the parameter estimates and a robust estimate of their variance-covariance matrix, without specific assumptions on the correlation structure between the transition times. Statistical inference, such as a test of the H_0, can be made based on the estimates and their variance-covariance matrix. This approach has been implemented in SAS (SAS, 2011).

The above has no direct connection to exposure models, apart from that $\mathbf{X}_i(t)$ may include exposure measures and can be fitted into any of the models. Indeed, the exposure dynamic and multiple events may go in parallel and can be modeled as such. For example, if $c_i^*(t)$ is independent of the states and a transition from SD to DP occurs at time t_a, then before t_a, $c_i^*(t)$ may appear in $h_{sp}(t, \mathbf{X}_i(t))$ and $h_{sd}(t, \mathbf{X}_i(t))$, and in $h_{pd}(t, \mathbf{X}_i(t))$ after t_a. It is also possible that a state transition may lead to exposure changes. For example, upon DP, some patients in the placebo arm may be switched to the active treatment, known as treatment switch-over. Therefore, $c_i^*(t)$ changes after t_a. This does not increase the technical difficulty of fitting the model if $c_i^*(t)$ can be calculated after t_a. However, if the switch-over depends on a patient's characteristics, confounding bias may be induced. But we leave this topic to Chapter 8.

5.4.2 Competing risks

As shown in the previous section, multiple events can occur to the same patient, and some event may terminate the patient from the study. Examples of such events are death due to different causes, or a severe AE that leads to the subject being withdrawn from a study. These types of events are known as competing events and have been studied intensively in survival analysis. In the three-state model (Figure 5.3), a patient may die with or without DP. The risks of death of the two types are competing risks, as only one can occur on one patient. Since there are multiple risks, we are interested in cause-specific risks. Ideally we would like to know the risk of one cause, with all the other causes removed. However, this is often not possible, due to the coexistence of the other risks. Letting T_k be the time to the event of type k, e.g., the time of death without DP, among all L types, $T = \min(T_1, ..., T_L)$ be the earliest event time, and K be the indicator of the occurred event type, the crude cause-specific risk of type k is defined as

$$h_k(t) = \lim_{\Delta \to 0} \frac{P(t < T < t + \Delta, K = k | T > t)}{\Delta}, \tag{5.32}$$

that is, the risk of having an event of type k, given no terminal event of any type happened before t. In contrast, the net risk of event k is defined as

$$h_k^n(t) = \lim_{\Delta \to 0} \frac{P(t < T < t + \Delta, K = k | T_k > t)}{\Delta}. \tag{5.33}$$

We can easily fit the model of the crude risk (5.32), since in reality the risk set at any time t is that of $T > t$. A Cox model can be applied to explore the relationship between covariates $\mathbf{X}_i$ and the risk of a specific event, e.g.,

$$h_{ki}(t) = h_{k0}(t) \exp(\mathbf{X}_i^T(t)\boldsymbol{\beta}_k). \tag{5.34}$$

Although one may also write (5.33) in the Cox model form, it is generally not possible to fit such a model, as the type specific risk set is not observed. Nevertheless, when the T_ks are independent, assuming no tie in $T_1, .., T_k$, the two risks are the same, since

$$
\begin{aligned}
h_k(t) &= f_T(t | K = k) / S_k(t) = f_k(t) \prod_{k' \neq k} S_{k'}(t) / \prod_{k=1}^{L} S_k(t) = f_k(t) / S_k(t) \\
&= h_k^n(t).
\end{aligned} \tag{5.35}
$$

To link competing risks to the multistate model, note that the transition times there follow the distributions defined by the net risks, and the competition is between events represented by the model, in the sense that if in (5.3) someone takes the route of SD-death without DP, then he would not be able to take the route SD-DP-death. The model may lead to independent competing risks between death and disease progression if t_{sd} is independent of t_{sp} and t_{pd}. The nonidentifiability of $h_k^h(t)$ is also reflected in fitting the multistate model, since for the example in Figure 5.3, it is not possible to fit the model with random effects in the transition risks. From here one can also see the importance of competing risks on the exposure effect, for example, if the exposure reduces the chance of DP, but also increases the risk of safety events such as death and severe AEs leading to treatment discontinuation. Then for those with higher exposure, the competing risk of safety events makes those with prolonged time to DP censored. Hence adjustment is needed for the estimation of treatment effects on the risk of DP. This is a topic for Chapter 8.

An important special case of competing risks is semicompeting risks, referring to the situation that one event time (e.g., death) is always observed but it may censor another event (e.g., disease progression). In this case, it is possible to further explore the relationship between the two events. For example, one may calculate the correlation between the two TTEs. For example, the risk of AE and death are semicompeting risks, as AE cannot censor death. In this case, one can measure the correlation between them by, e.g., Kendall's τ. Alternatively, the dependence between them may be represented by the conditional hazard of death, given the occurrence of AE.

5.5   Models for recurrent events

Some events, such as asthma attacks and epilepsy seizures, may occur multiple times. Modeling the relationship between the exposure and the risk of occurrence or recurrence provides important information for, e.g., dose selection and adjustment strategies.

5.5.1   Recurrent events as a counting process

We have introduced the counting process $N(t)$ to model the time-to-event. It is straightforward to extend it to let $N(t)$ be the number of occurred recurrent events to time t. The simplest yet commonly used is the homogeneous Poisson process with two key properties: 1) the number of events that occurred within a time interval follows a Poisson distribution and its mean is proportional to the length of the interval, i.e., $N(t) - N(s) \sim Poisson(\lambda(t-s)), t > s$, and 2) the gap time between two events follows exponential distribution. To model the impact of exposure on $N(t)$, one can follow the Cox regression for time to a single event and assume that the rate function, which is the probability of an event occurring to subject i at time t, follows

$$E(dN_i(t)) \equiv h_i(t) = h_0(t) \exp(\mathbf{X}_i^T \beta) \tag{5.36}$$

where $h_0(t)$ is the baseline rate function and the second term represents exposure and covariate effects.

5.5.2   Recurrent events in multistate models

A simple multistate model for recurrent events contains states of having had 1, 2, 3,... events. The gap time between events can be modeled as the transition time between the states of having k events to having $k + 1$ events. Let T_{ik} be the time to the kth event after having the $k - 1$th, one may assume the hazard of the kth event as

$$h_k(t) = h_{0k}(t) \exp(\mathbf{X}_i^T \boldsymbol{\beta}_k) \qquad (5.37)$$

where $h_{0k}(t)$ and $\boldsymbol{\beta}_k$ are state specific, but some components may be shared. When they are all shared, this model describes a renewal process with completely the same risk of having next event. This assumption is valid only for very special cases, such as the time to the next failure of a replaced machine part. In biostatistics models, the time to the next event often depends on the number of events that have happened. For example, gastric bleeding, a typical AE of some types of drug, may increase the risk of the recurrence of bleeding. The risk of recurrent myocardial infraction follows the same pattern. To avoid using many parameters while allowing $h_k(t)$ to change with k, one may assume

$$h_k(t) = h_0(t) \exp(\mathbf{X}_i^T \boldsymbol{\beta} + \beta_s \#(k) + \beta_c c(t)) \qquad (5.38)$$

where $\#(k)$ is a function of event number k. For example, if the log-baseline risk of recurrent events linearly increases with the number of events that have already happened, then using $\#(k) = k - 1$, the log-risk increases proportional to the number of previous events. Alternatively, one can also let the correlation between the transition times remain unspecified and use the approach of Wei, Lin and Weissfeld (1989) to obtain a robust variance-covariance matrix for the estimates of the β_ks.

Sometimes an event marks a state change. For example, an infection may last several days, so having an infection should be considered as a state. Therefore, the multistate model with state 0, 1, 2, 3 events may be extended to $0, 1, 1-, 2, 2-, ...$, where $k-$ denotes the state of no longer on the kth event state. This model needs another set of transition times from event to event free states, in addition to those from event free to event states. Therefore, one may need to specify several TTE models for these transition times. Under some assumptions, it may be simplified to a two-state (event on and off) model, possibly with a counter for number of events experienced, if the risk of on and off events has the form of (5.38).

5.5.3   Rate and marginal models

The multistate model approach requires specification of model structure and distribution of transition times. An alternative to modeling time to events is to model the average number of events $N(t)$ until a given time t as a function of time directly. Here we can consider $N(t)$ as a counting process with mean $m(t) = E(N(t))$, known as the mean function of $N(t)$. This approach has advantages in dealing with recurrent events. One may model $m(t)$ with a nonparametric, semiparametric or parametric model. To avoid the complex theory of stochastic process of continuous time, we consider discrete time $t = 1, 2, , 3, ...$, only and, without loss of generality, we use "day" as a shorthand for the time unit. Letting $l_i(t)$ be the indicator for subject i having an event at day t, we have

$$m(t) = E(N(t)), \quad l(t) = E(l_i(t)) = m(t) - m(t - 1). \tag{5.39}$$

Here $l(t)$ is often called the rate function, since when the time is continuous, $l(t)$ is the derivative of $m(t)$. To model recurrent events we can model either $m(t)$ or $l(t)$. We prefer modeling the latter as it is closer to the modeling of the hazard, although in this case, we model the marginal mean of occurrence, that is, we do not condition on the history. For a marginal semiparametric model for $l(t)$, we may use:

$$l(t|\mathbf{X}_i(t)) = E(l_i(t)|\mathbf{X}_i(t)) = l_0(t) \exp(\mathbf{X}_i^T(t)\boldsymbol{\beta}), \tag{5.40}$$

where $l_0(t)$ is a nonparametric marginal baseline function and $\exp(\mathbf{X}_i^T(t)\boldsymbol{\beta})$ is the parametric part with parameters $\boldsymbol{\beta}$. This model is very similar to the Cox model as they share the same type of model structure.

Model (5.40) can be fitted by an estimating equation (EE) approach . Letting C_i be the time of censoring for patient i and denoting $\delta_i(t) = I(C_i > t)$, the estimating equations for $l_0(t)$ and $\boldsymbol{\beta}$ can be written as

$$U_l(l_0(t), \boldsymbol{\beta},) = \sum_{i=1}^{n} \delta_i(t)(l_i(t) - l_0(t) \exp(\mathbf{X}_i^T(t)\boldsymbol{\beta})) = 0,$$

$$U_{\boldsymbol{\beta}}(l_0(t), \boldsymbol{\beta}) = \sum_{i=1}^{n} \sum_{t=1}^{C_i} \mathbf{X}_i(t)(l_i(t) - l_0(t) \exp(\mathbf{X}_i^T(t)\boldsymbol{\beta})) = 0.$$

$$\tag{5.41}$$

With discrete t and the nonparametric setting for $l_0(t)$, one can solve $l_0(t)$ from the first equation and substitute it into the second one and obtain:

$$U_{\boldsymbol{\beta}}(\boldsymbol{\beta}) = \sum_{i=1}^{n} \sum_{t=1}^{C_i} \mathbf{X}_i(t)[l_i(t) - \frac{\sum_{i=1}^{n} \delta_i(t) l_i(t)}{\sum_{i=1}^{n} \delta_i(t) \exp(\mathbf{X}_i^T(t)\boldsymbol{\beta})} \exp(\mathbf{X}_i^T(t)\boldsymbol{\beta})] = 0.$$

$$\tag{5.42}$$

This EE is also similar to that of the Cox model based on the partial likelihood. However, a key difference is that the baseline hazard cannot be completely eliminated, and has an impact on the estimation of $\boldsymbol{\beta}$.

Since $l_i(t)$ and $l_i(t')$ are not independent, for inference on $\boldsymbol{\beta}$, a robust estimate for $\text{var}(\hat{\boldsymbol{\beta}})$ is needed. The covariance matrix of $\hat{\boldsymbol{\beta}}$ depends on the joint distribution for $l_i(t)$s, which we do not specify in the marginal model. However, a robust variance estimate for $\hat{\boldsymbol{\beta}}$ is ready to use (Lawless and Nadeau,

1995). A more general robust covariance estimate was proposed by Wei et al. (1989) and has been implemented in SAS PROC PHREG (SAS, 2011).

This approach can also be used for modeling cumulative duration of events. The duration is equivalent to state occupancy in an on-off event model, and is generally difficult to estimate without a simulation. But the marginal model provides a straightforward way to model the relationship between the exposure and risk of having an event. To this end, we define $l_i(t)$ as the indicator that subject i is within an event episode. In this case $l(t) = E(l_i(t))$ is the mean event prevalence and $m(t) = \sum_{j=1}^{t} l(j)$ is the mean cumulative duration of events until time t. For fitting a model for $l(t)$, whether it is the mean incidence or prevalence function, the following pseudo event approach can generate data based on time to event and duration of event records and then use standard software to fit models such as (5.40) (Wang and Qartey, 2012c, 2013). For example, for an event episode starting at day j and ending at $j + s$, we generate $s + 1$ "pseudo" event records with $t = j + j', l_i(t) = 1, j' = 0, ..., s$. Here we follow the convention that the days that the event episode starts and ends are all counted as a whole day. The data then can be fed into SAS PROC PHREG using the counting process model formula. Note that this approach also allows exposure or covariate changes during an event, since $\mathbf{X}_i(t)$ values can be different among pseudo records during one event. Therefore, this approach also allows a dynamic exposure model to be included. One obvious drawback of this approach is the lack of computing efficiency due to the size of the pseudo dataset. However, it only leads to practical difficulty for very large and lengthy studies with long episode duration of events. In these cases, t may be discretized at a wider grid to reduce the number of records in the generated dataset.

After obtaining $\hat{\beta}$, one can replace it for β in (5.40) to obtain the Nelson estimate for $l(t|\mathbf{X}(t))$ for a given $\mathbf{X}(t)$:

$$\hat{l}(t|\mathbf{X}(t)) = \frac{\sum_{i=1}^{n} \delta_i(t) l_i(t)}{\sum_{i=1}^{n} \delta_i(t) \exp(\mathbf{X}_i^T(t)\hat{\beta})} \exp(\mathbf{X}_i^T(t)\hat{\beta}), \tag{5.43}$$

and an estimate for $L(t|\mathbf{X}(t))$ can be calculated based on $\hat{l}(t|\mathbf{X}(t))$. The estimate for $L(t|\mathbf{X}(t))$ and its pointwise confidence band based on a sandwich variance estimate for (5.43) have also been implemented in SAS PROC PHREG.

The following is an example using a dataset generated according to the scenario of therapeutic dose monitoring, as discussed in Chapter 1. The dataset is generated by the following SAS program, where the times AE occurred and resolved are generated alternately according to the drug exposure and subject random effects.

```
%let nsimu=50; %let cdose=log(1.3); %let bound=log(10);
%let base=-8; %let base2=-3; %let dose1=log(5); %let sigv=0.5;
%let sigu=0.3; %let sige=0.2; %let nsub=50; %let nrep=8;
%let beta=0.25; %let nday=14;
data simu;
   do sub=1 to &nsub;
```

```
   ui=rannor(12)*&sigu;
   vi=rannor(1)*&sigv;
   dose=&dose1;
     pevent=0;
   do i=0 to &nrep;
   lci=dose+ui;
   lce=lci+rannor(12)*&sige;
   conc=exp(lci);
   low=i*&nday; up=low+&nday;
     start=low;
     stop=low;
     do k=1 to 30 while (stop < up);
      if pevent in (0 2) then  lmd=exp(&base+&beta*exp(lci)+vi);
      else if pevent =1 then  lmd=exp(&base2-vi);
      ti=ranexp(12)/lmd;
      stop=start+ti;
      if stop le up then do;
         if pevent in (0 2) then event=1;
         else if pevent=1 then event=2;
         output;
         start=stop; pevent=event;
      end;
      else do;
         event=0; stop=up; output;
      end;
     end;
    if lce<&bound then dose=dose+&cdose;
    end;
  end; run;
```

In the data off- and on-event periods are indicated by pevent=0 or 2 and pevent=1, respectively. event=0 means censored. The program assumes that the exposure only affects the time to AE, but not the time AE is resolved (off-event). For both times of on- and off-event, a shared random effect, vi, is added to generate correlation between the recurrent event times, assuming that patients with a higher v_i would have a higher risk of AE and would also take longer to have the AE resolved. The following programs fit the rate models with the option of robust variance estimates.

```
proc phreg data=simu(where=(pevent in (0 2))) COVS(AGGREGATE);
  model (start, stop)*event(0)=conc;
  id sub; run;

proc phreg data=simu(where=(pevent in (1))) COVS(AGGREGATE);
  model (start, stop)*event(0)=conc;
  id sub; run;
```

The statistics and parameter estimates of the fitted model are

```
For time to event:
            Testing Global Null Hypothesis: BETA=0
```

Test	Chi-Square	DF	Pr > ChiSq
Likelihood Ratio	8.8217	1	0.0030
Score (Model-Based)	8.8725	1	0.0029
Score (Sandwich)	5.6290	1	0.0177
Wald (Model-Based)	8.7452	1	0.0031
Wald (Sandwich)	7.5404	1	0.0060

Analysis of Maximum Likelihood Estimates

Parameter	Parameter Estimate	Standard Error	StdErr Ratio	Chi-Square	Pr > ChiSq
conc	0.24278	0.08841	1.077	7.5404	0.0060

Test	Chi-Square	DF	Pr > ChiSq
Likelihood Ratio	0.1275	1	0.7210
Score (Model-Based)	0.1270	1	0.7215
Score (Sandwich)	0.1209	1	0.7281
Wald (Model-Based)	0.1269	1	0.7216
Wald (Sandwich)	0.1202	1	0.7288

and for time off event:

Analysis of Maximum Likelihood Estimates

Parameter	Parameter Estimate	Standard Error	StdErr Ratio	Chi-Square	Pr > ChiSq
conc	0.03560	0.10267	1.028	0.1202	0.7288

In the results the exposure effect estimates for time to event and time off event are similar to the true values 0.25 and 0 for to event and off event, respectively.

5.6 Frailty: Random effects in TTE models

To introduce the concept of frailty, we consider the Cox model

$$h_i(t) = h_0(t) \exp(\mathbf{X}_i^T \boldsymbol{\beta}) \tag{5.44}$$

in which covariates $\mathbf{X}_i$ represent all factors relating to the risk of subject i. Often there are unobserved factors that make subject i different from the others. As we did for mixed models introduced in the previous chapters, we can add a latent subject effect known as frailty in TTE models, in the hazard function, so that conditional on u_i:

$$h_i(t|u_i) = h_0(t)u_i \exp(\mathbf{X}_i^T \boldsymbol{\beta}). \tag{5.45}$$

A common choice for the distribution of u_i is gamma parameterized so that $u_i \sim gamma(\theta)$ with $E(u_i) = 1$ and $\text{var}(u_i) = \theta$ and frailty increases with the increase of θ. Since u_i is unobserved, one needs to derive the marginal model in order to use the ML approach. To this end one can derive conditional distribution from the hazard function, (5.45) then the marginal distribution can be derived by integrating the conditional distribution over the gamma distribution for u_i. The resulting marginal likelihood is complex (Klein, 1992), hence is omitted here. Unlike the partial likelihood, the marginal likelihood contains $h_{i0}(t)$ and $H_{0i}(t) = \int h_{0i}(\tau)d\tau$, which have to be specified in order to fit the model. For most other distributions of u_i, the marginal model does not have an analytical form.

Even if the marginal model is simple, frailty models can only be fitted reliably to recurrent event data to identify the frailty, just as a mixed effect model can only be fitted to repeated measurement data. To see the reason, suppose that there is no covariate and $h_0(t) = hu_i$, i.e., the baseline risk is constant but contains frailty u_i. However, those with higher risk (large u_i) will have an event in a short period, and those who did not have an event in the short period are likely to be the low risk ones. This leads to, on average, a higher risk at the beginning and a trend of reduction later on. Consequently, one cannot distinguish the case of time-varying baseline risk and a constant risk with frailty. This topic will be further discussed in Section 5.9.

Suppose that the risk of jth event on subject i is

$$h_{ij}(t) = h_{0j}(t)u_i \exp(\mathbf{X}_{ij}^T \boldsymbol{\beta}) \tag{5.46}$$

where $h_{0j}(t)$ is the baseline risk for the jth event and $\mathbf{X}_{ij}$ may include covariates for the jth event specifically, e.g., the previous event history. This model can be fitted with, e.g., R function survreg with the following codes:

```
rfit2a <- survreg(Surv(time, status) ~ conc + frailty.gaussian(id),
         dist="weibull", data=riskexp )
```

where frailty.gaussian(id) specifies a normally distributed frailty on a Weibull parametric model and id is subject identity. The same model can also be fitted with SAS proc NLMIXED and NONMEM using the built-in numerical integration function to calculate the marginal likelihood numerically. The following SAS program (adapted from SAS, (2011)) fits a model with a Weibull distribution conditional on u_i and $u_i \sim N(0, \sigma_u^2)$:

```
proc nlmixed data=mydata qpoints=8;
parms gamma=1 sigu=0.1;
bounds gamma sigu> 0;
linp = a + b ci+ui;
```

```
alpha = exp(-linp);
G_t = exp(-(alpha*time)**gamma);
g = gamma*alpha*((alpha*time)**(gamma-1))*G_t;
ll = (status=1)*log(g) + (status=0)*log(G_t);
model time ~ general(ll);
random ui ~ normal(0,sigu) subject=patient; run
```

where we have used the random statement to specify the frailty in the same way as to specify the random effects in an NLMM.

The ML approach can also be used for Cox models by replacing the conditional likelihood given u_i with a conditional partial likelihood function (5.6). One particular model is derived assuming $s_i = \log(u_i) \sim N(0, \sigma_s^2)$, and the joint log-likelihood for T_{ij} and s_i can be written as

$$l(\boldsymbol{\beta}, \mathbf{u}) = l_p(\boldsymbol{\beta}|\mathbf{s}) - \frac{\sum_{i=1}^n s_i^2}{2\sigma_s^2} - \frac{n}{2} \log(\sigma_s^2) + C. \qquad (5.47)$$

where $l_p(\boldsymbol{\beta}|\mathbf{s})$ is the partial likelihood function conditional on frailty $\mathbf{s}$. This is also called penalized partial likelihood since the second term acts to penalize large s_is. To fit the model, we still need to integrate out the s_is from (5.47). This can be done by a Laplace approximation (Therneau and Grambsch, 2000) and the algorithm has been implemented in software such as R library survival and SAS proc PHREG. The approximation can be used for other distributions for u_i, with an appropriate penalty term in (5.47). For R the following call fits such a model with log-normal frailty:

```
coxph(Surv(time,status) ~ exposure+ frailty(id, dist='gauss'),
                                                 data=mydata)
```

where in SAS proc PHREG, the statement RANDOM follows the same syntax as other procedures.

```
proc PHREG data=mydata;
model time*status(0)= exposure;
RANDOM id;
```

There are more choices for u_i in the R library but the log-normal distribution is the only one available in SAS. The following SAS codes fit the simulated data to frailty models for times "on" and "off" events.

```
proc phreg data=simu(where=(pevent in (0 2)));
  class sub;
  model (start, stop)*event(0)=conc;
  random sub/solution;
run;
proc phreg data=simu(where=(pevent in (1)));
  class sub;
  model (start, stop)*event(0)=conc;
  random sub/solution;
run;
```

Time to event

Cov Parm	REML Estimate	Standard Error		
sub	0.1769	0.2728		

Parameter	Parameter Estimate	Standard Error	Chi-Square	Pr > ChiSq
conc	0.24402	0.08549	8.1470	0.0043

Time off event

Cov Parm	REML Estimate	Standard Error
sub	0.2193	0.3324

Analysis of Maximum Likelihood Estimates

Parameter	Parameter Estimate	Standard Error	Chi-Square	Pr > ChiSq
conc	0.00568	0.11350	0.0025	0.9601

The β estimates are similar to those in the rate model. In addition, the estimated variances of the frailty are also given together with their SEs, although they are far from statistically significant.

The option "Solution" in the RANDOM statement requires estimation of the random effects. As vi is a shared frailty, it might be tempting to test the correlation between the predictions from the two models. However, only 30 patients had an estimate for vi, due to the lack of repeated events among other patients. Among the 30 patients, the correlation between the vis from the two models was 0.11 and the p-value for zero correlation (which indicates no sharing frailty) was 0.576. The reason behind this is the difficulty in the estimation of the random effects. It is only when the number of repeated events is large that the individual estimates for vi can approximate the true value. In common practical scenarios, this condition is rarely met.

In the general multistate models discussed in an earlier section, we have assumed that the risks of transition between the states are independent. This is often not true, since, e.g., the risks of disease related and non-disease related death are often correlated due to, e.g., age and general health conditions. It might be tempting to include frailty variables to reflect the correlation. However, most of the frailties, e.g., those which reflect the correlation between disease related and non-disease related deaths on the same subject, cannot be identified, since the two events cannot happen on the same subject. Frailties in sequential states, e.g., those in the risks of disease progression and death after progression, might be identified, but only under some additional conditions, although their correlation can always be estimated.

The frailty model can also be used to model multiple events of different types. Suppose there are K different types of events; the hazard function of a type k event may be specified as

$$h_{ik}(t) = h_{0k}(t)u_i \exp(\mathbf{X}_i^T \boldsymbol{\beta}_k) \qquad (5.48)$$

where $h_{0k}(t)$ and $\boldsymbol{\beta}_k$ are the baseline risk and parameters for the type k events and u_i is a shared frailty representing correlation between different events on the same subject i. To fit the model, one needs an indicator of event type then includes the event type-factor interaction terms in the model to estimate $\boldsymbol{\beta}_k$. However, the model may not be identifiable if the events censor each other, with the case of semicompeting risk as an exception.

5.7 Joint modeling of exposure and time to event

This section discusses similar topics as in Chapter 4, but here the response is TTE. In earlier sections we included exposure dynamic in the models as a time-varying covariate and estimated the parameters together with the parameters of the TTE model. However, often exposure measures are observed repeatedly at a set of time points $t_1, ..., t_J$, and using the exposure data can identify a more complex exposure dynamic and obtain better estimates of model parameters. For the exposure model we assume

$$\begin{aligned} c_i^*(t) &= m(d_i(t), \boldsymbol{\theta}_i) \\ c_{ij} &= c_i^*(t_j) + e_{ij}, \end{aligned} \qquad (5.49)$$

where $\boldsymbol{\theta}_i = \mathbf{X}_i\boldsymbol{\theta} + \mathbf{v}_i, \mathbf{v}_i \sim N(0, \Sigma_v)$, and $e_{ij} \sim N(0, \sigma_e^2)$ is the measurement error (ME). This model is the same as the second one in (4.1), except the notation $h(.)$ is changed to $m(.)$ to distinguish it from the risk function. This model, together with a TTE model,

$$h_i(t) = h_0(t) \exp(\beta c_i^*(t)), \qquad (5.50)$$

forms the classical ME model, that is, $c_i^*(t)$, which affects the risk directly and is measured with error. Therefore, this section follows the structure of Chapter 4 and discusses simultaneous modeling using the ML approach and sequential modeling (SM) using regression calibration approaches. Using exposure data in the modeling also allows the evaluation of the correlation between factors that affect the exposure and those affecting the risk of events. But we postpone this topic until Chapter 8, as it mainly deals with approaches dealing with confounding biases. Joint modeling of some special TTE models (e.g, those for recurrent events and interval censored data) is also left to these individual sections. For more details and a wide range of practical background of joint modeling TTE and longitudinal data, see Rizopoulos (2012).

5.7.1   Simultaneous modeling

Most simultaneous modeling (SM) approaches maximize a joint likelihood of
the TTE and the exposure data. The TTE part can use the likelihood or
the partial likelihood. But most SM approaches use a parametric TTE model
with the exception of Rizopoulos et al. (2009), where longitudinal models and
Cox models are fitted jointly. A major reason is that the partial likelihood
is difficult to specify in software such as proc NLMIXED and NONMEM.
Formally, we can write the joint likelihood function in the same form as (4.4):

$$L(\boldsymbol{\theta}, \boldsymbol{\beta}) = \prod_{i=1}^{n} \int_{\theta_i} \int_{\beta_i} L_t(\mathbf{t}_i | \boldsymbol{\beta}_i, \mathbf{c}^*(\mathbf{t}_i)) f_c(\mathbf{c}_i | \boldsymbol{\beta}_i, \sigma_e^2) dF(\boldsymbol{\theta}_i, \boldsymbol{\beta}_i | \boldsymbol{\theta}, \boldsymbol{\beta}, \boldsymbol{\Omega}) \quad (5.51)$$

where we use some general notation so that it can be used for multiple or recurrent events modeling. With this notation $\mathbf{t}_i$ is a set of times to events,
which may be times to multiple or recurrent events, and $L_t(\mathbf{t}_i | \boldsymbol{\beta}_i, \mathbf{c}^*(\mathbf{t}_i))$
is the likelihood function of the TTE part, in which $\mathbf{c}^*(\mathbf{t}_i)$, with density
function $f_c(\mathbf{c}_i | \boldsymbol{\beta}_i, \sigma_e^2)$, represents $c^*(t)$ at $\mathbf{t}_i$. For example, $L_t(t_i | c^*(t_i)) = f^{\delta_i}(t_i, c^*(t_i)) S^{1-\delta_i}(t_i, c^*(t_i))$, with all parameters omitted. Also we use $\boldsymbol{\beta}_i$ to
denote both fixed and random coefficients which we will introduce later, but
can be considered as a fixed coefficient β at the moment.

The implementation of the joint modeling approach now relies on maxi-
mizing the joint likelihood. To deal with the random effects in $\boldsymbol{\theta}_i$ and $\boldsymbol{\beta}_i$, we
can take the same approach as in Chapter 4 by using software such as SAS
proc NLMIXED and NONMEM, as both allow specifying the conditional like-
lihood, conditional on $\boldsymbol{\theta}_i$ and $\boldsymbol{\beta}_i$, then using an approximation or numerical
integration to evaluate (5.51). Here an additional issue is the calculation of
$H_i(t)$. A few situations under which $H_i(t)$ can be calculated analytically have
been given in earlier sections. They are 1) constant baseline risk in combination with a wide range of $c^*(t)$ functions, 2) a wide range of baseline risk
functions (e.g., for gamma and Weibull distributions) in combination with
piecewise constants $c_i^*(t)$, and 3) a wide range of $c^*(t)$ and baseline functions
but with an additive risk model.

For joint modeling with Cox models, Rizopoulos et al. (2009) used the
Laplace expansion of the joint likelihood and the expectation-maximization
(EM) algorithm. Based on this algorithm, an R library JM was developed, with
its details and applications in Rizopoulos (2012). Although in principle it can
also be implemented in a general purpose software, e.g., proc NLMIXED, the
major problem is the lack of an efficient way to program the partial likelihood, since at one failure time, data from all subjects who are still at risk
have to be available. To allow for some flexibility in the baseline risk distribution, the piecewise constant baseline risk has been widely used, which can
be programmed in, e.g., proc NLMIXED. Although one can solve the corresponding estimating equation to eliminate parameters for the baseline risk,
the programming becomes difficult due to the same issue as when using the
Cox regression. Therefore, one has to estimate these parameters explicitly.

When the exposure is driven by a complex popPK model, it is more convenient to calculate $H(t)$ numerically. NONMEM has the facility to model the popPK and TTE model jointly. The following is a sample program for a simple popPK model (adapted from Holford's course notes for advanced pharmacometrics at

http://holford.fmhs.auckland.ac.nz/teaching/pharmacometrics/advanced),

but it can be extended to any other popPK models specified in terms of ODE (linear or nonlinear). The idea is to calculate $c_i^*(t)$ and $H_i(t)$ by solving the following ODEs with initial conditions $H_i(0) = 0$ and $c_i^*(0) = 0$,

$$\frac{dc_i^*(t)}{dt} = -Cl_i/V_i c_i^*(t)$$
$$\frac{dH_i(t)}{dt} = \exp(\beta_0 + \beta c_i^*(t)) \tag{5.52}$$

where Cl_i and V_i are random coefficients in the popPK model with variance components estimated in NONMEM as usual.

```
$DES
DCP=A(1)/V
DADT(1)=-CL*DCP
DADT(2)=EXP(BETA0+BETA*DCP)
$ERROR
CP=A(1)/V
CUMHAZ=A(2) ; cumulative hazard
IF (DV.EQ.0) THEN ; censored
Y=EXP(-CUMHAZ)
ENDIF
IF (DV.EQ.1) THEN ; event
HAZ=EXP(BETA0+BETA*CP)
Y=EXP(-CUMHAZ)*HAZ
ENDIF
```

Here DADT(k) is the kth ODE for A(k) (A(1) $= c_i^*(t)$ and A(2) $= H_i(t)$), HAZ and CUMHAZ are $h_i(t)$ and $H_i(t)$, respectively, and DV is an event indicator. The $ERROR part specifies the likelihood function (5.51). One can also find a very interesting link between the dynamics of $H(t)$ and $c^*(t)$ in the above lecture notes. It can also be programmed in R, with an ODE solver.

As an example for simultaneously fitting the models without calculation of $H(t)$ by numerical integration, consider the situation of exposure change related to dose change and we assume that $c_i^*(t)$ changes instantly with dose change. We assume model

$$c_i^*(t) = \theta_0 + \theta d_i(t) + v_i \tag{5.53}$$

where $c_{ij} = c_i^*(t_j) + e_{ij}, j = 1, ..., J$ are observed, and $d_i(t)$ is piecewise constant within intervals $t_1, ..., t_j, ...t_J$. For the TTE model we assume

$$h_i(t) = h_0 \exp(\beta c_i^*(t) + u_i) \tag{5.54}$$

where we add a random effect u_i, and assume that u_i, v_i have a joint normal distribution. Conditional on u_i, this leads to an exponential distribution with piecewise constant risk, hence $H_i(t)$ can be calculated analytically. The constant baseline risk is not an essential restriction and can be replaced with a piecewise function. The following SAS program generates the paths of exposure and TTE for 200 patients in a TDM trial and fits the exposure and TTE data simultaneously. indexSAS

```
%let nsimu=3000; %let cdose=log(1.3); %let bound=log(10);
%let base=-6; %let base2=-3; %let dose1=log(5); %let sigv=0.5;
%let sigu=0.3; %let sige=0.2; %let nsub=200; %let nrep=8;
%let beta=0.25;  %let nday=14;
data simu;
  do sub=1 to &nsub;
    fevent=0; retain fevent;
    latent=rannor(12)*0;
    ui=(latent+rannor(12))*&sigu/2;
    vi=(latent+rannor(12))*&sigv/2;
    dose=&dose1;
      pevent=0;
    do i=0 to &nrep;
    lci=dose+ui;
    lce=lci+rannor(12)*&sige;
    conc=exp(lci);
    low=i*&nday; up=low+&nday;
      start=low;
      stop=low;
      do k=1 to 30 while (stop < up);
       if pevent in (0 2) then  lmd=exp(&base+&beta*exp(lci)+vi);
       else if pevent =1 then  lmd=exp(&base2-vi);
       ti=ranexp(12)/lmd;
       stop=start+ti;
       if stop le up then do;
          if pevent in (0 2) then event=1;
          else if pevent=1 then event=2;
          output; fevent=event;
          start=stop; pevent=event;
       end;
       else do;
          event=0; stop=up; output;
       end;
      end;
     if lce<&bound then dose=dose+&cdose;
     end;
  end; run;

data first; set simu; if fevent=0; run;

proc nlmixed data=first(keep=lci lce dose start stop event sub)
                            qpoints=4 ebopt;
```

```
      parm b0=-5 beta=0.25 a=0 b=1 sige=0.02 a11=0.3 a12=0 a22=0.3;
      bounds sige a11 a22 >0;
            lci=a+b*dose+vi;   resc=lce-lci;
        if (abs(resc) > 1E50) or (sige < 1e-20) then llw = 1e-20;
        else  llw = -0.5*(resc**2 / sige  + log(sige)); ll=1e-20;
      hi=exp(b0+beta*exp(lci)+ui);
      Ht=hi*(stop-start);
      ll=event*log(hi+1e-20)-Ht;
  *  put sub start stop event lce ll llw;
      model stop ~general(ll+llw);
      random vi ui ~normal([0,0],[a11, a12, a22]) subject=sub; run;
```

Parameter	Estimate	Standard Error	t Value	Pr > |t|
b0	-5.9949	0.3881	-15.45	<.0001
beta	0.2603	0.04650	5.60	<.0001
a	-0.04873	0.05010	-0.97	0.3319
b	1.0189	0.02479	41.10	<.0001
sige	0.04166	0.002475	16.83	<.0001
a11	0.02296	0.003660	6.27	<.0001
a12	0.01541	0.01506	1.02	0.3075
a22	0.08719	0.1816	0.48	0.6317

The results shows that β is estimated rather well, with the estimated value 0.26 comparing with the true value 0.25. The parameters a and b for the PK part are well estimated, as expected. The variance of u_i is estimated much worse than that of v_i, due to the difficulty identifying the frailty. Note that the likelihood for TTE can be written in the form of $L_t(\mathbf{t}_i|\boldsymbol{\beta}_i, \mathbf{c}^*(\mathbf{t}_i))$, as u_i can be considered as a part of $\boldsymbol{\beta}_i$, as it is the random baseline risk. The program takes advantage of built-in multi-normal random effects for vi and ui in proc NLMIXED.

5.7.2 Sequential modeling and regression calibration

To estimate β, a very intuitive approach is the two-stage method that fits model (5.49) at the first stage. In the second stage the fitted model is used to predict $c_i^*(t)$ as $\hat{c}_i^*(t) = m(t, d_i(t), \hat{\boldsymbol{\theta}}_i)$, which is used as a known time-varying covariate to fit the exposure–risk model. The second stage model can be fitted using the MLE or a semiparametric approach, depending on the nature of the exposure–risk model. An important advantage of this approach is its simplicity to applied statisticians using common software. Formally, we may write

$$\hat{c}_i^*(t) = c_i^*(t) + e_i(t) \tag{5.55}$$

where $e_i(t) = \hat{c}_i^*(t) - c_i^*(t)$. The simplest SM approach replaces $c_i^*(t)$ in the TTE model with $\hat{c}_i^*(t)$ and ignores $e_i(t)$. However, the major issue is that ignoring $e_i(t)$ makes the SEs of the estimates in the second stage often too optimistic, and more importantly may lead to inconsistent estimates for β. The problem is similar to that of regression calibration (RC) in Chapter 4, but for TTE models these issues are often more complex.

To see where complexity lays and how to avoid it, we follow the second stage modeling strategy in Chapter 4. For example, for $\hat{c}_i^*(t) = 1 - \exp(-\hat{K}_i t)$ and constant baseline risk,

$$H_i(t) = \int_0^t \lambda_0 \exp(-\beta(1 - \exp(-\hat{K}_i \tau)))d\tau \qquad (5.56)$$

needs numerical calculation. Again this can be implemented in NONMEM, with individual predicted $\hat{K}_i$s based on the first stage model fitting. This approach can also be used when the popPK model is given by ODEs without an analytical solution. One can provide population parameters in the popPK model, then resolve the ODEs of the popPK model together with the ODE for $H(t)$ to provide prediction for θ_i and $c_i^*(t)$. The population parameters in the popPK model, however, are not estimated again. Therefore, this is still a two stage approach and has the pros and cons of such approaches, namely, the stability in the estimation but it cannot take the variability of first stage estimation into account.

One may take the first stage variation into account by using a computing intensive approach such as bootstrapping to generate bootstrapped samples for $c_i^*(t)$ and using them in the second stage to fit the Cox model to obtain bootstrapped estimates for β. In this way the variation in $\hat{c}_i^*(t)$ is carried into the bootstrapped estimates and can be evaluated, e.g., to form the confidence interval for β. Here we use the Bayesian bootstrap, which can be considered as a smoothed version of the ordinary bootstrap (Appendix A.8). The following is a SAS program to fit the power model to the TDM example data and generate bootstrapped predictions $c_{ib}^*(t), b = 1, ..., B$ for a large B.

```
data sub;
  set tdm(where=(group=1));
  keep sub start; run;

data wei;  set sub;
  do simu=1 to 100;
    wei=-log(ranuni(12));
    output;
  end; run;

proc sort data=wei; by sub start;run;

data all;
  merge tdm0 wei;
  by sub start; run;

proc sort data=all; by simu sub start;run;
```

```
proc mixed data=all;
  class sub dose;
  model lconc=ldose/ outp=pred;
  random sub;
  weight wei;
  by simu; run;

data cox; set pred;
  pconc=exp(pred); run;

proc phreg data=cox;
  model (start,stop)*event(0)=pconc;
  by simu; run;
```

We use $B = 100$ bootstraps to generate $\hat{\beta}_b, b = 1, ..., 100$. Summary statistics for them are

N	Mean	Std Dev	Minimum	Maximum
100	0.0250844	0.0292990	-0.0544430	0.1069432

As a comparison, the standard two stage approach gives

Parameter Estimate	Standard Error	Chi-Square	Pr > ChiSq	Hazard Ratio
0.02580	0.00759	11.5513	0.0007	1.026

Although $\hat{\beta} = 0.0258$ is very close to the mean of $\hat{\beta}_b$s, 0.0251, the SE of 0.00759 is much smaller than the SD of the $\hat{\beta}_b$s. Therefore, using this SE for statistical inference is misleading.

Consider the RC approach for the Cox regression. Given $h_i(t|e_i) = h_0(t)\exp(\beta c_i^*(t)) = h_0(t)\exp(\beta(\hat{c}_i^*(t) + e_i))$, the key step for the RC is to find the marginal hazard function,

$$h_i(t|\mathbf{c}_i) = h_0(t)E(\exp(\beta(c_i^*(t)))|\mathbf{c}_i, T_i > t), \qquad (5.57)$$

conditional on exposure measures $\mathbf{c}_i$ from subject i. For rare events $E(\exp(\beta(c_i^*(t)))|\mathbf{c}_i, T_i > t) \approx E(\exp(\beta(c_i^*(t)))|\mathbf{c}_i)$ (Prentice, 1982). Therefore, one can directly replace $c_i^*(t)$ with $\hat{c}_i^*(t)$. When the events are not rare, Clayton et al. (1991) proposed the risk set RC that, for each failure time T_i, fits the exposure model on those still at risk, i.e., those in R_i, to get $c_i^*(T_i)$. However, practical experiences show the difference between the risk set RC and the simple RC using all subjects to fit the exposure model is often small.

Heterogeneity in $e_i(t)$ may also lead to inconsistency of the β estimate and may need an adjustment. Note that $e_i \sim N(0, \text{var}(e_i))$ leads to $K = \exp(\text{var}(e_i)/2)$. This suggests that if one can model the heterogeneity as

$$\text{var}(e_i) = v(c_i^*(t), \gamma) \qquad (5.58)$$

one may adjust for the effect of heterogeneity by including $v(c_i^*(t), \gamma)$ as a covariate.

5.8   Interval censored data

Although right censoring is the most common type of censoring, interval censoring (IC) is also common (Sun, 2006). An interval censored TTE refers to the fact that we only know subject i had an event within a time interval (L_i, R_i), but not the exact time of occurrence. IC includes right censoring as a special case with $R_i = \infty$, observed event time with $L_i = R_i = T_i$. One example of IC is the time to disease progression (DP) in many types of diseases. For solid tumors, DP is defined according to tumor size changes, which can only be measured infrequently. Therefore, time to DP is always interval censored between the time of the first assessment that classifies the DP and the previous assessment time. To estimate parameters with IC data, one may still use the ML approach. Letting $S(t|\mathbf{X}_i, \boldsymbol{\beta})$ be the survival function with $\mathbf{X}_i$, and subject i had an IC event (L_i, R_i), we can write $P(L_i < T_i \leq R_i) = S(L_i) - S(R_i)$. The likelihood function can be written as

$$L(\boldsymbol{\beta}) = \prod_{i=1}^{n}(S(R_i|\mathbf{X}_i, \boldsymbol{\beta}) - S(L_i|\mathbf{X}_i, \boldsymbol{\beta})) \tag{5.59}$$

which can be maximized to estimate $\boldsymbol{\beta}$. Note that, although $L(\boldsymbol{\beta})$ holds for $R_i = \infty$, the contribution of an observed event time $L_i = R_i = T_i$, if any, to $L(\boldsymbol{\beta})$ needs to specify by the density function as in (5.4).

If the TTE follows a parametric distribution, the ML approach can be carried out similarly as for right censored data, as long as $S(.|\mathbf{X}_i, \boldsymbol{\beta})$ is given. However, if it follows a semiparametric model such as the Cox model, the baseline risk function $h_0(t)$ cannot be eliminated by the partial likelihood approach. Let $s_0, ..., s_J$, be all the ordered values of $(L_i, R_i), i = 1, ..., n$. Since one can only determine the value of $h_0(t)$ at times s_js, and they are all needed to estimate $\boldsymbol{\beta}$, we can assume that $h_0(t)$ takes value a_j when $t = s_j$ and 0 elsewhere. Consequently,

$$H_0(t) = \int_{\tau=0}^{t} h_0(\tau)d\tau = \sum_{j:s_j<t} a_j. \tag{5.60}$$

Then $S(R_i|\mathbf{X}_i, \boldsymbol{\beta})$ can be written as

$$S(R_i|\mathbf{X}_i, \boldsymbol{\beta}) = \exp(-H_0(t)\exp(\boldsymbol{\beta}^T\mathbf{X}_i)) \tag{5.61}$$

and the a_js have to be estimated together with $\boldsymbol{\beta}$. See Sun (2006, Chapter 6).

For implementation, SAS proc LIFEREG can fit parametric models to interval censored data by maximizing the likelihood (5.59). The following call fit a Weibull model to data with censoring intervals (lower, upper) and exposure conc:

```
proc lifereg data = ic;
  model (lower upper) = conc/ d=weibull;
run;
```

Neither SAS PHREG nor R function coxph (.) can fit a Cox model to interval censored data, although some specialized SAS macros and R libraries are available.

IC may also occur on recurrent events. For example, one may be interested in the effect of an anti-asthma treatment on reducing the risk of asthma attack. However, often only the number of attacks in a period, e.g., a week, rather than the time of each attack, is recorded. In a clinical trial patients may be asked how many asthma attacks occurred between the previous and the current visits. The exact time of occurrence may not be recalled and may not be important. Therefore, the times of the events are interval censored between the intervals. In this case, IC leads to longitudinal count data (Sun 2006, Chapter 9). In this case, one can use a simple longitudinal model to fit the data yet still capture important features of the data. Letting y_{ij} and c_{ij} be the number of events and the exposure between intervals j and $j+1$, we may assume that

$$y_{ij} \sim Poisson(\lambda_{ij})$$
$$\lambda_{ij} = \exp(\mathbf{X}_{ij}^T\beta + u_i). \tag{5.62}$$

In addition, we can also allow for extra variation in y_{ij} for given u_i, either using a negative binomial distribution or an overdispersion parameter representing the variance inflation. This model can easily be fitted with, e.g., proc GENMOD or proc GLIMIXED. See Section 3.6 in Chapter 3 for details. In fact, one may also consider y_{ij} as a counting process and λ_{ij} as the rate function. It is also possible to fit the data with proc PHREG, with the option allowing for dependency of events within the same patients. In principle, if we assume that the underlying mechanism of events is a multistate model with on and off states, it is possible to calculate λ_{ij}. But, in general, the multistate model may not lead to a time constant λ_i, even with the simple relationship to $\mathbf{X}_{ij}$ as in (5.62). Therefore, model (5.62) should be considered as an empirical model for interval censored data.

Interval censored data can also be modeled jointly with exposure data. To this end, we maximize the joint likelihood of the Poisson model (5.62) and an exposure model. Here we use a simulated dataset generating exposure data from a power model

$$\log(c_{ij}) = \log(d_{ij}) + u_i + e_{ij} \tag{5.63}$$

with the following program:

```
%let nsimu=50; %let nsub=100; %let cdose=log(1.3);
%let bound=log(10); %let base=-3; %let dose1=log(5);
%let sigv=0.5; %let sigu=0.3; %let sige=0.2;
%let nsub=5; %let nrep=8; %let beta=0.25;

data simu0 simu1;
  do simu=1 to &nsimu;
```

```
do group=0 to 1;
do sub=1 to &nsub;
   ui=rannor(1)*&sigu;
   vi=rannor(1)*&sigv;
   nn=0;
   dose=&dose1;
   do i=1 to &nrep;
   lci=dose+ui;
   lce=lci+rannor(12)*&sige;
   trend=0;
   if group=1 then lmd=exp(&base+&beta*exp(lci)+vi+trend);
   else lmd=exp(&base+vi+trend);
   yi=ranpoi(1,lmd);
   if group=0 then output simu0;
   else if group=1 then output simu1;
   if lce<&bound then dose=dose+&cdose;
     end;
  end;
 end;
end; run;
```

The following SAS program fits the joint models of (5.63) and (5.62) by maximizing the joint likelihood with u_i as frailty shared with the exposure model:

```
proc nlmixed data=simu(drop=ui vi) qpoints=8 ebsubsteps=100 ebsteps=150
   ebssfrac=0.8 ebopt ftol=1e-12 gtol=1e-12;
  parms base=-3 beta=1 a=0 b=1 sigu=1 sigv=1  siguv=0 sige=1;
  bounds sigu sigv sige >0;
    logci=a+b*dose+ui;
    resc=log(tci+0.0001)-logci;
    if (abs(resc) > 1E50) or (sige < 1e-12) then llw = -1e20;
    else  llw = -0.5*(resc**2 / sige  + log(sige));
    lambda=exp(base+beta*exp(logci)+vi);
   lly = yi*log(max(lambda,0.00001)) - lambda;
  model yi ~general(lly+llw);
   random ui vi ~ normal([0,0],[sigu,siguv,sigv]) subject=sub;
run;
```

which generates the following results:

Parameter	Estimate	Standard Error	DF	t Value	Pr > |t|
base	-3.7550	0.1474	198	-25.47	<.0001
beta	0.3329	0.01203	198	27.67	<.0001
a	0.08036	0.03909	198	2.06	0.0411
b	0.9712	0.01465	198	66.30	<.0001
sigu	0.08622	0.009239	198	9.33	<.0001
sigv	0.2545	0.04903	198	5.19	<.0001
siguv	0.02125	0.01514	198	1.40	0.1620

| sige | 0.04256 | 0.001611 | 198 | 26.41 | <.0001 |

The result shows reasonably good estimates for almost all parameters, including the variance of the frailty.

5.9 Model identification and misspecification

Model identification is a more important issue for TTE data than it is for other endpoints, particularly for some common types of TTE models. We consider two situations when model identifiability becomes problematic in ER modeling for TTE data. One is that the effect of exposure variation on population level cannot be distinguished from the unspecified baseline hazard $h_0(t)$ in the Cox model. Suppose that the exposure at time t for subject i can be written as

$$c_i^*(t) = c_0(t) + c_i'(t) \tag{5.64}$$

where $c_0(t)$ is the common time profile of the exposure for all subjects and $c_i'(t)$ represents variations between subjects. In this case, we have

$$h(t) = h_0(t) \exp(\beta c_i^*(t)) = h_0(t) \exp(\beta c_0(t) + \beta c_i'(t)). \tag{5.65}$$

But the Cox regression will estimate $h_0(t) \exp(\beta c_0(t))$ as the baseline hazard and $h_0(t)$ and $\exp(\beta c_0(t))$ are distinguishable only when $c_0(t)$ is known. Consequently, β is only estimated with $c_i'(t)$. Although it can be estimated as long as $c_i'(t)$ exists, the estimate could be very poor if this component is small. If $c_0(t)$ forms a major part of variation in $c_i(t)$, the impact of measurement error in it will also be strong and severely dilute the treatment effect.

The nonidentifiability problem is obvious in the Cox model. But the problem may also exist even if a parametric model is fitted. The difference is that in a parametric model, the baseline hazard part is not completely separated from the part of exposure driven hazard, but for the same reason, a part of the common trend cannot be distinguished from the baseline risk and the shape of TTE distribution reflects the risk changes of natural course.

The other situation where identification becomes problematic is due to confounding between the baseline function and the risk time profile change as a result of frailty. Consider an example where the conditional risk function is $h(t|u_i) = u_i a(t)$ with u_i following a unit mean gamma distribution (hence we set the scale and shape parameters all equal to γ). This leads to risk function

$$h(t) = a(t)/(1 + A(t)/\gamma) \tag{5.66}$$

where $A(t) = \int_0^t a(\tau)d\tau$. It shows that a time constant frailty causes risk time profile changes. Deriving the corresponding density function involves some lengthy calculation (Aalen et al., 2008, Chapter 6), hence is omitted.

Frailty also causes changes in the risk ratio in a proportional risk model. Suppose that two conditional risk functions are risk proportional with $h_1(t|u_i) = u_i a(t)$ and $h_2(t|u_i) = u_i r a(t)$, with u_i following the same distribution as above; then the ratio of the corresponding risk functions is

$$h_2(t)/h_1(t) = r \frac{1 + A(t)/\gamma}{1 + rA(t)/\gamma}, \tag{5.67}$$

hence the risk ratio also becomes time varying and the model with frailty is no longer risk proportional.

Often in a clinical trial or an observational study there is a control group representing the "baseline" disease process, in the context of TTE modeling, $h_0(t)$. It is apparently reasonable to hope that with the control group, one may be able to eliminate the baseline and frailty in order to identify the exposure effects. This is, however, generally not possible. Consider an example that $h(t, c(t)) = u_i \exp(c(t))$ with u_i following a unit mean gamma distribution. Can we estimate the exposure effect of $c(t)$ on $h(t)$ by the difference in log-hazard between the active and placebo? Figure 5.4 shows the estimate as log-risk with a nonparametric estimate and the true $h(t)$ with different shape parameter values. It demonstrates that even with this simple model, the true treatment effect cannot be identified in the presence of frailty. This is because the risk sets of the two groups become more and more different over time, if treatment effects change the risk. Therefore, the two groups are not comparable at all times.

5.10   Random sample simulation from exposure–risk models

5.10.1   Basic simulation approaches for time-to-events

Time-to-event samples that follow some standard parametric distributions can be generated easily. If T_i has distribution function $F(t)$, and $F^{-1}(u)$ can be calculated easily, a straightforward way to generate T_is is to generate a uniform distribution $U_i \sim U(0,1)$, the uniform distribution function within $(0,1)$, then $T_i = F^{-1}(U_i) \sim F(t)$. One can also use the survival function $S(t)$ to generate $T_i = S^{-1}(U_i) \sim F(t)$, since $S^{-1}(u)$ and $F^{-1}(u)$ are symmetric on $(0,1)$. Note that for the same U_i, $F^{-1}(U_i)$ and $S^{-1}(U_i)$ follow the same distribution and are also negatively correlated. This approach can be used to generate TTE data under timeconstant exposures efficiently as long as the survival function under the exposure has an analytical inverse function.

Generating TTE samples from a Cox model with an underlying parametric distribution corresponding to the baseline risk needs more careful consideration. We start from Cox models with constant exposures, then turn to those

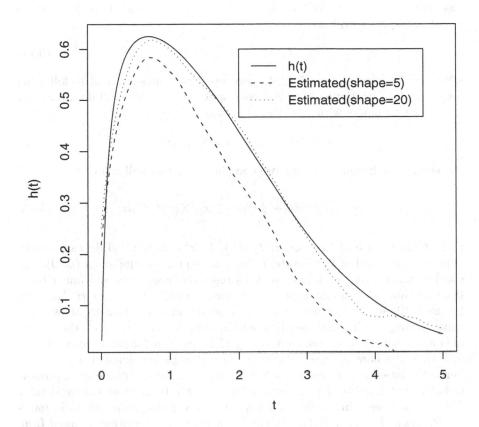

FIGURE 5.4
True and estimated exposure effects with placebo adjustment: constant baseline hazard with gamma frailty (shape = 5 and 20).

with time-varying exposures. Consider the Cox regression model

$$h(t, \boldsymbol{\beta}) = h_0(t) \exp(\mathbf{X}^T \boldsymbol{\beta}). \tag{5.68}$$

The survival function of this model is

$$S(t|\mathbf{X}) = \exp[-H_0(t) \exp(\mathbf{X}^T \boldsymbol{\beta})]. \tag{5.69}$$

When $H_0(t)$ is known, and its inverse function can be found, a sample T following the Cox model can be generated as (Bender et al., 2005)

$$T = H_0^{-1}[-\log(U) \exp(-\mathbf{X}^T \boldsymbol{\beta})]. \tag{5.70}$$

One can choose from some standard parametric distributions for the baseline risk. For example, the Weibull distribution with scale and shape parameters s and γ, respectively, has

$$H_0^{-1}(t) = (s^{-1}t)^{1/\gamma}, \tag{5.71}$$

which can be inserted into (5.70) to generate a random sample of Ts following a Cox model with the baseline risk corresponding to a Weibull distribution.

For a Cox model with time-varying covariate

$$h(t, \boldsymbol{\beta}) = h_0(t) \exp(\mathbf{X}^T(t)\boldsymbol{\beta}), \tag{5.72}$$

the simulation becomes much more complex. One can still write

$$S(t|\mathbf{X}) = \exp[-\int_0^t h_0(\tau) \exp(\mathbf{X}(\tau)^T\boldsymbol{\beta})]d\tau, \tag{5.73}$$

and, if the inverse of function $H(t) = \int_0^t h_0(\tau) \exp(\mathbf{X}(\tau)^T\boldsymbol{\beta})d\tau$ can be calculated, one could still generate T from an equation similar to (5.70). For special time-varying covariates such as piecewise constants and linear functions of time, the simple approach in Austin (2012) based on $H^{-1}(u)$ can be used. Zhou (2001) suggested a monotone transformation to allow for a time-varying baseline risk function within each interval in which the covariates are constant. If a transform $s = g(t)$ is used on t with a constant hazard, then the baseline risk within an interval is proportional to $dg^{-1}(t)/dt$, hence the baseline risk can be made time varying, yet the property of proportional risk still holds. This approach has recently been proposed by Hendry (2013). For generating a piecewise exponential variable with risk λ_k in interval $(t_k, t_{k+1}), k = 1, ..., K, t_1 = 0$, the following is an algorithm adapted from Zhou (2001). Start from $k = 1$ and $Y_0 \sim Exp(\lambda_1)$, then repeat the following steps.

- If $Y \le t_{k+1}$, then output Y and stop.

- Else $Y \leftarrow t_k + \lambda_k(Y - t_k)/\lambda_{k+1}$.

- $k \leftarrow k + 1$.

The following is a fully vectorized R program for implementing this algorithm, which generates a large number of samples simultaneously, and the loop in it runs until K regardless if an individual $Y < t_k + 1$. If it is programmed in a software that does not support vectorized operations, the iteration should stop when an individual $Y \le t_{k+1}$.

```
nsimu=10000
cut=1:10
rate=c(10/cut,11)
rate=rep(1,11)
yv=rexp(nsimu)/rate[1]
```

```
for (i in 1:length(cut)){
  ti=cut[i]
  yv=ifelse(yv<ti,yv,ti+(yv-ti)/rate[i+1]*rate[i])
}
```

Note that if the transformation $g(t)$ is used, the intervals should also be transformed.

However, in most cases numerical calculation is needed to evaluate $H^{-1}(u)$ even when $H(t)$ has a simple analytical form. Consider the following simple case in which $h_0(t) = a$ (i.e., t follows an exponential distribution), $c(t) = 1 - \exp(-Kt)$, and $h(t) = h_0(t)c(t)$. The latter is a simple function describing the process of exposure reaching the steady state. Under these conditions one can derive

$$
\begin{aligned}
h(t) &= a(1 - \exp(-Kt)) \\
H(t) &= \int_0^t a(1 - \exp(-K\tau))d\tau = a(t + (\exp(-Kt) - 1)/K) \\
S(t) &= \exp(-a(t + (\exp(-Kt) - 1)/K)).
\end{aligned}
\tag{5.74}
$$

It is still not easy to sample directly from $S(t)$, as $S^{-1}(t)$ is not an explicit function. Although one can calculate it by numerically solving the implicit function

$$
S(T_i) = U_i \tag{5.75}
$$

to generate T_i for a given U_i, there are more efficient algorithms than this one, as we will show later on.

For general time-varying covariates, a permutation algorithm (Sylvestre and Abrahamowicz, 2008) can be used. But it is inefficient to generate data of a large sample size.

5.10.2 Rejection-acceptance algorithms

The rejection-acceptance sampling technique is commonly used to generate samples from a distribution which is not easy to sample from. The idea is to generate samples from a distribution close to the target distribution, then reject those unlikely to follow the target distribution. Specifically, to sample from $f(t)$, we find, $g(t)$ which is easy to sample from with $f(t) \le Cg(t)$, where C is a constant; then we can

- Generate $Y_i \sim g(t)$

- Generate $U_i \sim U(0, Cg(t))$

- If $U_i \le f(Y_i)$, then accept Y_i, else regenerate Y_i and U_i

There is also an equivalent version in which one generates $U_i \sim U(0,1)$ then accepts Y_i if $U_i \leq C^{-1}f(Y_i)/g(t)$. Applying it to the example above, we note that

$$
\begin{aligned}
f(t) &= a(1 - \exp(-Kt))\exp(-a(t + (\exp(-Kt) - 1)/K)) \\
&\leq a(1 - \exp(-Kt))\exp(-a(t - 1/K)) \\
&\leq a\exp(-at)\exp(a/K) \qquad\qquad (5.76)
\end{aligned}
$$

and we find $g(t) = a\exp(-at)$ is the density function of the exponential distribution with parameter a, and $C = \exp(a/K)$. The following algorithm can be used to generate T:

- Generate $T^* \sim Exp(a)$

- Generate $U \sim U(0, a\exp(-aT^*)\exp(a/K))$

- If $U \leq f(T^*)$, then $T = T^*$; otherwise reject T^* and go back to the first step

Its efficiency depends on the ratio a/K. When the time to reach the steady state is comparable to the time to event or even longer than it, this algorithm may be quite inefficient. Nevertheless, compared to the time needed for model fitting, the lack of efficiency is acceptable in most practical situations.

This approach can also be used for some other $h(t)$s. Suppose that $h(t) = B(t)$ where $B(t)$ is a sum of spline functions; then

$$
H(t) = \int_0^t B(\tau)d\tau, \qquad\qquad\qquad (5.77)
$$

which, in principle, can be derived analytically. When $H^{-1}(t)$ is difficult to calculate, the rejection method can be used. Since $f(t) = B(t)\exp(-H(t))$, if $N < B(t) < M$, then

$$
f(t) \leq (M/N)N\exp(-tN), \qquad\qquad (5.78)
$$

and T can be generated with the rejection-acceptance method based on an exponential distribution with parameter N. The efficiency of the generator depends on the ratio M/N. This approach is inefficient if the lower bound N is very low relative to the upper bound M.

5.10.3   Generating discrete event times

In clinical trials it is often sufficient to record the event time in days, hence using discrete time may allow considerable simplification of the simulation. In the following we assume that $t = 1, 2, \ldots$, so that we can write

$$
h_i(t) = h_0(t)\exp(\mathbf{X}_i^T(t)\boldsymbol{\beta}) \qquad\qquad (5.79)
$$

as an approximation to the hazard at day t. A very general approach for discrete time sample generation follows a similar idea to the conditional logistic regression model. That is, for a subject at risk at time t, the event at time t follows a binary distribution with probability $h(t)$. Here we assume that day is a sufficiently small time unit so that $h(t) < 1$ for all t. Otherwise, a smaller unit may be used. With this approach we can generate events from the hazard function directly, hence it is straightforward even when the corresponding $f(t)$ and $S(t)$ have no analytical form and are difficult to calculate. $h(t)$ can be an analytical function or empirical distribution if the simulation is to mimic a specific practical scenario.

The following algorithm implements this approach:

- for each subject set censor indicator $C_i = 0$

- do $t = 1, 2, \ldots$, until the end of the study

    - while $C_i = 0$, generate $y_i(t) \sim bin(h_i(t))$
    - when $y_i(t) = 1$, set event time to t
    - when a censoring event occurs or when $y_i(t) = 0$, set $C_i = 1$

- next t

- repeat for all subjects

Note that if a Cox model is to be fitted, only data at time points when an event occurs are needed. Dropping the other data may substantially reduce the size of the final dataset. On the other hand, one may want to keep all the records when a parametric model will be used for analysis. The only shortcoming is the lack of efficiency when the follow-up time is very long. In this case, one may increase the interval length for discretizing the time. For example, in a trial with a five year follow-up period, the time to event may be sufficiently accurate if calculated in weeks.

Since this approach only needs the hazard function, it can be easily used for a complex model specified by $h_i(t)$. For example, to generate TTE from a frailty model with $h_i(t) = h_0(t) u_i \exp(\mathbf{X}_i^T \boldsymbol{\beta})$, we generate u_i from a desired distribution then, use the same algorithm. The same approach can also be used for a model $h_i(t) = h_0(t) u_i \exp(\beta c_i(t))$ with time-varying exposure $c_i(t)$.

This approach can be made more efficient for rare event simulation by simulating the number of events at each t for a set of subjects sharing the same covariates and exposure levels.

5.10.4 Simulation for recurrent and multiple events

Simulation for transition times based on a multistate model seems complex, but it follows the same rules as for single TTE simulation. For example, at state k, suppose that the risk of having the event of moving to state k' is

$$h_{kk'}(t) = h_{0kk'}(t) \exp(\beta_{kk'} c(t)) \tag{5.80}$$

where $h_{0kk'}(t)$ is the specific baseline risk for transition between the two states. Then, given function $c(t)$, the transition time can be generated accordingly. Other state dependent covariates can be added without additional difficulties. One special issue is whether one should restart the clock for $h_{0kk'}(t)$, i.e., use $h_{0kk'}(t - t_k)$ where t_k is the time of reaching state k. However, this is a model specification issue rather than a simulation issue, as the approaches stated here can cope with both situations.

The following algorithm generates recurrent events from a two-state (event "on" and "off") model for piecewise constant exposure within time intervals $t_j, t_{j+1}, j = 1, ..., J$. Let $\lambda_{kk'j}$ be the risk of transition from k to k', calculated based on the exposure in interval j.

- set state = "off"

- For $j = 1, .., J$, do while $(T < t_{j+1})$

    - set $T = t_j$.
    - if state = "on" then generate $t \sim \exp(\lambda_{10j})$
      else generate $t \sim \exp(\lambda_{01j})$
    - set $T = T + t$. If $T < t_{j+1}$, then switch the state, back to the previous step,
      else let $T = t_{j+1}$ and go to the next j

- Collect all Ts and the state each time the state changes

A SAS code implementing this algorithm can be found in the appendix.

A very simple algorithm can generate recurrent events on a discrete time scale, based on the dependence of current hazard on the history. Suppose that we denote if a patient is having an AE at day t by an indicator y_t with $y_t = 1$ for event occurrence and 0 otherwise, Then between the days the event started and resolved, there is a series of "1"s. The continuation of the "1"s means when a patient is having an AE today, he is more likely to have the AE tomorrow than to have no event. Therefore, we may assume that

$$h(t, y_{t-1}) = h_0(t) \exp(\beta c(t) + \gamma y_{t-1}), \tag{5.81}$$

where γ controls the duration the patient stays in the "on" state. The term γy_{t-1} can be replaced by a more complex function multiple lagged $y(t)$. One will find that this approach in fact uses a dynamic generalized linear model described in the next chapter. This algorithm can be easily implemented by adapting the algorithms in Section 5.10.3.

5.10.5   Simulation for TTE driven by complex dynamic systems

Often behind the TTE, there is a latent process, and the event occurs when the process hits a threshold. This assumption forms the first hit model (FHT)

(Aalen et al., 2008, Chapter 10), which provides simple hypothetical mechanisms for the TTE. The latent process may be abstract but could also be real. Latent general health status may be considered as an abstract process, which is 1 for perfect health and drops to 0 when a patient dies. There may also be real processes, such as CD4 cell counts in HIV/AIDS, driving the TTE. An HIV patient is diagnosed with AIDS if the CD4 count is less than $200/mm^3$. Therefore, the time from HIV infection to AIDS may have the HIV dynamic as a latent process, and the event occurs when the CD4 count hits $200/mm^3$. In oncology, disease progression (DP) for solid tumors is indicated if the tumor size increase is more than 20% and is also more than 5 mm (RECIST V1.1). Hence, the time to DP is driven by a tumor dynamic model.

As an example, we consider the relationship between time-to-DP and tumor dynamics in cancer patients. Consider the tumor growth model (Claret, 2009) as the latent process

$$\frac{dY(t)}{dt} = (K_L - K_D C(t))Y(t) \tag{5.82}$$

where $Y(t)$ and $C(t)$ are the tumor size and exposure at time t, and K_L and K_D are constants. This model can easily be solved if the exposure is constant, but may only have numerical solutions for an arbitrary $C(t)$. Since we often only measure the tumor size periodically at $t_j, j = 1, ..., r$, e.g., by MRI or CT scans. For simplicity we assume that the t_js are at equal distance Δ apart. At time t_j, we observe y_j which satisfies

$$\log(y_j) = \int_0^{t_j} (K_L - K_D C(t))dt + \varepsilon_j \tag{5.83}$$

where ε_j is a measurement error. This leads to

$$\log(y_j) - \log(y_{j-1}) = \Delta K_L - \int_{t_j}^{t_{j-1}} K_D C(t)dt + \varepsilon_j. \tag{5.84}$$

Therefore, y_j is a random walk on the log-scale with drift $\Delta K_L - \int_{t_j}^{t_{j-1}} K_D C(t)dt$. A random walk is a series of random variables that behaves like walking randomly. It has the form $x_{j+1} = x_j + K + e_j$, which goes from x_j to x_{j+1} by a random step e_j and a fixed step K, called draft. Let $y_{min}(j) = \min(y_1, ..., y_j)$, the time to disease progression is the first t_j when $y_j - y_{min}(j) > 5$ and $y_j/y_{min}(j) > 1.2$ (20% increase) are observed. Therefore, generating the time to DP is essentially a generation of the latent process. The simulation provides a useful tool to evaluate empirical models when fitting to TTE that has clearly defined algorithm.

The time to death or time to AIDs for HIV infected patients may have a more complex mechanistic latent process, as the the virus dynamic model is complex. Consider a dynamic model for an HIV study (Wu et al., 2008) where drug exposures were measured:

$$\frac{dT}{dt} = \lambda - d_T T - (1 - \gamma(t))kTV$$

$$\frac{dT^*}{dt} = (1 - \gamma(t))kTV - \delta T^*$$

$$\frac{dV}{dt} = N\delta T^* - cV \tag{5.85}$$

and where T, T^* and V are the uninfected and infected T-cells and free virons, respectively. $\gamma(t) = (1 + \phi/(\sum_{k=1}^{K} w_k C_k A_k(t)/IC_{50}^k(t)))^{-1}$ is the treatment effect (which is an Emax model with $E_{max} = 1$). $C_k, A_k(t)$ and $IC_{50}^k(t)$ are the PK exposure, adherence and the exposure achieving 50% inhibition for the kth regimen at time t, respectively, all measured periodically during the study. To simulate the TTE, we may use a numerical approach to solve the equations from time $t = 0$ to $T(t) < T_0$ for a given threshold T_0.

Using the mechanism but without specifying a threshold, we can link the latent process in a normal way. We may assume that $T(t)$ is associated with the risk of death and $h(t) = h_0(t)\exp(m(T(t), \beta))$ represents the relationship between the exposure, the virus dynamics and TTEs. However, even though one can solve $T(t)$ from the dynamic model then calculate $H(t)$ numerically, it is still not easy to sample from $S(t) = \exp(-H(t))$. Crowther and Lambert (2013) proposed using a Gaussian quadrature to calculate $H(t)$ and a numerical root finder to generate the event time $S^{-1}(u)$ in general situations. But applying this approach here is inefficient, as a large number of function evaluations are needed for generating a single TTE. A more efficient way is to use Gaussian quadratures in combination with a rejection-acceptance algorithm, with $H(t)$ calculated with a Gaussian quadrature. Since $H(t)$ is often bounded from below, it is easy to find an appropriate $g(t)$ that leads to a reasonably efficient rejection algorithm. For example, with an Emax model in $h(t)$: $h(t) = \exp(\beta_0 + E_{max}/(1 + EC_{50}/C(t)))$, and $N \leq h(t) \leq M$. With a positive E_{max}, $N = \exp(\beta_0)$ and $M = \exp(\beta_0 + E_{max})$ for any $C(t)$. Therefore, $f(t) \leq M\exp(-Nt) = N\exp(-Nt)(M/N)$, hence we can use $g(t) = N\exp(-Nt)$. As $a = M/N = \exp(E_{max})$, when $E_{max} = 1$, $a \leq \exp(1) = 2.72$, i.e., on average, one sample is generated by every 2.72 simulations.

As an example, suppose that $T(t)$ is related to the risk of death by $h(t) = \exp(\beta_0 + m(T(t)))$, where $m(T(t))$ may be an Emax model. The rejection algorithm for the HIV dynamic model can be implemented as follows. Since $T(t)$ does not have a closed form, we need to solve the ODEs, which can be done together with the numerical integration for $H(t)$. To implement this approach, we add a dummy ODE

$$\frac{dH}{dt} = \exp(\beta_0 + m(T(t))) \tag{5.86}$$

to the dynamic model (5.85). Then $f(t)$ is given as $f(t) = \exp(\beta_0 + m(T(t)) - $

$H(t)$), where $T(t)$ and $H(t)$ are the solutions to the first equation in (5.85) and equation (5.86). Then the same rejection-acceptance algorithms as discussed before, with N and M determined in the same way, can be implemented.

5.10.6   Example programs

The following R program simulates TTE data using the binary distribution with $c(t) = \exp(-t) - \exp(-2t)$. Censoring time is 5 and between 0 to 5, 1000 intervals of equal size are used. log(etime/ntime) converts the log-baseline risk h0=-1 to that per interval.

```
etime=5
ntime=1000
ttime=(1:ntime)/ntime*etime
EC50=0.15
Emax=1
h0=-1
h0adj=h0+log(etime/ntime) #width=5year/1000
Yi=NULL
for(i in 1:5000){
 ct=(exp(-ttime))-exp(-2*ttime)
 hi=pmin(0.9999,exp(h0adj+Emax/(1+EC50/ct)))
ei=rbinom(ntime,1,hi)
tti=min(which(ei==1))
Yi=c(Yi,min(ntime+1,tt1)) #min(NA)=Inf
}
Yi=Yi/1000*etime
```

This is a fully vectorized algorithm, which is more efficient in R (but not in SAS) than using an additional loop for each individual to generate ei until ei=1 occurs.

The following R program uses the RA algorithm together with a numerical integration algorithm to generate TTE data with an Emax model and $c(t) = \exp(-t) - \exp(-2t)$.

```
h0=-1
EC50=0.15
Emax=1
ht=function(t){
 ct=(exp(-t))-exp(-2*t)
 exp(h0+Emax/(1+EC50/ct))
}
M=exp(h0+Emax)
N=exp(h0)
nsimu=10000
  yv=rexp(nsimu,N)
  un=runif(nsimu,max=M*exp(-N*yv))
Yi=NULL
for(i in 1:nsimu){
```

```
  Ht=integrate(f=ht,lower=0, upper=yv[i])$value
  if (un[i]<exp(-Ht)*ht(yv[i])) Yi=c(Yi,yv[i])
}
```

where the R function intergate(.) calculates $H(t)$ for calculating $f(t)$ for the
RA algorithm.

A simulation algorithm for HIV dynamic model:

```
parameters <- c(a =0.3, lmd=33,d=0.3,k=1,del=0.5,N=1,c=0.2,u=0.5)
 state <- c(T = 1, Ts = 0, V = 1,H=0)
gm=function(t) 0.5 #assume constant

h0=-1
EC50=1
Emax=1
#set upper and lower boundaries for h(t)
M=exp(h0+Emax)
N=exp(h0)

 HIV<-function(t, state, parameters) {
 with(as.list(c(state, parameters)),{
 # similar to $DES in NONMEM
 dT <- lmd - d*T-(1-gm(t))*k*T*V
 dTs <- (1-gm(t))*k*T*V -del*Ts
 dV= N*del*Ts-c*V
 dH= exp(h0+Emax*T/(T+EC50)) #h(t)
 list(c(dT, dTs, dV,dH))})
 }
 Yi=NULL
   yv=rexp(nsimu,N)
  un=runif(nsimu,max=M*exp(-N*yv))
for(i in 1:nsimu){
  out=ode(y = state, times = c(0,yv[i]), func = HIV,
    parms = parameters,method="lsoda")
  T=out[2,2] #col2-5 are T, Ts, V and H
  h=exp(h0+Emax*T/(T+EC50))
  Ht=out[2,5]
  if (un[i]<exp(-Ht)*h) Yi=c(Yi,yv[i])
}
```

where $H(t)$ is also calculated by the ODE solver function ode(.), together with
T, Ts and H.

5.10.7   Sample size estimation using simulations

As an application of the simulation approaches, here we use these simulation
tools for sample size calculation for estimation of exposure effect, e.g., log-
RR, when the exposure has a complex distribution. Sample size calculation
for parametric models and Cox models for comparing two or more groups

has been well developed, hence the reader is referred to a standard text such as Schoenfeld (1983). In general, sample size calculation for exposure–risk modeling is rather complex, since the information matrix for model parameters cannot be easily calculated. The following simulation tools provide an easy way for applied statisticians and modelers. The basic idea is to use simulation in combination with model fitting to estimate the information matrix with a large sample simulation. Then the sample size needed to achieve a given precision (e.g., given by a target SE) can be determined accordingly. The approach described here can be applied almost directly to sample size calculation for other type of models.

We start from a simple situation of no covariates in a Cox model, which is often sufficient for sample size calculation, and that the exposure $c_i(t)$ can be simulated. We would like to determine the sample size that gives SE_t, a target SE size, for $\hat{\beta}$, assuming a baseline hazard function. Although a trial-and-error approach using multiple simulations with different sample sizes can be used to determine the right sample size, a much more efficient way is to run a single simulation of a much larger sample size so that random variations in the SE estimate are almost eliminated. Suppose that we have run a simulation of sample size n_l and have obtained SE_l, from which we can calculate the information matrix (which is a scalar here) $I_l = SE_l^{-2}$ of n_l samples. The contribution of one sample is then $I_u = I_l/n_l = SE_l^{-2}/n_l$. From I_u we can calculate that the SE of sample size n is $1/\sqrt{I_u n}$. Therefore, to obtain SE_t, the sample size needed is $n = n_l SE_l^2/SE_t^2$. As this n is based on asymptotic properties, we may still need to assess the small sample properties by running multiple simulations with the sample size n and adjusting if necessary. But the approach based on the asymptotic variances helps determine the starting point for the multiple simulations. This formula can be derived directly based on the fact that the asymptotic SE^{-2} is proportional to the sample size. However, the derivation via the information matrix is a more general approach.

This basic approach can be extended to more complex situations. The following is a typical scenario in drug development when deciding exposure measures for later phase trials, where an active and a placebo arm are included. It is often needed to determine from how many patients in the active arm exposure measures should be taken. If the modeling strategy is to exclude the placebo arm, the previous approach can be used directly. But if the placebo arm is included as having zero exposure, the approach has to be modified. In this case, the total information is

$$I = n_0 I_{u0} + n_1 I_{u1} \tag{5.87}$$

where n_k and I_{uk} are the sample sizes and the unit (of a single patient) information matrix in arms $k = 0$ (placebo) and $k = 1$ (active), respectively. We would like to decide the number of exposure measures in the active arm to obtain an estimate with a given SE_t value. For given $n_0 I_{u0}$ and I_{u1}, the sample size to obtain a given SE_t value is $n_1 = (SE_t^{-1} - n_0 I_{u0})/I_{u1}$. Therefore, the task becomes how to calculate $n_0 I_{u0}$ and I_{u1} based on simulations. This can be done by running two simulations with different sample sizes n_{1a} and n_{1b} for the active arm and obtaining SE_a and SE_b, respectively. Then $n_0 I_{u0}$ and I_{u1} can be solved from

$$n_0 I_{u0} + n_{1a} I_{u1} = SE_a^{-2}$$
$$n_0 I_{u0} + n_{1b} I_{u1} = SE_b^{-2} \qquad (5.88)$$

which leads to $I_{u1} = (SE_a^{-2} - SE_b^{-2})/(n_{1a} - n_{1b})$ and $n_0 I_{u0} = SE_b^{-2} - n_{1b} I_{u1}$. Note that we only need to estimate $n_0 I_{u0}$ as a single value.

Another extension of this approach is for multiple parameters to allow for time-varying effects or covariates in the model. These can be spline functions to increase model flexibility, which is important when the placebo arm is included in the analysis, or covariates must be adjusted for, particularly when the exposure is related to these covariates. The same approach can be applied to the information matrix of multiple parameters $\boldsymbol{\beta}$. The determinant of the information matrix is inversely proportional to the volume of the confidence region, hence is a general measure for the precision of multiparameter estimates. Considering the situation with a placebo arm in the model, we can use the same approach of running two simulations with sample sizes n_{1a} and n_{1b}. Writing (5.88) in terms of the variance-covariance matrices

$$n_0 \mathbf{I}_{u0} + n_{1a} \mathbf{I}_{u1} = \text{var}(\hat{\boldsymbol{\beta}}_a)^{-1}$$
$$n_0 \mathbf{I}_{u0} + n_{1b} \mathbf{I}_{u1} = \text{var}(\hat{\boldsymbol{\beta}}_b)^{-1}, \qquad (5.89)$$

where $\text{var}(\hat{\boldsymbol{\beta}}_a)$ and $\text{var}(\hat{\boldsymbol{\beta}}_b)$ are the variance-covariance matrices from the two simulations, one can solve these equations to give $\mathbf{I}_{u1} = (\text{var}(\hat{\boldsymbol{\beta}}_a)^{-1} - \text{var}(\hat{\boldsymbol{\beta}}_b)^{-1})/(n_{1a} - n_{1b})$ and $n_0 \mathbf{I}_{u0} = \text{var}(\hat{\boldsymbol{\beta}}_b)^{-1} - n_{1b} \mathbf{I}_{u1}$. If the target is to make $|\text{var}(\hat{\boldsymbol{\beta}})| \leq K$, then the required sample size for the exposure measures can be solved from the equation

$$|n_0 \mathbf{I}_{u0} + n_1 \mathbf{I}_{u1}| = K^{-1} \qquad (5.90)$$

numerically. Since there may not be any integer n_1 that satisfies equation (5.90), to do this efficiently, one can treat n_1 as a continuous variable and use a root finder to find the solution to (5.90) then round the value to the nearest integer. The following code defines an R function and calls a root finder:

```
obj=function(n1) det(n0*Iu0+n1*Iu1)-1/K
uniroot(f=obj, interval=c(lower,upper))
```

where one provides $n0$, $Iu0$ and $Iu1$ and K, as well as the lower and upper boundaries for the root search.

The approach is quite universal, as it can be used for almost any type of TTE model and model fitting procedure. When a non-ML approach is used to estimate $\boldsymbol{\beta}$, one can still consider $(\text{var}(\hat{\boldsymbol{\beta}}))^{-1}$ as the information matrix and use the approach directly. This approach only uses the assumption that asymptotically $\text{var}(\beta) \propto 1/n$, which is satisfied by almost all estimates in this book. For multiple parameter models, there are a number of alternatives to the determinant of the information matrices for measuring the precision of $\hat{\beta}$, e.g., the trace and a linear combination of the information matrix. These measures can be found in standard texts for optimal experimental designs such as Atkinson and Donev (1992).

5.11 Comments and bibliographic notes

There are a number of excellent books on TTE data analysi; among those that the author has used extensively are Fleming and Harrington (1991), Kalbfleisch and Prentice (2002), and Aalen et al. (2008). Cook and Lawless (2007) focus on recurrent events. Putter et al. (2007) give a tutorial on competing risk and multistate models. The use of parametric models for TTE data analysis has a long history, while a landmark in the semiparametric approach is the Cox model (Cox, 1972). Numerous works have been published on the use of the Cox model for TTE data analysis, in particular in the biomedical fields.

Joint modeling is currently a very active research area with a large number of publications in the past ten years. No attempt is made to review the literature. The reader is referred to a recent monograph (Rizopoulos, 2012) for details and references.

6

Modeling dynamic exposure–response relationships

Until now we have assumed that the effect of exposure on the response is instantaneous. However, this assumption is rarely true, although sometimes reasonable as an approximation. In pharmacometrics, in most cases, the PD change lags behind the PK change that caused it, a phenomenon known as hysteresis. In this case, it is important to investigate and to model the temporal relationship between PK and PD. Ignoring hysteresis may induce bias in the estimation of exposure–response relationships. Even when hysteresis does not cause significant bias, if not dealt with properly, it may increase the residuals, hence reduce the accuracy/power of statistical inference. Often the temporal relationship itself may also be important, e.g., to predict the time of onset of an adverse event or the efficacy response to treatment. Csajka and Verotta (2006) give a comprehensive review of PKPD modeling , in particular, hysteresis in PKPD relationships. There are mainly two ways to model the temporal relationship between the exposure and response. One assumes that the response model is instantaneous, but it depends on a latent "exposure", which follows a dynamic model that links to the exposure we observe. The other is based on a dynamic response model describing the natural course of the disease process, e.g., tumor growth. The exposure affects the disease process by changing its natural course.

Although the modeling of the dynamic exposure–response relationship is our main interest, the dynamics of exposure is an integrated part of the dynamic system. Figure 6.2 shows the system with three dynamic subsystems, the dose–exposure part has been discussed in Chapter 3. The two other subsystems represent (1) exposure effect accumulation: the dynamic of drug effects, and (2) disease processes: the dynamic of disease with or without exposure. The dynamic of exposure effect cumulation and disease process may not be separable.

The exposure dynamic may also be affected by the exposure and response via a feedback loop in, e.g., a trial with therapeutic dose monitoring (TDM) where the dose is adjusted according to the individual exposure level. Dose adjustment may also be made according to clinical outcomes. These are exposure dynamics with more complex mechanisms than that under a fixed dosing scheme and will be discussed in Chapter 9.

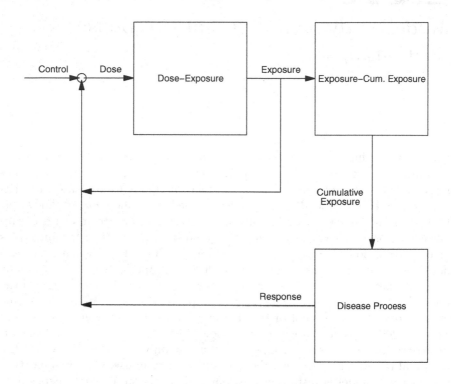

FIGURE 6.1
An illustration of the dynamic system between dose, exposure, cumulative exposure (e.g., an effect compartment), and disease process. The boxes represent dynamic subsystems with a temporal relationship between the input and the output.

6.1 Effect compartment models

A simple way to deal with data showing noninstantaneous exposure–response relationships is to assume that the exposure (in the form of drug or chemical) has to reach a hypothetical site to cause the response. For example, to introduce hysteresis in PKPD relationships, a widely accepted approach is to assume that the drug enters a hypothetic effect compartment, where the drug induces instantaneous effects. The hysteresis is reflected in the time needed for the change in the central compartment to propagate to the effect compartment. The effect compartment model is an extension of the well developed PK compartment models, often supported by the pharmacological mechanism. A combination of an effect compartment model and an instantaneous response model such as the Emax model can describe nonlinear relationships with hysteresis. The effect compartment model approach has been widely used in practice. One example, among many others, is the modeling of the PK/QT relationship by Minematsu et al. (2001).

First we assume that the effect compartment is a linear system described by

$$c^*(t) = K_{in} \int_0^t \exp(-K\tau)c(t - \tau)d\tau \qquad (6.1)$$

where $c^*(t)$ and $c(t)$ are the concentrations in the effect compartment and the central compartment in a population PK model, and K_{in} and K are rate constants of input and elimination of the drug to and from the effect compartment. Only for a few types of population PK models, (6.1) can be calculated analytically. One is the linear compartmental model with constant coefficients. A simple but typical one is the one-compartment model after an IV bolus injection, which gives

$$c(t) = C_0 \exp(-K_e t) \qquad (6.2)$$

where C_0 is determined by dose and parameters such as the volume of distribution and K_e is the elimination rate from the central compartment. Taking it into (6.1), we obtain

$$c^*(t) = \frac{C_0 K_{in}}{K - K_e}(\exp(-K_e t) - \exp(-Kt)). \qquad (6.3)$$

In general, if the PK model is a linear compartmental model with constant parameters, the concentration $c(t)$ is then a linear combination of exponential and polynomial functions, hence the concentration in the effect compartment $c^*(t)$ always has an analytical form. However, the derivation of $c^*(t)$ is often complex. We recommend the use of software such as MATLAB® to derive it automatically when the PK model is more complex than a one-compartment model with first-order absorption with oral doses or a two-compartment model with IV injection.

When there are repeated measurements for the exposure, we also need to use an NLMM for the effect compartment. Taking model (6.3) as an example, $c^*(t_{ij})$ for subject i can be written as

$$c_{ij}^* = \frac{C_0 K_{ini}}{K_i - K_{ei}}(\exp(-K_{ei}t) - \exp(-K_i t)) \qquad (6.4)$$

where K_{ini}, K_{ei} and K_i are random parameters for subject i, as in the standard NLMM. However, in practice, the effect compartment is not observable, hence the unknown parameters can only be determined when fitting the PKPD model. Replacing c_{ij} in model (3.16) with c_{ij}^*, we obtain

$$y_{ij} = g(c_{ij}^*, \boldsymbol{\beta}_i) + e_{ij}. \tag{6.5}$$

Note that, in addition to the parameters in (3.16), extra parameters are in c_{ij}^* and need to be estimated. Model (6.5) is nonlinear even when (3.16) is linear, hence is more difficult to fit.

6.2   Indirect response models

The other type of model, the indirect response model (Csajka and Verotta, 2006) is based on another assumption that there is a baseline biomedical process of concern, and the exposure acts to modify the baseline process by, for example, inhibiting/promoting the input or elimination process. The PD process may be a biopharmaceutical process or a disease process such as tumor growth or the regulation of glucose. These models can be combined with non/semiparametric approaches to reduce the dependence of model assumptions if appropriate.

Let $R(t)$ be the response at time t. The following are two simple but typical indirect response models to describe the dynamic relationship between exposure $C(t)$ and $R(t)$:

$$
\begin{aligned}
\frac{dR(t)}{dt} &= K_{in}(1 \pm g(C(t))) - K_{out}R(t) \\
\frac{dR(t)}{dt} &= K_{in} - K_{out}(1 \pm h(C(t)))R(t)
\end{aligned}
\tag{6.6}
$$

where $g(C(t))$ and $h(C(t))$ are functions of exposure $C(t)$ with $g(0) = h(0) = 0$. The exposure effect can be either inhibited $(-)$ or stimulated $(+)$ and may act on K_{in} or K_{out}. When $C(t) = 0$, both models in (6.6) degenerates to

$$\frac{dR(t)}{dt} = K_{in} - K_{out}R(t), \tag{6.7}$$

which describes the natural process of the response without exposure. Note that without exposure, $R(t)$ will finally reach the level of K_{in}/K_{out}, derived from setting $dR(t)/dt = 0$. The same approach can be used on models (6.6) when $C(t)$ is constant or tends to a constant level. A common model for $g(C(t))$ and $h(C(t))$ is the Emax model

$$g(C(t)) = E_{max}/(EC_{50}/C(t) + 1) \tag{6.8}$$

on which random error and coefficients can be added. However, even when $C(t)$ is the output of a simple model such as (2.5), the model (6.6) may not have an analytical solution. Therefore, $R(t)$ has to be calculated via an ODE solver, which can be found in either NONMEM or R. The following R program calculates the solution of the above model with three parameter settings, using the deSolve(.) library.

```
yini <- c(y1 = 0, y2 = 1)
OneComp <- function(t,y,parms){
     with(as.list(c(parms,y)),{
        dy1 = 1 - Emax*y[2]/(EC50+y[2]) -kout*y[1] #response dynamic
        dy2 = - theta*y[2] #Central comp. with elimination rate ke=theta
list(c(dy1, dy2))
})
}
times <- seq(0, 10, by = 3/24)

out1 <- ode(func = OneComp, times = times,y = yini,
 parms = c(Emax=1.5,EC50=0.3,theta=0.3,kout=1), method = "ode45")
out2 <- ode(func = OneComp, times = times,y = yini,
 parms = c(Emax=1.5,EC50=0.6,theta=0.3,kout=1), method = "ode45")
out3 <- ode(func = OneComp, times = times,y = yini,
 parms = c(Emax=1.5,EC50=0.6,theta=0.9,kout=1), method = "ode45")
```

Solutions with the three parameter settings are plotted in Figure 6.2. There are only two $C(t)$ curves, as only K_e affects it. This program is for the purpose of illustration of solving ODE systems with the R library. Since for this model, y[2] is not affected by y[1], a computationally more efficient approach is to solve y[2] analytically and only use deSolve(.) for y[1].

An advantage of using an ODE solver is that fitting much more complex models than (6.6) is also feasible. For example, the natural course model (6.7) can be nonlinear; K_{in} and/or K_{out} may also be a time varying function. However, although using complex models does not greatly increase the programming complexity, it does increase computing intensity and reduce computational stability.

6.3 Disease process models

Disease process models are the basis of modeling exposure effects on disease outcomes following a complex time pattern even without the exposure. Some simple disease processes may be modeled by indirect response models, but in many situations a dynamic model dedicated to disease processes is needed. This section introduces a number of models for modeling different diseases.

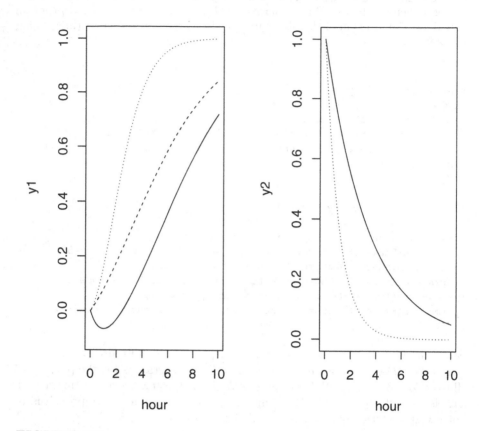

FIGURE 6.2
Solution to the system of ODEs consists of a central compartment, an indirect response model (6.6) and an Emax model for exposure effects, with $K_e = 0.3, EC_{50} = 0.3$ (solid line), $K_e = 0.6, EC_{50} = 0.3$ (dash line) and $K_e = 0.6, EC_{50} = 0.9$ (dot line) using the deSolve(.) library. Left panel: response; right panel: concentration in the central compartment.

6.3.1   Empirical models

Empirical models always play an important role in the situation where mechanistic disease models are not available or are difficult to use. These models describe the disease process by specifying the changes in measurements for disease outcomes, e.g., tumor sizes over time. Linear time trends with random coefficients, e.g.,

$$y_i(t) = \beta_{0i} + \beta_{1i}t + e_i(t), \tag{6.9}$$

are frequently used. This model can be easily extended to using spline functions to increase flexibility. Fitting these models is easy with software for linear mixed models, as described in Chapter 3. It can also be combined with exponential functions, e.g., with

$$y_i(t) = \beta_{0i} + \beta_{1i}t + \beta_{2i}\exp(-\beta_{3i}t) + e_i(t), \tag{6.10}$$

which offers a component that decays to zero naturally. A technical difficulty is that $y_i(t)$ is not linear with respect to β_{3i}, hence the model (6.11) cannot be fitted with linear mixed model software.

Extending to other types of outcomes is straightforward. For example, if $y_i(t)$ is a binary indicator of a disease state, a logistic model can be used with the linear predictor (see Chapter 3) replaced by the right hand side of (6.9). If $y_i(t)$ is a count, e.g., of asthma attacks in time interval t, a Poisson or negative binomial model can be adapted to include a time trend.

To use a disease model in ER modeling one needs to link the model to the exposure as the output of either the effect compartment or the indirect response model. It is obvious that the exposure can be directly added to the right hand side of the two models as an additional term. But this setting leads to an instantaneous ER model. If the exposure has an impact on the rate of change, e.g.,

$$y_i(t) = \beta_{0i} + \beta_{1i}t + \beta_{2i}\exp(-(c_i(t)\beta_{3i}t) + e_i(t), \tag{6.11}$$

then the model is an empirical indirect response model specifying a dynamic relationship between $y_i(t)$ and $c_i(t)$.

6.3.2 Slow and fast subsystems

Although the combination of exposure and disease process systems is often complex, sometimes it can be simplified if one system (often the exposure dynamic) is much faster than the other. For example, suppose that the systems for the central and effect compartments are the fast and slow systems, respectively:

$$\begin{aligned} \frac{dc(t)}{dt} &= h(c(t), c^*(t)) \\ \frac{dc^*(t)}{dt} &= \epsilon h^*(c(t), c^*(t)) \end{aligned} \tag{6.12}$$

where $h(c(t), c^*(t))$ and $h^*(c(t), c^*(t))$ are functions of comparable value range, ϵ is a small number so that it makes the second in model (6.12) the slow system (i.e., with much lower rate $dc^*(t)/dt$ than $dc(t)/dt$). When the first one is much faster than the second one in the sense that it reaches steady state while $c^*(t)$ has almost not changed, an approximation to the solution of the system can be obtained in two stages. First, when $t < t_0$ with a given t_0,

$$\lim_{\epsilon \to 0} c^*(t) = c^*(0) = 0, \tag{6.13}$$

where $c^*(0)$ may also be set to a non-zero value. Under this condition, the system is simplified to

$$\frac{dc(t)}{dt} = h(c(t), 0) \tag{6.14}$$

for which we assume a solution $c(t) = \psi_0(t)$ exists. In the second stage, we replace $c(t)$ with $\psi_0(t)$ in the second system

$$\frac{dc^*(t)}{dt} = \epsilon h^*(\psi_0(t), c^*(t)), \tag{6.15}$$

then solve this system to obtain $c^*(t)$.

Now we consider a more specific system consisting of models for the central and effect compartments and the response:

$$\begin{align}
\frac{dc(t)}{dt} &= h(d(t), \boldsymbol{\theta}) \\
\frac{dc^*(t)}{dt} &= -Kc^*(t) + K_{in}c(t) \\
y(t) &= \beta_0 + E_{max}/(1 + EC_{50}/c^*(t))t \tag{6.16}
\end{align}$$

where the third one is a linear disease process model with growth rate $E_{max}/(1 + EC_{50}/c^*(t))$ controlled by latent exposure $c^*(t)$. Note that, since there is no $c^*(t)$ in the first submodel, one can solve the system sequentially starting from the first one. However, we show that the system can be simplified as long as the first one is much faster than the rest. In fact, the first model may be rather complex and may not be easily solved. Consider the case of a single dose which leads to solution $c(t) = \psi_0(t)$ that decays to 0 much quicker than the second system, then

$$\begin{align}
c^*(t) &= \int_0^t \exp(-K(t - \tau))K_{in}\psi_0(\tau)d\tau \\
&= K_{in}\exp(-Kt)\int_0^t \exp(K(\tau))\psi_0(\tau)d\tau. \tag{6.17}
\end{align}$$

As long as the $c(t)$ dynamic is much faster than the elimination rate K, there exists a t_0 such that

$$\begin{align}
\exp(K\tau) &\approx 1 & \tau \leq t_0 \\
\exp(K\tau)\psi_0(\tau) &\approx 0 & \tau > t_0. \tag{6.18}
\end{align}$$

Using these approximations, we can write

$$\int_0^t \exp(K(\tau))\psi_0(\tau)d\tau \approx \begin{cases} U(t_0) & t > t_0 \\ U(t) & t < t_0 \end{cases} \tag{6.19}$$

where $U(t) = \int_0^t \psi_0(\tau)d\tau$ is the area under the curve (AUC) of $c(t)$, which may be calculated analytically for a simple system, or non-parametrically for a complex system. Note that as $\psi_0(t) \to 0$ quickly, $U(t) = U$, a constant when $t > t_0$, for not small t_0 (relative to the $c(t)$ dynamic), hence

$$c^*(t) \approx K_{in} \exp(-Kt)U \qquad (6.20)$$

is a good approximation to (6.17). Taking $c^*(t)$ into the third equation leads to

$$y(t) = \beta_0 + E_{max}/(1 + EC_{50}/(\exp(-Kt)U))t, \qquad (6.21)$$

where we omitted K_{in}, as it is absorbed into EC_{50}. This simplification supports a simple approach to using PK parameter AUC as a summary exposure in the effect-compartment model when the $c(t)$ sub-system is a much faster system, as Wu (2006) used in the HIV dynamic models. Further improvement to the simple approximation is possible by using the perturbation method to dynamic systems that consist of fast and slow subsystems. Due to the complexity of these approaches, the reader interested in the development of this area is referred to texts such as O'Malley (1991).

6.3.3 Semimechanistic models

Some models may be considered semimechanistic since there is no clear pharmacological mechanism to support it, but they may consists of hypothetical compartments to represent important quantities of measurement. For example, the Positive and Negative Syndrome Scale (PANSS) is a common measure for the symptoms of schizophrenia, a severe mental illness. To model the disease process of schizophrenia the following model has been used (Reddy, 2013),

$$d\text{PANSS}(t)/dt = K_{in} - K_{out}(1 + \beta C_e)\text{PANSS}(t), \qquad (6.22)$$

where C_e is a constant exposure ($C_e = 0$ for placebo). The model assumes that the treatment affects the elimination rate of PANSS. The model also assumes the existence of $dPANSS(t)/dt$, although PANSS is in fact on a discrete scale. This assumption, however, does allow practical model fitting using the solution $\text{PANSS}(t) = \text{PANSS}(0) \exp((K_{in} - K_{out}(1 + \beta C_e))t)$. A similar approach was used for modeling the Unified Parkinsons Disease Rating Scale (UPDRS) scores.

6.3.4 Modeling tumor growth and drug effects

A number of empirical models have been used to model tumor growth and treatment effects on it. Wang et al. (2009) used the following model for non-small cell lung cancer (NSCLC). Let $y(t)$ be the tumor size at time t, and the model assumes a linear growth and an exponential decay:

$$y(t) = y(0) \exp(-St) + Kt \qquad (6.23)$$

where S and K are decay and growth parameters. Another model (Claret et al., 2009) incorporated treatment effects

$$
\begin{aligned}
dy(t)/dt &= K_g y(t) + K_d Dose \exp(-\lambda t) y(t) \\
&= (K_g + K_d Dose \exp(-\lambda t)) y(t).
\end{aligned}
\tag{6.24}
$$

In this model $Dose \exp(-\lambda t))$ can be considered as the solution to an exposure kinetic model.

Houk et al.'s (2009) model replaces the dose in model (6.25) with concentration $c(t)$:

$$
dy(t)/dt = (K_g + K_d c(t) \exp(-\lambda t)) y(t).
\tag{6.25}
$$

This model is in fact an indirect response model with exposure affecting the rate of tumor decay. The solution to (6.25) can be written as

$$
y(t) = y(0) \exp(K_g t + K_d \int_0^t c(\tau) \exp(-\lambda \tau) d\tau).
\tag{6.26}
$$

When $c(t)$ is considered piecewise constant, e.g., $c(t) = c_j, t \in (t_j, t_{j+1})$, the solution can be written as

$$
y(t_{j+1}) = y(t_j) \exp(K_g \Delta_j + K_d c_j [\exp(-\lambda t_j) - \exp(-\lambda t_{j+1})]/\lambda).
\tag{6.27}
$$

where $\Delta_j = t_{j+1} - t_j$. Taking the log-transformation on both sides, we obtain on the log-scale

$$
\begin{aligned}
\log(y_{j+1}) &= \log(y_j) + K_g \Delta_j + K_d c_j [\exp(-\lambda t_j) - \exp(-\lambda t_{j+1})]/\lambda \\
&= \log(y_j) + K_g \Delta_j + K_d c_j \exp(-\lambda t_j)(1 - \exp(-\lambda \Delta_j))/\lambda.
\end{aligned}
\tag{6.28}
$$

Hence the log-tumor growth rate has a constant and an exposure dependent component, and the latter decays with a rate λ.

6.4   Fitting dynamic models for longitudinal data

We have described a number of dynamic systems. To fit longitudinal data to these models, for those with an analytical solution, the approaches given in Chapter 3 can be applied directly. For complex models without an analytical solution, one has to combine the algorithm for fitting NLMM with that solving the dynamic system numerically. Both NONMEM and R provide options to implement them in a straightforward manner. The implementation in R uses an ODE solver interface to the NLMM model fitting library nlme. NONMEM uses the same solver and similar model specification as in R. For example, to fit the models:

$$\frac{dR(t)}{dt} = E_{maxi}/(EC_{50}/C(t) + 1) - K_{out}R(t)$$

$$\frac{C(t)}{dt} = -\theta C(t) \tag{6.29}$$

where $E_{maxi}|u_i \sim N(E_{max} + u_i, \sigma_e)$ is the only random coefficient in the models, the following program, similar to that in Chapter 3, can be used.

```
bdata=bdata[order(bdata$Subject,bdata$Time),]
gdata=groupedData(conc~Time|Subject,bdata)

OneComp <- list(DiffEq=list(
        dy1dt = ~ Emax*y2/(EC50+y2) -kout*y1 ,     #Absoprtion dynamic
        dy2dt = ~ - theta*y2, #Central comp. with elimination rate ke
    ObsEq=list(
        c1 = ~ y1,          #No measure for absorption
        c2 = ~ y2,  #Observe concentration c2=amount y2/volume
    Parms=c("Emax","EC50","theta"),
    States=c("y1","y2"),
    Init=list(0,0))  #Initial states for the ODE

moximodel=nlmeODE(OneComp,gdata) #generate function for nlme() call

jk=nlme(conc~moximodel(ka,ke,Cl,Time,Subject)  ,fixed=ka+ke+CL~1,
    random=pdDiag(Cl~1), start=c(Emax=1,EC50=0.5,theta=1), data=gdata,
    control=list(returnObject=TRUE,msVerbose=TRUE),
    verbose=TRUE)
```

The call to the R function nlme(.) follows the same way as shown in Chapter 3, except the model specification is via the moximodel, which calls the ODE solver.

6.5 Semiparametric and nonparametric approaches

6.5.1 Effect compartment model with nonparametric exposure estimates

Sometimes, one may prefer to base the analysis on fewer model assumptions. For example, when the PK model is not well established, one may prefer to use a nonparametric method to describe the concentration time profile in the system. A simple approach is to use linear interpolation to approximate $c(t)$

within $t_{j+1} \leq t \leq t_j$ as

$$
\begin{aligned}
c(t) &= c_j + \frac{c_{i+1} - c_j}{\Delta_j}(t - t_j) \\
&\equiv a_j + b_j t
\end{aligned}
\tag{6.30}
$$

where $\Delta_j = t_{j+1} - t_j, b_j = (c_{j+1} - c_j)/\Delta_j$, $a_j = c_j - t_j b_j$, and c_j is the concentration measured at t_j. Taking (6.30) into the effect compartment model (6.1) and using the well known formula for integration by parts, $\int x \exp(ax)dx = \exp(ax)(ax - 1)/a^2$, we obtain for $0 < t \leq t_1$, with $c(0) = 0$,

$$
c^*(t) = K_{in}K^{-2}[Kt - 1 + \exp(-Kt)].
\tag{6.31}
$$

The $c(t)$ in other intervals can be calculated consecutively. See Wang (2005) for details. When sampling times are well designed, a moderate number of PK samples provide a good approximation to a moderately complex PK profile (Wang, 2001). In this case, since $c(t)$ is a piecewise linear function, it is possible to calculate the effect compartment concentration analytically. The class of approximation tools for PK profiles allowing analytical effect compartment concentrations includes spline functions and polynomials. This algorithm can be implemented in SAS proc NLMIXED with some simple programming (Wang, 2005). Although the calculation is elementary, it is recommended to use MATLAB or similar software to derive it and generate programming codes automatically, particularly when high order polynomials are involved.

6.5.2   Nonparametric dynamic models and lagged covariates

In this section we explore the relationship between the dynamic model based approach and the nonparametric lagged covariate method (Diggle et al., 2002) for modeling delayed exposure effects. Let y_{ij} and c_{ij} be the response and exposure from subject i at time t_j, respectively. A linear model with the exposure as lagged covariates can be written as

$$
y_{ij} = \beta c_{ij} + \sum_{r=1}^{R} \beta_r c_{ij-r} + u_i + \varepsilon_{ij},
\tag{6.32}
$$

where we have omitted other covariates for simplicity. Coefficients $\beta_r, r = 1, ..., R$, represent the contribution of the delayed exposure effect (hysteresis) to the present response, and hence the existence of hysteresis can be tested by the null hypothesis $H_0 : \cap_{r=1}^{R} \beta_r = 0$. This model is easy to fit as long as all necessary data are observed, for example, from a longitudinal study with repeated measures at fixed intervals at equal distances. Note that the test for the null hypothesis may not work under some situations. One example is when hysteresis is short and the time intervals between the t_js are wide. Specifically, the method may not detect a constant delayed effect with $c^*(t) = c(t - \delta)$ if the time intervals are wider than δ.

The lagged covariate model can be considered as a discrete approximation to a model based on the effect compartment model (6.1). With sufficiently dense time intervals $t_1, t_2, ...$, (6.1) can be approximated as

$$c^*(t) = K_{in} \int_0^t \exp(-K\tau)c(t-\tau)d\tau \approx K_{in} \sum_{t_r < t} \exp(-Kt_r)c(t-t_r). \quad (6.33)$$

It is clear that β_r is an estimate for $\exp(-Kt_r)$. In fact, the lagged covariate model can be considered as a nonparametric approximation to a linear model in combination with a morel general linear system,

$$c^*(t) = K_{in} \int_0^t h(\tau)c(t-\tau)d\tau, \quad (6.34)$$

for the effect compartment without assuming a constant rate K. In this case, β_r is an estimate for $h(t_r)$.

Next we apply the lagged covariate approach to the moxifloxacin data. To this end, we use the fitted popPK model, as described in Section 3.4. The prediction of PK profiles based on the fitted model can also be found there. To create lagged concentrations as covariates for each response, we predict the PK profiles with a number of different time lags as the covariates. For this example, we have used unequal lag intervals (0.5, 1, 3, 6, 10) to cover a large range of time intervals with a relatively small number of lag times. Some earlier QT measures may have some lagged concentrations set to zero. The lagged concentrations were included in an LMM with random effects on all their coefficients. The fitted model is summarized below.

```
    AIC      BIC     logLik
 3779.202 3933.431 -1853.601
```

```
Random effects:
 Formula: ~PHKRSL1N + pk05 + pk1 + pk3 + pk6 + pk10 | id
 Structure: General positive-definite, Log-Cholesky parametrization
             StdDev       Corr
(Intercept) 4.276355e+00 (Intr) PHKRSL pk05   pk1    pk3    pk6
PHKRSL1N    2.498573e-03 -0.226
pk05        1.381401e-03  0.051  0.048
pk1         1.614406e-03 -0.035 -0.620 -0.645
pk3         7.838476e-07 -0.071 -0.112 -0.682  0.584
pk6         1.773276e-03 -0.173  0.209  0.117 -0.285 -0.281
pk10        2.191611e-03  0.092  0.027 -0.025  0.078  0.170 -0.544
Residual    6.323127e+00
```

```
Fixed effects: dd_tavg ~ PHKRSL1N + pk05 + pk1 + pk3 + pk6 + pk10
                Value Std.Error  DF  t-value  p-value
(Intercept) 2.3008091 1.2376942 470 1.858948  0.0637
PHKRSL1N    0.0029934 0.0005968 470 5.015411  0.0000
pk05        0.0004095 0.0005101 470 0.802786  0.4225
pk1         0.0006924 0.0004593 470 1.507614  0.1323
```

```
pk3           -0.0000001 0.0000002 470 -0.451544  0.6518
pk6            0.0004219 0.0004832 470  0.873148  0.3830
pk10           0.0007972 0.0005246 470  1.519473  0.1293
 Correlation:
           (Intr) PHKRSL pk05    pk1     pk3     pk6
PHKRSL1N -0.625
pk05        -0.127 -0.226
pk1          0.013 -0.179 -0.636
pk3          0.041 -0.068 -0.145  0.353
pk6         -0.368  0.302  0.083 -0.218 -0.222
pk10        -0.334  0.260  0.057  0.012  0.018 -0.329
```

```
Standardized Within-Group Residuals:
       Min          Q1         Med          Q3          Max
-3.76389458 -0.52315789  0.00844584  0.52550998  3.76154683
```

```
Number of Observations: 536
Number of Groups: 60
```

The distribution of predicted individual coefficients from the model can be found in Figure 6.3. Note that, if the linear effect compartment model holds, the coefficients should follow the shape of $\exp(-Kt)$. However, for the QTcF data this seemed not to be the case, as there was no exponential decay in the exposure effects. This suggests that the temporal relationship between moxifloxacin concentration and its QT effect may be more complex than a linear effect compartment model with constant rate can describe.

6.6 Dynamic linear and generalized linear models

The effect compartment and indirect response models are all based on differential equations which describe the response dynamic and its relationship with the exposure. The model fitting heavily relies on solving the differential equations, either analytically or numerically. Under some situations where repeated measures on the response, e.g., the tumor size or a biomark value, are taken at equally spaced time intervals $t_1, ..., t_J$, with $t_j - t_{j-1} = \Delta, j = 1, ..., J$, some simple models can also be used to describe the dynamic relationship. The dynamic linear model (DLM) is among the simplest, yet has a wide range of applications. Letting y_{ij} be the jth response measure from subject $i, i = 1, ..., n$, we assume that

$$y_{ij} = \rho y_{ij-1} + \beta c_{ij} + u_i + \varepsilon_{ij} \tag{6.35}$$

where parameter ρ controls the dependence of y_{ij} on y_{ij-1}; the other parts are the same as in an ordinary repeated measurement model.

Apparently fitting this model could be done by the LS method, treating u_i as a fixed parameter. However, to do so one needs to condition on y_{i0} for the

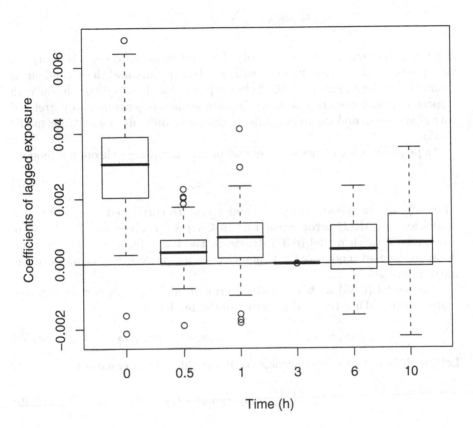

FIGURE 6.3
Distribution of predicted coefficients of lagged moxifloxacin exposures in the
linear mixed model with lagged exposures for QTcF change from baseline.

model of y_{i1}. In most situations in ER modeling with fixed (and often small) J, the LS estimates are inconsistent when $n \to \infty$. When treating u_i as an random effect, however, the MLEs for all model parameters are consistent, hence it is a practical choice. How to treat y_{i0}s in the model is sometimes an important issue. See Hsiao (2003) for details.

When u_i is considered random, this model can be written as a marginal model

$$y_{ij} = \rho^j y_{i0} + \beta \sum_{j'=1}^{j} \rho^{j-j'} c_{ij'} + \varepsilon_{ij}^* \tag{6.36}$$

as a function of whole exposure history. The first term represents the marginal time profile of the system dynamic without the exposure and the second one is a sum of cumulative exposure effects in the past. Model (6.36) is not linear with respect to ρ and the error term ε_{it}^* is variance heterogeneous and correlated within subjects, making fitting this model more difficult than fitting model (6.35).

Dependence on the history may also occur on ε_{ij}, e.g., through a model:

$$\varepsilon_{ij} = \rho \varepsilon_{ij-1} + \varepsilon_{ij}^* \tag{6.37}$$

where ε_{ij}^* are independent, but ε_{ij} and ε_{ij-1} are correlated via ρ. This is a model for the AR(1) error introduced in Chapter 3, given here to illustrate the similarity with model (6.35). However, there is a fundamental difference between the two models, as this model does not introduce dynamics in the mean response.

The model (6.35) without random terms u_i and ε_{ij} is in fact an approximation to an ODE. To see this, we write the model as

$$(y_{ij} - y_{ij-1})/\Delta = [(\rho - 1)y_{ij-1} + \beta c_{ij}]/\Delta. \tag{6.38}$$

Letting $\Delta \to 0$, this model becomes ordinary differential equation

$$\frac{dy_i(t)}{dt} = \eta y_i(t) + \beta^* c_i(t) \tag{6.39}$$

where $\eta = \lim_{\Delta \to 0} (\rho - 1)/\Delta$ and $\beta^* = \lim_{\Delta \to 0} \beta/\Delta$. These limits exist when $y_i(t)$ (without the random components) is sufficiently smooth. Ignoring the random terms avoids technical complexity of introducing a stochastic differential equation.

A more general DLM contains a hidden dynamic part

$$\begin{aligned} \xi_{ij} &= \rho \xi_{ij-1} + \beta c_{ij} + u_i \\ y_{ij} &= \xi_{ij} + \varepsilon_{ij} \end{aligned} \tag{6.40}$$

in which the first one is a dynamic model and the second is an observation model, as ξ_{ij} can only be observed through y_{ij}. DLMs have been used for prediction in engineering for a long time and the most significant method using these models is the Kalman filter. But model (6.40) is much more difficult to fit than model (6.35). One may be tempted to write the former into the form of the latter by replacing ξ_{ij} in the first one in (6.40) with $y_{ij} - \varepsilon_{ij}$ and getting

$$
\begin{aligned}
y_{ij} &= \rho y_{ij-1} - \rho \varepsilon_{ij} + \beta c_{ij} + u_i + \varepsilon_{ij} \\
&= \rho y_{ij-1} + \beta c_{ij} + u_i + \varepsilon_{ij}^*
\end{aligned}
\tag{6.41}
$$

where $\varepsilon_{ij}^* = -\rho \varepsilon_{ij-1} + \varepsilon_{ij}$. Although this model looks like model (6.35), it cannot be fitted with lagged y_{ij}, as ε_{ij}^* and y_{ij} are correlated. This issue can also be looked at from the measurement error aspect, as (6.40) can be considered as a measurement error model in which we only observe ξ_{ij} via y_{ij} with measurement error ε_{ij}. The measurement error for ξ_{ij} does not cause bias, but that in ξ_{ij-1} does. Recent developments on fitting these models have been focused on Bayesian approaches since, unlike model (6.35), model (6.40) cannot be fitted with linear mixed model software, and the performance of the MLE is unsatisfactory (Hsiao, 2003).

Another useful extension of the DLM is introducing time varying coefficients, yet still keeping the linear structure. For example, if the exposure effect is time varying, then it may be described by a dynamic system of its own,

$$
\beta_j = \gamma \beta_{j-1} + b,
\tag{6.42}
$$

where γ and b are parameters to be estimated. Subject effects can also be introduced into these parameters. When $|\gamma| < 1$, β_j tends to its asymptote $b/(1 - \gamma)$. One may choose to write β_j as an explicit function of j and β_0 without adding the DLM for β_j. But the system is no longer linear.

As an application, a linear dynamic model can be a simple alternative to model (6.28) for tumor size per cycle and has nonzero asymptotes. Let $Z_k = \log(Y_k)$ be the log-size at the end of treatment cycle k,

$$
Z_k = \rho Z_{k-1} + K - g(C_k),
\tag{6.43}
$$

where K is a constant and C_k is the cumulative treatment effect in cycle k. This model is more general than (6.28) as in which $\rho = 1$. With constant treatment $C_k = C$ and $\rho < 1$, $Z_k \to (K - g(C))/(1 - \rho)$, i.e., tumor size $Y_k \to \exp((K - g(C))/(1 - \rho))$ when the equilibrium between the natural growth and treatment effects is reached. Therefore, we find ρ plays a key role in determining the long term outcome. This is derived by assuming $Z_k = Z_{k-1} = Z$ in the model, which becomes $Z = \rho Z + K - g(C_k)$, then solving Z from it, a similar approach to finding equilibrium from an ODE by setting the derivatives to zero. Note that these approaches only work when Z_k tends to a constant, i.e., the system is stable. When $\rho = 1$, it is a random walk on the log scale again, while $\rho > 1$ leads to unstable growth; both have no equilibrium states.

As another example, we consider a crossover clinical trial that compares the FEV1 on patients after they were given two active treatments (a) and (c) and a placebo (p) with a crossover design. FEV1 was measured hourly for 8 hours following each treatment and a baseline measure was also taken 11 hours before application of the treatment from 24 patients. The dataset has been used in multiple publications, including Littell et al. (2002). Figure 6.4 gives individual curves of FEV1 measures for all subjects, together with their means. There is a small placebo effect, as the placebo curve has a slight increase after

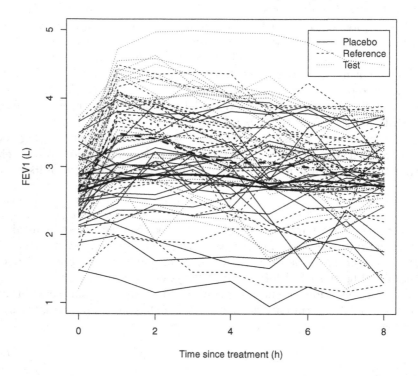

FIGURE 6.4
Mean (bold lines) and individual time profile of FEV1 measures after three treatments.

the treatment. Both active treatments show a significant initial increase at the first hour, then the treatment effects reduce gradually afterwards. Therefore, treatment effects may be measured by the initial increase and also the rate of reduction. From the individual curves, one can also find patterns of correlation between consecutive measures.

We first fit a model to FEV1, i.e, y_{ijk} from patient i under treatment k at hour j on the log-scale:

$$\log(y_{ijk}) = \rho \log(y_{ij-1k}) + \beta_k I(j = 1) + u_i + \varepsilon_{ijk} \tag{6.44}$$

where β_k is the effect of treatment k at 1 hour only, as it is only given at 0 hour. Note that using $\log(y_{i0k})$ as the lagged dependent variable for $\log(y_{i1k})$ is not restrictively correct, due to the much longer than 1 hour interval the post treatment measures have, but it used this way here for illustration. This setting assumes a direct pike effect on the first observation to this dynamic system. Direct effects on later hours can be tested by adding, e.g., $\beta_{2k}I(j=2)$. The model is fitted with codes, where logy and loglag are $\log(y_{ijk})$ and $\log(y_{ij-1k})$ respectively,

```
proc mixed data=fev1;
 class drug hour1 hour2 patient;
 model logy=loglag drug*hour1/solution;
 random patient;
 estimate 'a vs p' drug*hour1 -1 1 0 0 -1 1;
 estimate 'c vs p' drug*hour1 -0 0 -1 1 -1 1;
run;
```

which results in fixed effect estimates and comparisons of the two active treatments with the placebo.

Effect	DRUG	hour1	Estimate	Standard Error	t Value
Intercept			0.3738	0.03966	9.43
loglag			0.6722	0.03051	22.03
DRUG*hour1	a	0	-0.03488	0.02289	-1.52
DRUG*hour1	a	1	0.2065	0.02948	7.00
DRUG*hour1	c	0	-0.01760	0.02340	-0.75
DRUG*hour1	c	1	0.2736	0.02948	9.28
DRUG*hour1	p	0	-0.05162	0.02235	-2.31
DRUG*hour1	p	1	0		

Label	Estimate	Standard Error	DF	t Value	Pr > |t|
a vs p	0.1897	0.03168	546	5.99	<.0001
c vs p	0.2396	0.03198	546	7.49	<.0001

Both active treatments showed significant effects at 1 hour. There is also a strong dependency on the previous measure.

One may suspect if the treatments also have effects on the time course (represented by ρ), e.g., lowering the rate of decay. This can be tested by adding treatment effect on loglag.

```
*effect on rate change;

proc mixed data=fev1;
 class drug hour1 patient;
 model logy=loglag loglag*drug drug*hour1 /solution;
 random patient;
run;
```

and the results show no effect on it (the term loglag*DRUG).

```
             Type 3 Tests of Fixed Effects

                  Num     Den
Effect            DF      DF     F Value    Pr > F

loglag             1      544    476.68     <.0001
loglag*DRUG        2      544      0.79     0.4552
DRUG*hour1         5      544     53.81     <.0001.
```

One can also test if there is a random effect on ρ by adding loglag in the random statement:

```
*random dynamic?;

proc mixed data=fev1;
 class drug hour1 patient;
 model logy=loglag loglag*drug drug*hour1 /solution;
  random intercept loglag/subject=patient;
run;
```

but the results showed no random effect too.

```
Covariance Parameter Estimates

Cov Parm       Subject      Estimate

Intercept      PATIENT      0.007296
loglag         PATIENT             0
Residual                    0.01043
```

Finally, we add a term to represent direct effects at 2 hours:

```
*drug effect later (hour2);
proc mixed data=fev1;
 class drug hour1 hour2 patient;
 model logy=loglag drug*hour1 drug*hour2/solution;
 random patient;
  estimate 'a vs p' drug*hour1 -1 1 0 0 1 -1;
 estimate 'c vs p' drug*hour1  0 0 -1 1 1 -1;
 estimate 'a vs p 2h' drug*hour2 -1 1 0 0 1 -1;
 estimate 'c vs p 2h' drug*hour2 0 0 -1 1 1 -1;
run;
```

but although the effects are overall significant, there is no significant difference between the actives and the placebo.

Effect	Num DF	Den DF	F Value	Pr > F
loglag	1	543	413.52	<.0001
DRUG*hour1	3	543	86.51	<.0001
DRUG*hour2	3	543	3.92	0.0087

Estimates

Label	Estimate	Standard Error	DF	t Value	Pr > |t|
a vs p	0.1899	0.03162	543	6.01	<.0001
c vs p	0.2367	0.03190	543	7.42	<.0001
a vs p 2h	0.02590	0.03169	543	0.82	0.4140
c vs p 2h	0.02159	0.03168	543	0.68	0.4958

In general, it seems the first model we fit is reasonable.

6.7 Testing hysteresis

A proper assessment for hysteresis is advisable before fitting a complex dynamic model to deal with hysteresis. Even when the hysteresis is statistically significant, sometimes it may not affect the PKPD modeling significantly enough to warrant modeling it in the PKPD model. When a number of repeated measures are taken at several time points for both PK and PD, hysteresis may be identified by plotting the PKPD data and linking the points in the order of observation times, if the random error is not large (Csajka and Verotta, 2006). Hysteresis is indicated by an anti-clockwise loop, showing different responses at the same concentration level in the ascending and descending PK phases. The anti-clockwise curves provide an initial view on hysteresis. Intuitively, one may think that the larger the area in the curve, the longer the hysteresis. However, the area within the circle does not quantify the extent of hysteresis. The rate constant in the effect compartment is another quantitative measure of hysteresis with a clear meaning. We will not make a recommendation on which measure of hysteresis should be used, since it is only appropriate to choose it on a case by case basis. In the following, we will discuss approaches to testing hysteresis in a general framework.

As the effect compartment (6.1) is the source of hysteresis, a natural approach to testing the existence of hysteresis is to test if this compartment is necessary. Intuitively, the larger the K (here we assume $K_{in} = K$) the less the hysteresis, since with a sufficiently large K, the effect compartment becomes almost instantaneous. This suggests testing hysteresis with hypotheses

$$H_0 \quad : \quad c^*(t) = c(t)$$

$$H_a \quad : \quad c^*(t) = K \int_0^t \exp(-K\tau)c(t-\tau)d\tau, \tag{6.45}$$

where $K < \infty$. It can be verified that if $c(t)$ is absolutely continuous on a closed set, then $c^*(t) \to c(t)$ when $K \to \infty$. Therefore, we reject H_0 by showing $K < \infty$ with an appropriate statistical test procedure. To this end a natural approach is the likelihood ratio test. This test compares the likelihood function values evaluated at H_0 and at H_a, hence fitting a model for $c(t)$ together with (6.1) is needed. To simplify the model fitting under H_a, the profile likelihood approach (see Appendix A.5) can be used to test hypothesis (6.45). The key parameter for hysteresis is K, hence we can calculate the likelihood function for a large number of fixed K values, then plot the profile likelihood curve for statistical inference on K. However, the calculation of $c^*(t)$ under H_a may not be easy, and the profile likelihood approach may not be accurate when K and other parameters closely depends on each other.

Although the score test does not need model fitting under H_a, it cannot be used to test H_0 as $K \to \infty$ (Wang and Li, 2014c). However, instead of using the effect compartment model, we can use a very simple model as an instrument to perform the score test. Note that to control the type I error of the test, it is sufficient that the model under H_0 is correct. We can consider the simplest model

$$c^*(t) = c(t - \delta), \tag{6.46}$$

where δ represents a constant delay between the changes in the central and effect compartments. Therefore, we can test hysteresis by testing $H_0 : \delta = 0$ vs $H_a : \delta > 0$. A more general instrumental model can also be introduced as

$$c_i^*(t) = c_i(t - \delta_i), \tag{6.47}$$

where $c_i^*(t)$ and $c_i(t)$ are concentrations from subject i, and $\delta_i = \delta_0 \exp(s_i)$ is the ith subject's hysteresis with a fixed component δ_0 and a random component $\exp(s_i)$. However, since it is more difficult to test both together, we will only use (6.46). This model is rarely true, but as an instrumental model, it yields a simple score statistic for δ. It is easy to find that $\partial c^*(t)/\partial \delta|_{\delta=0-} = \dot{c}^-(t)$, the left derivative of $c(t)$ at time t.

We illustrate the approach with a simple ER model

$$y_j = g(c^*(t_j), \boldsymbol{\beta}) + \varepsilon_j. \tag{6.48}$$

The score statistic can be written as

$$S(\boldsymbol{\beta}) = \sum_{j=1}^{J} \dot{c}^-(t_j)\frac{\partial g}{\partial c}(c(t_j), \boldsymbol{\beta})(y_j - g(c(t_j - \delta), \boldsymbol{\beta}))|_{\delta=0}. \tag{6.49}$$

Under some technical conditions, $S(\beta)$ follows a normal distribution with a zero mean and a variance that can be derived (Wang and Li, 2014c). The score test based on the instrumental model does not have an issue with fitting the model with an effect compartment, but still requires extra programming for the calculation of $S(\beta)$ and its variance.

The score statistic for a mixed model can also be derived, but the calculation is complex. To further simplify it, we propose using an approximation for the instrumental model:

$$c^*(t) = c(t - \delta) \approx c(t) - \delta \dot{c}^-(t). \tag{6.50}$$

Note that H_0 is equivalent to $\delta = 0$, so the approximation does not affect the type I error. To illustrate the implementation, we take the linear longitudinal model

$$y_{ij} = c^*_{ij}\beta_i + u_i + \varepsilon_{ij} \tag{6.51}$$

as an example. To test H_0, we fit model

$$y_{ij} = c_{ij}\beta_i + \dot{c}^-_{ij}\eta + \mathbf{Z}^T_{ij}\gamma_i + w_i + \varepsilon_{ij}, \tag{6.52}$$

where $\dot{c}^-_{ij} = \dot{c}^-(t_{ij})$ and $w_i = -u_i\delta\dot{c}^-_{ij}$. That is to add $\dot{c}^-_{ij}$ as a covariate. Then H_0 is tested by testing $H^*_0 : \eta \equiv -\beta\delta = 0$. Although H^*_0 is equivalent to H_0 only when $\beta \neq 0$, this restriction does not limit practical use of the test since it is only when $\beta \neq 0$ that the test for hysteresis is meaningful. The power may depend on how good the approximation in (6.50) is. In practice, $\dot{c}^-_{ij}$ may be derived from a compartmental PK model. For example, with model (2.5), $\dot{c}^-_{ij}$ is proportional to $(-K_e \exp(-K_e t) + K_a \exp(-K_a t))|_{t=t_{ij}}$. Alternatively, it can also be estimated by a linear interpolation approach. The major advantage of this test over the others is the easiness in implementation using common statistical software. This test can also be applied to other models. For example, we can test $H_{0a} : \eta = 0$ for hysteresis in a logistic or beta regression model

$$\mathrm{logit}(E(y_{ij})) = c_{ij}\beta_i + \dot{c}^-_{ij}\eta + \mathbf{Z}^T_{ij}\gamma_i. \tag{6.53}$$

Alternatively, the same approach can be used for generalized estimating equation approaches. The details are very similar to using GLMM, hence are omitted here.

We apply the score test procedure to the moxifloxacin and QT example described in the introduction, using both the analytical derivatives and one using the finite difference approximation. For the approximation, the slope of linear interpolation to the left of each time point is used. To calculate the analytical one, the concentration data from each subject are fitted to the open one compartment model with first order absorption (6.2). The derivative of concentration at each time point is calculated from the fitted PK model for each individual. Then we fit a linear mixed model to the ΔQTcF repeated measures with the corresponding concentrations and their analytical or approximate derivatives as fixed effects and subject as a random effect. Hysteresis is tested via the coefficient of $\dot{c}^-_{ij}$. With the analytical derivatives the χ^2

statistic (with 1 degree of freedom) is 16.19 and with the approximate one it is 12.45; both are very significant. The profile likelihood test can also be used and it yields a χ^2 of 9.0, consistent with the other two tests. Therefore, all three tests suggest the existence of hysteresis between the PK and PD measures, although the linear model fits the data well.

The test based on instrumental models is closely related to the lagged covariates model. Hysteresis can be tested based on this model since the contribution of lagged covariates represents the hysteresis. It has been shown (Wang and Li, 2014c) that under some special situations the score test for a lagged covariate is equivalent to the instrumental model based test using an approximation to $\dot{c}_{ij}^-$. However, in general cases, the instrumental model based test is more powerful.

6.8   Comments and bibliographic notes

Many real examples for how dynamic models play a central role in ER modeling can be found in papers in the *Journal of Pharmacokinetics and Pharmacodynamics*, as well as in Bonate (2006) and Ette and Williams (2007). As typical disease process models, tumor growth models with different levels of complexity vary from empirical ones to those derived from biological and pharmacological mechanisms. Bonate (2011) gives a nice summary and can serve as a starting point. Collaboration with clinical and pharmacological experts is the key to successful dynamic modeling of the ER relationship.

Discrete time dynamic models have been used in statistics as empirical models for longitudinal and time series data. As ways to describe a dynamic system with a simple model, they seem underused in general. A statistical introduction can be found in Fitzmaurice et al. (2006). For more technical details the reader is referred to Hsiao (2003) and econometrics literature specializing in this area.

7

Bayesian modeling and model–based decision analysis

This chapter deals with a few different but closely related topics: Bayesian modeling, meta analysis and model based decision analysis (DA). DA is introduced in a Bayesian framework, following texts on this topic, e.g., Parmigiani et al. (2009), and includes some contents of DA with multiple objectives, in particular, applications in treatment benefit assessment. The meta analysis part includes mixed treatment comparison with an attempt to deal with some special topics in a general framework.

7.1   Bayesian modeling

7.1.1   An introduction to the Bayesian concept

In recent years we have seen a rapid development in Bayesian modeling and inference as well as their application in a wide range of areas, including exposure–response modeling. Bayesian inference is based on a framework which is significantly different from the frequentist one we have used in the previous chapters. In the Bayesian framework, parameters in a model are considered as random variables, in contrast to the frequentist assumption that they are fixed but unknown values. The Bayesian framework also allows us to use prior information to describe the parameters quantitatively via prior distribution. This involves elicitation of prior information into probability distribution to be used for fitting models and Bayesian inference.

So far we have used mainly the MLE and EE approaches for ER modeling. Suppose that we have observed $\mathbf{y} = (y_1, ..., y_n)$ generated from a linear model $y_i = \beta_0 + \beta_1 c_i + \varepsilon_i$ with $\varepsilon_i \sim N(0, \sigma^2)$. Denote $\boldsymbol{\xi} = (\beta_0, \beta_1, \sigma^2)$; we can specify the likelihood function based on the distribution $f(\mathbf{y}|\boldsymbol{\xi})$, then use MLE to estimate $\boldsymbol{\xi}$. This approach is implicitly based on the frequentist assumption that $\boldsymbol{\xi}$ are fixed unknown values. Note that here we treat c_i as fixed for simplicity. In contrast, the core of Bayesian analysis is based on the following coherent logic of information update:

$$Prior\ knowledge \rightarrow Data \rightarrow Model fitting \rightarrow Knowledge\ update \quad (7.1)$$

Since in Bayesian framework parameters are random, the main focus of Bayesian inference is to find the distribution of the parameters that reflects updated knowledge after observing the data. In the Bayesian framework prior knowledge is given by the prior distribution of the parameters $f(\xi)$. For example, an expert may believe that β_0 should follow a normal distribution $N(0, 10)$ without knowing the data. Two key tasks in Bayesian modeling are prior specification, which we will discuss in detail later, and calculating the posterior distribution $f(\xi|\mathbf{y})$ based on data $\mathbf{y}$, model $f(\mathbf{y}|\xi)$ and prior $f(\xi)$. The latter one in principle can be based on the well known Bayesian theorem

$$
\begin{aligned}
f(\xi|\mathbf{y}) &= \frac{f(\mathbf{y}|\xi)f(\xi)}{f(\mathbf{y})} \\
&= \frac{f(\mathbf{y}|\xi)f(\xi)}{\int f(\mathbf{y}|\xi)f(\xi)d\xi}
\end{aligned}
\tag{7.2}
$$

where $f(\mathbf{y})$ is the marginal distribution of $\mathbf{y}$ derived from the prior and the model. The posterior distribution $f(\xi|\mathbf{y})$ is the key outcome of Bayesian analysis derived from the model and data, and is the basis of Bayesian inference. With some simple models and priors, $f(\xi|\mathbf{y})$ can be calculated analytically. But in most cases, it has to be calculated via intensive simulations, such as Markov chain Monte Carlo (MCMC), as described later.

7.1.2 From prior to posterior distributions

The first key step of Bayesian modeling is to specify the prior distribution $f(\xi)$ for model parameters. The prior distribution is often specified in a parametric form $f(\xi|\xi_0)$, with hyper-parameters ξ_0. Therefore, the task of prior specification can be simplified to determining the hyper-parameter values after the function $f(\xi|\xi_0)$ is determined. Although there are many choices in terms of prior distribution, most commonly used are conjugate priors that make the posterior distribution belonging to the same family. For example, if $y \sim Bin(n, p)$ with prior p following $Beta(a, b)$ prior, then $f(p|y) \sim Beta(a + y, b + n)$. Other model-conjugate prior combinations are normal-normal, Poisson-gamma and multinomial-Dirichlet pairs. One practical advantage of using a conjugate prior is that, for some simple situations, the posterior distribution can be derived analytically. This advantage is not as important as in the past, since most models do not have a conjugate prior, and widely used Bayesian software in combination with modern computer power can easily calculate the posterior distribution. Nevertheless, using conjugate priors is still a common choice.

Sometimes a nonparametric prior may also be needed when the model or a part of it is nonparametric, e.g., an nonparametric exposure–response curve, or the baseline risk in Cox regression. Another example of using a nonparametric prior is the Bayesian bootstrap (Rubin, 1972), as described in the appendix. An application to ER modeling can be found in Wang et al. (2002).

Ideally, the prior distribution should reflect the information and knowledge prior to observing the data in the analysis. Priors of this nature are called informative priors. However, often noninformative priors that carry no information are preferred. One reason is to make analyses as objective as possible, hence the analysis uses the Bayesian framework but is not influenced by prior information or knowledge. Another reason is that often there is no prior information or knowledge, or they are too vague to quantify. Sometimes it is not easy to specify noninformative priors, as it may not be clear if a prior is truly noninformative, even in some very simple cases. For example, a prior of equal probability for any possible parameter value (uniform prior) seems to be noninformative. However, if the parameter is transformed on a log-scale, then the prior on the log-scale is no longer uniform, hence its noninformativeness is not transformation invariant. This uniform prior is improper (i.e., it is not a density function) if the parameter range is not finite. This might not be a problem for Bayesian inference as long as the resulting posterior distribution is proper. Nevertheless, improper priors do cause issues in some situations, such as model selection.

There are priors derived for a model that minimize the information it carries into the posterior distribution, known as reference priors. The most commonly used is the so-called Jeffreys prior,

$$f(\boldsymbol{\xi}) \propto |\mathbf{I}(\boldsymbol{\xi})|^{1/2}, \tag{7.3}$$

where $|\mathbf{I}(\boldsymbol{\xi})|$ denotes the determinant of the information matrix $\mathbf{I}(\boldsymbol{\xi})$. It can be easily checked that this prior is transformation invariant, so it does not have the problem of uniform priors. Details about reference priors and Bayesian analysis based on these priors involve some technical complexity. They can be found in, e.g., Robert (2001). In practice, standard software such as SAS and R have the option to use Jeffreys prior, hence its implementation is an easy task.

To avoid the problem of improper uniform priors, one may specify a prior distribution that carries a very small amount of information, but is also easy to handle. For example, $f(\beta) \sim N(1, 1)$ is a strong prior suggesting β being around 1. However, $f(\beta) \sim N(1, 10000)$ carries almost no information, although the prior is still centered around 1. An apparently informative prior but with huge variability is known as a diffused prior and has been widely used in practice. However, a very variable prior may make model fitting unstable, particularly when the sample size is small.

Informative priors are useful, as they represent the expert's best knowledge on the ER relationship, which is invaluable for decision making when data are sparse. The process of transferring the knowledge (or belief) to the prior is called prior elicitation . A common way to elicit a prior is to let the expert specify some statistics and to use a parametric distribution to approximate them. For example, if a coefficient is believed to have mean $= 2$ and a 90% chance of being within the range of 1 and 3, then a normal prior distribution with the property should have mean 2 and $SD = \Phi^{-1}(0.95) = 0.61$. Note that in this case the form of prior distribution, i.e., normality, is also a part of the prior, which is often suggested by the data analyst rather the expert. Other distributions can also be specified to meet the same criteria, and they may behave quite differently when put into a model. The issue is that, although the mean and range are specified correctly, different distributions extrapolate them into different global specifications, which may not reflect the expert's knowledge. Therefore, using informative priors carries considerable risk, especially when the sample size is small.

A Bayesian approach should be robust to prior misspecification in the sense that when the sample size becomes large, the impact of the prior distribution becomes weak and finally diminishes. However, when using a Bayesian approach in practice, we are more interested in its small sample properties. A measure of the impact of a prior on modeling is the effective sample size (ESS) of the prior (Morita et al., 2008). Consider a situation of using a vague or noninformative prior and observing k samples. Suppose that one can derive the posterior distribution based on the vague prior and the data. If an informative prior gives the same amount of information, e.g., in terms of the posterior variance-covariance matrix of parameters, then k can be considered as the ESS of the informative prior. The ESS can be found easily for some specific situations, in particular for conjugate priors. For example, for $y_i \sim N(\mu, 1)$ with prior $\mu \sim N(\mu_0, \sigma^2)$, the ESS is $1/\sigma^2$. In most situations there is no exact equivalence between the prior and a dataset of a certain size, but the ESS can be defined as the sample size that minimizes the determinant of the difference between the information matrices of the posterior distribution from the data and vague prior, and of the informative prior distribution.

Closely related to ESS is the the pseudo data prior (Whitehead and Williamson, 1998), which is specified in the form of data, but reflects prior knowledge about model parameters. Sometimes this comes naturally from the prior elicitation process. For example, a physician may have treated three patients with a treatment and believe that if the three patients are given a 1 mg dose, it is likely two thirds of them may have an AE. Then a record in the pseudo data prior will be two events from three trials under a 1 mg dose. Approximately the ESS of this pseudo prior is 3 from a binary distribution with $p = 2/3$. The whole pseudo dataset may consist of multiple records at different dose levels. The use of a pseudo prior is simply fitting the prior data together with observed data to the model. The resulting parameter estimates are for the posterior modes, hence this approach may be a good choice if one only needs the posterior mode, e.g., for a simulation study to assess dose escalation algorithms, as it is simple and more efficient than using prior distributions in combination with an MCMC approach to obtain the posterior distribution. The pseudo data may not correctly reflect the uncertainty of prior information. For example, to specify a complex prior, one may need 10 records in the pseudo data, but the expert feels that the amount of information, after taking uncertainty into account, should be equivalent to ESS = 3. A practical way to adjust it is to weight the records properly so that the contribution of the prior relative to observed data reflects the uncertainty. In the example above, the weight for the prior data should be 3/10, while observed data have weight 1.

Prior specification is closely related to model structure. Take the Emax model

$$y_i = E_0 + E_{max}/(1 + EC_{50}/c_i) + \varepsilon_i \tag{7.4}$$

as an example. Here the parameters are E_0, E_{max}, EC_{50} and $\text{var}(\varepsilon_i) = \sigma^2$. For convenience we denote $\beta = (E_0, E_{max}, EC_{50})$. Note that one needs to specify prior distribution for all parameters, including σ^2, to find the posterior distribution (7.2). In general, for model

$$y_i = g(c_i, \beta) + \varepsilon_i \tag{7.5}$$

one often uses the priors

$$\begin{aligned} \beta &\sim N(\beta_0, \Sigma) \\ \sigma^2 &\sim IG(a, b), \end{aligned} \tag{7.6}$$

where $IG(a, b)$ is an inverse gamma distribution, in which parameters β_0, Σ, a and b are hyper-parameters we have to supply as a part of the prior distributions. To reduce the impacts of the prior selection, often the hyper-parameters are selected to make the distribution as wide as possible to diffuse the prior. For example, $a = 0.001$ gives a quite variable distribution for σ^2, but it may also cause problems during MCMC, particularly with small sample sizes.

For linear models

$$y_i = \mathbf{X}_i^T \beta + \varepsilon_i \tag{7.7}$$

the priors

$$\begin{aligned} \beta|\sigma^2 &\sim N(\beta_0, \sigma^2\Sigma) \\ \sigma^2 &\sim IG(a, b), \end{aligned} \tag{7.8}$$

where $x|y \sim F$ means that x conditional on y follows distribution F, lead to posterior distribution in a closed form. Note that the prior for β is specified as distribution conditioning on σ. Letting $S_x = \sum_{i=1}^n \mathbf{X}_i^T \mathbf{X}_i$, $\hat{\beta}$ be the LS estimate for β, $\Sigma^* = (S_x + \Sigma^{-1})^{-1}$, and s^2 be the residual sum squares, the posterior distributions are

$$\begin{aligned} \beta|\sigma^2 &\sim N(\Sigma^*(S_x\hat{\beta} + \Sigma^{-1}\beta_0), \sigma^2\Sigma^*) \\ \sigma^2 &\sim IG(a + n/2, b + (s^2 + (\hat{\beta} - \beta_0)^T\Sigma^*(\hat{\beta} - \beta_0))/2). \end{aligned} \tag{7.9}$$

They have the same form as their priors. With these posteriors, one can easily generate posterior samples by first taking samples for σ^2, then sample for β, plugging the σ^2 sample as the parameter in the β distribution.

If we replace the prior for β by Zellner prior (Zellner, 1986)

$$\beta|\sigma^2 \sim N(\beta_0, \sigma^2 S_x^{-1}/n_0) \tag{7.10}$$

the posterior distributions have the simpler form

$$\beta|\sigma^2 \sim N((\hat{\beta} + n_0\beta_0)/(n_0 + 1), \sigma^2 S_x^{-1}/(n_0 + 1))$$
$$\sigma^2 \sim IG(a + (n - p)/2, b + (s^2 + n_0(\hat{\beta} - \beta_0)^T S_x(\hat{\beta} - \beta_0)/(n_0 + 1))/2)$$

$$(7.11)$$

where p is the dimension of β. This prior has a simple relationship with EES, as it means that the prior information on β is equivalent to n_0 "copies" of the observed data. This is reflected in the posterior distribution, as its mean is the weighted average of the LS and prior mean based on the observed and "pseudo" sample sizes and the inverse variances are proportional to the total sample size. One may specify a weak prior by specifying a prior equivalent to, say, 0.1 copy of the observed information.

For NLMM

$$y_{ij} = g(c_{ij}, \beta_i) + \varepsilon_{ij} \qquad (7.12)$$

with $\beta_i \sim N(\beta, \Sigma)$, one may use priors

$$\begin{aligned} \beta &\sim N(\beta_0, \Sigma) \\ \sigma^2 &\sim IG(a, b) \\ \Sigma &\sim IW(\rho, \mathbf{R}) \end{aligned} \qquad (7.13)$$

where $IW(\rho, \mathbf{R})$ is the inverse Wishart distribution with degree of freedom ρ and variance-covariance matrix $\mathbf{R}$. These priors are commonly used for technical convenience, but there is a range of alternatives.

Extending the model-prior combination to GLMM is also straightforward. Supposing that

$$g^{-1}(y_{ij}) = \beta_i^T \mathbf{X}_i \qquad (7.14)$$

with $\beta_i \sim N(\beta, \Sigma)$, we can use

$$\begin{aligned} \beta &\sim N(\beta_0, \Sigma) \\ \Sigma &\sim IW(\rho, \mathbf{R}). \end{aligned} \qquad (7.15)$$

Sometimes there may be extra parameters in the conditional distribution for y_{ij}, for example, the overdispersion parameter. In general, for a parameter representing the magnitude of variation, a prior for variance σ^2 such as the inverse gamma distribution can be used.

For a parametric survival model, prior specification is similar to that for GLM or a nonlinear model. But for a Cox model additional consideration is needed for its unspecified baseline function $h_0(t)$. In the Bayesian framework, the baseline function does not cancel out. Therefore, one has to specify a prior distribution for $h_0(t)$, then calculate the likelihood from

$$S(t) = \exp(-H_0(t)\exp(\beta \mathbf{X}_i)) \qquad (7.16)$$

and $f(t) = -dS(t)/dt$ for Bayesian analysis. The main difficulty is that $h_0(t)$ is a functional and needs a more complex prior than a parameter needs. In order to calculate $H_0 = \int_0^t h_0(\tau)d\tau$ without numerical integration, a common choice is a piecewise increment prior for H_0 within the intervals formed by failure times, which is equivalent to piecewise constant priors for $h_0(t)$. Piecewise constant priors can be implemented in SAS proc PHREG, with the following command as an simple example:

```
BAYES CPRIOR=UNIFORM PIECEWISE=(NINTERVAL=10 PRIOR = UNIFORM);
```

where CPRIOR specifies the prior for the coefficient β and PIECE-WISE=(NINTERVAL=10 PRIOR = UNIFORM) specifies a piecewise constant baseline function with 10 intervals and a uniform prior for the log-hazards for each interval, that is, we assume that $h_0(t) = h_k, k = 1, ..., K$ when t is in the kth interval and $h_k \propto C$. Several other options are available (SAS, 2011).

An alternative Bayesian approach to the above one for fitting a Cox model is to treat the partial likelihood as an ordinary likelihood and derive the posterior likelihood for β as

$$p(\beta|\mathbf{y}) \propto L_p(\mathbf{y}|\beta)p_0(\beta) \qquad (7.17)$$

where $L_p(\mathbf{y}|\beta)$ is the partial likelihood (5.6) and $p_0(\beta)$ is the prior distribution for β. Note that the prior for the baseline risk is not needed, as it does not exist in $L_p(\mathbf{y}|\beta)$. Although this approach seems very simple, its implementation is not easy, as the calculation of $L_p(\mathbf{y}|\beta)$ (5.6) requires identification of the risk set at each failure time. An example for implementing this approach can be found in examples for SAS proc MCMC (SAS, 2011).

We end this section with a simple example to contrast the difference between the frequentist and Bayesian approaches. Suppose we have the following data with three subjects given three dose levels and the response is the occurrence of an event.

```
data logit;
  input dose event;
cards;
 10 0
 20 1
 30 1;
```

We would like to estimate the coefficient for dose, as it represents the dose–response relationship. But with these data, it is not estimable with the logistic regression, since the low dose level had no event and the two higher ones both had an event. This situation is known as complete separation. Nevertheless proc GENMOD still reports convergency and gives an estimate of 7.63 with a huge SE.

```
Analysis Of Maximum Likelihood Parameter Estimates
```

Parameter	DF	Estimate	Standard Error	Wald 95% Confidence Limits	
Intercept	1	-114.565	4.4905E8	-8.801E8	8.8012E8
dose	1	7.6275	27195412	-5.33E7	53302037
Scale	0	1.0000	0.0000	1.0000	1.0000

Next we fit a model with diffused prior $N(0, 1000)$. (options nbi=5000 nmc=5000 thinning=5 will be discussed later.)

```
proc genmod data=logit;
  model event=dose/d=b ;
  bayes nbi=5000 nmc=5000 thinning=5 cprior=normal(var=1000);
run;
```

which gives Bayesian posterior

```
Posterior Summaries
```

Parameter	N	Mean	Standard Deviation
Intercept	1000	-34.9541	17.5928
dose	1000	2.5290	1.2744

```
Posterior Intervals
```

Parameter	Alpha	Equal-Tail Interval		HPD Interval	
Intercept	0.050	-71.3520	-6.3441	-69.1895	-5.8979
dose	0.050	0.4350	5.0684	0.3022	4.8904

with a more reasonable posterior mean and SD of β and the lower equal-tail and the 95% highest probability density (HPD) interval larger than zero.

As the sample size is very small, we expect a strong impact of the prior. With var=500 we have

```
Posterior Summaries
```

Parameter	N	Mean	Standard Deviation
Intercept	1000	-25.7519	13.1703
dose	1000	1.9152	0.9798

| | | Posterior Intervals | | | |
Parameter	Alpha	Equal-Tail Interval		HPD Interval	
Intercept	0.050	-55.2237	-3.6299	-53.4359	-2.6741
dose	0.050	0.3442	3.9477	0.3437	3.9418

One can see the estimate is quite sensitive to the prior variance, but both lower bounds of the 95% equal-tail and highest probability density (HPD) intervals (Bayesian counterparts of confidence intervals, Section 7.1.4) of the coefficient for dose are larger than zero. Therefore, although the prior information is toward the null, it may help to ascertain a positive ER relationship, which is uncertain in the simple logistic regression.

7.1.3 Bayesian computing

7.1.3.1 Markov chain Monte Carlo

Bayesian inference is based on the posterior distribution, hence its calculation has been the key issue in implementation of Bayesian approaches. However, nowadays, this calculation can be done much easier and quicker, even for quite complex models, than twenty years ago, thanks to the development of efficient MCMC algorithms and powerful modern computers. Although $p(\xi|Y)$ has no closed form for most model and prior combinations, sampling from $p(\xi|Y)$ with MCMC is still an efficient approach.

MCMC uses Markov chains in simulations to obtain posterior samples from posterior distributions that are difficult or impossible to sample directly, including distributions with no analytical form. Two samplers are commonly used for MCMC and are implemented in common statistical software. When sampling from a multidimension distribution is not feasible, the Gibbs sampler is a standard tool to use, if the distribution of an individual element, conditional on the others, is easy to sample from. For example, if we would like to sample $\mathbf{y} = (y_1, ..., y_k)$ from $p(y_1, ..., y_k)$, using the Gibbs sampler, we first sample y_1^* from $p(y_1|y_2, ..., y_k)$ where $y_2, ..., y_k$ are initial values, then y_2 from $p(y_2|y_1^*, y_3, ..., y_k)$. After obtaining $(y_1^*, ..., y_k^*)$, we sample y_1 from $p(y_1|y_2^*, ..., y_k^*)$. After a large number of steps (known as burn-in in MCMC), under some technical conditions on $p(y_1, ..., y_k)$, the samples $(y_1^*, ..., y_k^*)$ approximately follow distribution $p(y_1, ..., y_k)$. To use the Gibbs sampler, the conditional distributions should be in a simple form.

Under some situations, even a conditional distribution may not be easy to sample from. The Metropolis–Hastings (MH) algorithm provides a feasible way to sample from a wide range of distributions. Supposing that we would like to sample from a distribution $p(\beta)$, the algorithm starts from an initial value, which may be the estimate from a fitted model using a non-Bayesian approach. The algorithm iteratively samples from a "proposal" distribution $q(\beta^*|\beta)$. At step k after obtaining sample $\beta^{(k)}$, β^* is sampled from the proposal distribution $q(\boldsymbol{\beta}^*|\boldsymbol{\beta}^{(k)})$. Then

$$r = \frac{p(\boldsymbol{\beta}^*)q(\boldsymbol{\beta}^{(k)}|\boldsymbol{\beta}^*)}{p(\boldsymbol{\beta}^{(k)})q(\boldsymbol{\beta}^*|\beta^{(k)})} \tag{7.18}$$

is calculated and $\boldsymbol{\beta}^*$ is taken as $\boldsymbol{\beta}^{(k+1)}$ with probability $\min(1, r)$, and $\boldsymbol{\beta}^{(k+1)} = \boldsymbol{\beta}^{(k)}$ with probability $1 - \min(1, r)$. It can be shown that with a large number of iterations, $\boldsymbol{\beta}^{(k)}$ is approximately $p(\boldsymbol{\beta})$ distributed. The proposal distribution should satisfy some technical conditions to avoid the algorithm running in circle in a region and to ensure that β^k explores all the parameter space. The proposal distribution might be symmetric, i.e., $q(\boldsymbol{\beta}^*|\boldsymbol{\beta}^{(k)}) = q(\boldsymbol{\beta}^{(k)}|\boldsymbol{\beta}^*)$. For example, normal proposal distribution $\boldsymbol{\beta}^* \sim N(\boldsymbol{\beta}^{(k)}, \boldsymbol{\Sigma})$ is symmetric. An algorithm using a symmetric proposal distribution is known as a random walk Metropolis algorithm.

Although with MCMC Bayesian computing is feasible to program, in practice, using readily made software is much more convenient. For some specific models such as GLM or survival models, the MCMC procedure has been integrated into SAS proc GENMOD, LIFEREG and PHREG, so the user only needs to specify options for the prior and the MCMC algorithm. There are other MCMC packages such as WinBugs with an R-interface. One practical advantage of SAS proc MCMC is that it has a similar structure for random effects as in proc NLMIXED, hence it is easier to program mixed models than in other software. Proc MCMC has the same random statement to specify random effects in the same way as in proc NLMIXED. The PRIOR statement is to specify prior distribution for the parameters. The PARMS statement declares model parameters (except the hyper-parameters, for which values are specified directly in the distribution function) and assigns initial values. See Section 7.1.5 for details. NONMEM can also be used for Bayesian analysis, with fewer options than SAS and R.

7.1.3.2   Other computational approaches

There are other Bayesian (or quasi-Bayesian) approaches that do not need a MCMC algorithm. One is the Bayesian bootstrap (BB) proposed in (Rubin, 1981) which has also been considered as a smoothed version of the standard bootstrap, or a weighted likelihood approach. The approach is extremely simple in the case of using a noninformative prior. Suppose that we have data pairs $y_i, c_i, i = 1, ..., n$ to fit an ER model. The BB approach simply weights each pair with a weight following a Dirichlet distribution, $w_i \sim D(1, ..., 1)$, to fit a model, say, $y_i = g(\boldsymbol{\beta}, c_i) + \varepsilon_i$ as usual. Repeating this for B times, each with a new set of weights, one gets $\hat{\boldsymbol{\beta}}_b, b = 1, ..., B$, which, according to Rubin, are samples from the posterior distribution of $\boldsymbol{\beta}$ when we treat the individual data pair as skeletons with prior probabilities of occurrence following the Dirichlet distribution with all parameters equal to 1. The approach can be further simplified by taking $w_i \sim Exp(1)$ independently. This approach, although much simpler, is equivalent to the original BB, in view of the construction of the Dirichlet distribution from independent exponential variables. The BB method is easy to implement as long as fitting the model with weighted data is technically easy. This includes GLM models, linear mixed models, GLMM fitted with GEE approaches, and nonlinear models with nonsparse data.

There is an obvious link between BB and the standard bootstrap (SB), which is equivalent to using integer (including 0) weights to fit the model in the same way as BB. Therefore, BB can also be considered as a smoothed version of SB. One technical advantage of BB is that it is also more stable than SB, as it ensures each pair having a contribution to each $\hat{\beta}_b$, while for SB, some pairs may be assigned a zero weight and have no contribution to the estimate, which may cause difficulty in fitting the model due to the reduction of sample size.

In some situations approximations can substantially reduce the computing burden simulation based approaches need. Recall the robustness of a Bayesian approach which shows that when sample size is sufficiently large, the posterior distribution of the model parameter tends to the same asymptotic normal distribution the MLE follows. This suggests that with a sufficiently large sample size, it is possible to approximate the posterior distribution based on the normal distribution. This approach was used frequently before MCMC algorithms were well developed and computer power made MCMC reasonably quick. We will not explore in this direction, as its implementation is not as straightforward as MCMC using general or specific software.

7.1.4   Bayesian inference and model-based prediction

Bayesian inference for model parameters of interest can be mostly based on the posterior distribution. After obtaining $p(\theta|Y)$ with observed data Y, inference for θ is straightforward. For example, we can calculate $P(\theta > \theta_0|Y)$, or the 95% credible interval which can be defined as the narrowest (HPD) interval (C_1, C_2) satisfying $P(C_1 \leq \theta \leq C_2|Y) = 0.95$, or that with equal 2.5% tails. One may also test hypothesis $H_0 : \theta < \theta_0$ according to $P(\theta < \theta_0|Y)$ and reject H_0 if $P(\theta < \theta_0|Y) < \alpha$. This approach is similar to the frequentist test procedure. But an important difference is that the type I error of the frequentist test refers to the rejection rate of repeating the experiment under H_0, while $P(\theta < \theta_0|Y)$ is simply the probability that H_0 is true, given the data and model. Similarly, $P(C_1 \leq \theta \leq C_2|Y)$ is the probability of $C_1 \leq \theta \leq C_2$, given the data and model. Therefore, the Bayesian framework leads to more a intuitive interpretation of modeling results.

The Bayesian point estimate for θ can also be derived from the posterior distribution. In fact, the posterior mean, median and mode all can be estimated from it. If a criterion for point estimate can be specified, an optimal one can be found too. For example, the posterior mean is the optimal estimate that minimizes the posterior mean squared error of the prediction, in the sense of Bayesian decision analysis, as discussed later.

In the frequentist framework, although one can predict further outcomes with estimated parameters, it is often difficult to quantify the uncertainty and variability of the prediction, as they typically depend on true parameter values. Bayesian prediction can be easily based on the posterior distribution. Letting y_0 be the outcome in the future, its distribution can be written as

$$p(y_0|Y) = \int p(y_0|\beta)p(\beta|Y)d\beta \qquad (7.19)$$

and inference on y_0 can be made in the same way as for β. As the MCMC approach gives the posterior samples $\beta_1, .\beta_s.., \beta_S$, rather than $p(\beta|Y)$, we may calculate (7.19) by

$$\hat{p}(y_0|Y) = \sum_{s=1}^{S} p(y_0|\beta_s). \qquad (7.20)$$

Alternatively, one can generate posterior samples for y_0 from $p(y_0|\beta_i)$ so that $p(y_0|Y)$ is characterized by these samples.

Bayesian model selection is also straightforward. Supposing that there are two possible models M_1 and M_2 and their prior probabilities are $P(M_1)$ and $P(M_2)$ and under model M_k response Y has distribution $p(Y|\theta_k, M_k)$, one can calculate the odds of model M_1 being true as

$$\frac{P(M_1|Y)}{P(M_2|Y)} = \frac{p(Y|M1)}{p(Y|M2)} \frac{P(M_1)}{P(M_2)} \qquad (7.21)$$

where $p(Y|M1)/p(Y|M2)$ is known as the Bayesian factor, and, e.g., $p(Y|M1)$ can be calculated as $p(Y|M1) = \int p(Y|\theta_1, M_1)p(\theta_1|M_1)d\theta_1$. Note that the approach applies when models M_1 and M_2 are completely different, as are the parameters. In contrast, the likelihood ratio test for model selection requires nested models, i.e, either M1 is included in M2, or vice versa. Therefore, the Bayesian approach also has an advantage for model selection.

7.1.5 Argatroban example by Bayesian analysis

Here we revisit the argatroban example to fit the same models with Bayesian approaches. To start we use the sequential approach to fit the Emax model for the relationship between APTT and predicted concentration. For simplicity we reuse the frequentist concentration prediction to concentrate on the ER model. The following proc MCMC program is adapted from the proc NLMIXED program in Section 4.2.3.

```
proc mcmc data=pred nbi=5000 nmc=3000 outpost=postout seed=23
                                            init=random;
ods select Parameters REParameters PostSummaries;
array theta[3] emax ec50 e0;
array mu0[3] (0 0 0);
array Sig0[3,3] (1000 0 0 0 1000 0 0 0 1000);
prior theta ~ mvn(mu0, Sig0);
```

```
prior sigu ~ igamma(0.01,scale=0.01);
 prior var_y ~ igamma(0.01, scale=0.01);
 parms  theta {20 5 5} var_y {4} sigu {1};
    emaxi=exp(emax+u1);
    ec50i=exp(ec50);
    pred=E0+(emaxi-E0)*conc/(conc+ec50i);
  random u1 ~ normal(0,var=sigu) subject=subject;
model pd ~ normal(pred, var=var_y);
run;
```

In the program diffused priors are used for all the parameters and the MCMC starting values for some parameters are based on the fitted model by the frequentist approach. The following shows summary statistics of the posterior samples. The means are similar to those of the frequentist estimate, but some SDs are considerably different from the SEs of the frequentist ones.

Parameter	Mean	Standard Deviation
emax	4.8384	0.1205
ec50	7.6948	0.1220
e0	30.7843	0.0888
var_y	21.3648	2.4320
sigu	0.0925	0.0317

However, this fitting as been shown not converged properly. The following program fits a model with a slightly different parameterization and with additional random effects on the baseline.

```
proc mcmc data=pred nbi=10000 nmc=3000 outpost=postout seed=23
                                          init=random;
ods select Parameters REParameters PostSummaries;
array theta[3] emax ec50 e0;
array mu0[3] (0 0 0);
array Sig0[3,3] (1000 0 0 0 1000 0 0 0 1000);
prior theta ~ mvn(mu0, Sig0);
prior sigu ~ igamma(0.01,scale=0.01);
 prior var_y ~ igamma(0.01, scale=0.01);
 parms  theta {20 5 5} var_y {4} sigu {1};
    emaxi=exp(emax+u1);
    ec50i=exp(ec50);
    pred=E0+(emaxi-E0)*conc/(conc+ec50i);
  random u1 ~ normal(0,var=sigu) subject=subject;
model pd ~ normal(pred, var=var_y);
run;
```

The posteriors samples for the model parameters are summarized as follows.

Posterior Summaries

| | Standard | Percentiles |

Parameter	Mean	Deviation	25%	50%	75%
emax	4.8846	0.1188	4.8038	4.8787	4.9665
ec50	7.7193	0.1185	7.6394	7.7106	7.7984
e0	30.6840	0.0932	30.6036	30.6947	30.7614
var_y	21.5298	2.4275	19.8539	21.3968	23.0193
sigu	0.0851	0.0291	0.0646	0.0805	0.1001

and their densities are plotted in Figure 7.1. The convergence of MCMC is satisfactory.

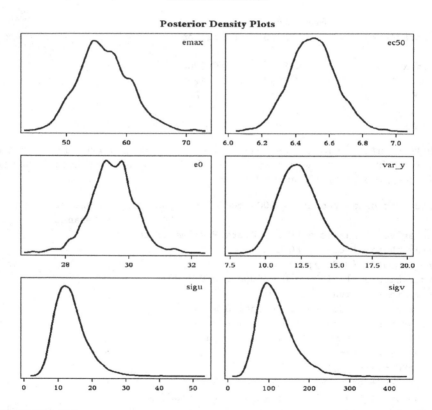

FIGURE 7.1

Posterior distribution of the Emax model parameters using proc MCMC.

The proc NLMIXED program for the joint modeling can be adapted to fit the joint models with proc MCMC. Again we use the results from proc NLMIXED to determine initial parameter values here.

```
proc mcmc data=arg nbi=20000 nmc=3000 thin=5 outpost=postout
    seed=123 init=random PROPCOV=QUANEW;
ods select Parameters REParameters PostSummaries;
array beta[3] emax ec50 e0;
array theta[2] theta1 theta2;
array mu0[3] (0 0 0);
array S[2,2] (1 0 0 1);
array theta0[2];
array Sigma[2,2];
array theta00[2] (0 0);
array Sig0[3,3] (1000 0 0 0 1000 0 0 0 1000);
array Sigb[2,2] (1000 0 0 1000);
prior beta ~ mvn(mu0, Sig0);
prior theta0 ~ mvn(theta00, Sigb);
prior Sigma ~ iwish(4, S);
prior sigu ~ igamma(0.01,scale=0.01);
prior sig ~ igamma(0.01,scale=0.01);
 prior var_y ~ igamma(0.01, scale=0.01);
 parms   theta0 {-5 -2} beta {20 5 5} sig {1} Sigma var_y {4} sigu {1};
cl=exp(theta1);
v=exp(theta2);
pk=(dose/cl)*(1-exp(-cl*t2/v))*exp(-cl*(1-t1)*(time-tinf)/v);
var=(sig**2)*(pk**(2*0.22)));
if ind=1 then pred=pk;
else if ind=0 then do;
   emaxi=exp(emax+u1);
   ec50i=exp(ec50);
   pred=E0+(emaxi-E0)*pk/(pk+ec50i);
   var=var_y;
end;
  random u1 ~ normal(0,var=sigu) subject=subject;
  random theta ~mvn(theta0,Sigma) subject=subject;
model resp ~ normal(pred,var=var);
run;
```

The summary of posterior distribution for the parameters is as follows.

Posterior Summaries

Parameter	Mean	Standard Deviation	25%	Percentiles 50%	75%
theta01	−5.4194	0.0723	−5.4642	−5.4183	−5.3734
theta02	−1.9020	0.0470	−1.9334	−1.9040	−1.8735
emax	9.1851	0.0569	9.1463	9.1765	9.2121
ec50	6.6917	0.1381	6.6074	6.7166	6.7978
e0	30.0969	0.5443	29.6991	30.0835	30.4841
sig	22.9332	0.7897	22.3485	22.9465	23.4893
Sigma1	0.1741	0.0427	0.1429	0.1661	0.1972
Sigma2	0.00937	0.0206	−0.00422	0.00849	0.0215

Sigma3	0.00937	0.0206	−0.00422	0.00849	0.0215
Sigma4	0.0631	0.0187	0.0504	0.0594	0.0721
var_y	16.5058	1.9074	15.1683	16.2539	17.6840
sigu	22.8149	5.4977	18.9758	21.6388	26.1234

However, diagnostic plots (not shown here) suggested that not all the Markov chains were properly converged. Nevertheless, comparing the Bayesian posterior means of the parameters with the estimates using proc NLMIXED, there is no substantial difference. The proc MCMC run with the setting above took about ten times CPU time as that for the proc NLMIXED run. Hence one may ask what is the added value for the Bayesian approach. In terms of statistical inference, the most substantial one is that the Bayesian approach does not require asymptotic assumptions, which is essential for the inference based on fitted models using proc NLMIXED. The distribution of posterior samples is ready to use for estimation of credible intervals, hypothesis testing and decision analysis that we will introduce later.

7.2   Bayesian decision analysis

7.2.1   Decision analysis and ER modeling

The purpose of decision analysis (DA) is to find the best action based on available information to optimize some criteria reflecting the goal of the decision maker. For example, DA can be applied to the following tasks of ER modeling and using an ER model to control the exposure level and/or its outcomes.

- To find the optimal estimation of dose-exposure–response relationships and to predict the outcome based on certain criteria.

- Dose selection to achieve target exposure or response level, e.g., to find the optimal argatroban dose to control APTT level at the normal range.

- Dose adjustment for individual patients to maximize tumor size reduction yet to control the risk of severe AEs.

- Optimal dose selection based on a phase II trial to maximize the probability of success in a phase III trial. This example is typical for using DA for drug development.

- To find stopping rules for searching individual optimal treatment regimens with multiple n-of-one trials to evaluate individuals' response by switching treatments for each patient. One needs to judge when information for a patient is sufficient so no further switching is needed.

- To find the optimal dose to balance the benefit and risk of drug exposure with appropriate measures for the benefit and risk.

The first one is a classical task for DA in Bayesian inference, as one can formulate estimation and prediction as a decision making task and find solutions that satisfy certain criteria, e.g., to minimize the estimation or prediction error. Some of them require balancing between two objectives, hence criterion selection is more complex and involves determining the utility of outcomes. This will be covered later in the chapter.

DA is to select the optimal action based on given criteria and models, in the Bayesian framework, as introduced in the earlier part of this chapter. Three elements in DA are

- the action to be determined,

- the criterion for optimization,

- the outcome determined by the model and action.

Specifying the criterion for optimization based on the task and model is a key step. To consider these tasks more specifically, we consider the following dose-exposure model:

$$y(d, \boldsymbol{\xi}) = g(d, \boldsymbol{\beta}) + \varepsilon \qquad (7.22)$$

where $\varepsilon \sim N(0, \sigma^2)$ is a random error term, and $\boldsymbol{\xi} = (\boldsymbol{\beta}, \sigma)$ is the set of all the parameters. Suppose that our task is to find a dose to deliver the response as close to the target response y_t as possible. We may use squared error $(y_t - y(d, \boldsymbol{\beta}))^2$ to measure the difference between them. But even when $\boldsymbol{\beta}$ is known, $y(d, \boldsymbol{\beta})$ is random due to ε. To eliminate the random term, we may take the average over ε and use a loss function defined as

$$L(d, \boldsymbol{\xi}) = E_\varepsilon[(y_t - y(d, \boldsymbol{\beta}))^2] = (y_t - g(d, \boldsymbol{\beta}))^2 + \sigma^2, \qquad (7.23)$$

hence, in terms of dose selection, the criterion is equivalent to $L(d, \boldsymbol{\xi}) = (y_t - g(d, \boldsymbol{\beta}))^2$. ε does not need to be normally distributed. However, if σ^2 is not constant but a function of dose, then we need to take this part into account as well.

Suppose that instead of achieving a specific response level we would like the response to be within a certain range (y_L, y_H), hence we select d to maximize $P(y_L < y < y_H)$, then the criterion

$$
\begin{aligned}
P(y_L < y < y_H) &= P(y_L < g(d, \boldsymbol{\beta}) + \varepsilon < y_H) \\
&= P(y_L - g(d, \boldsymbol{\beta}) < \varepsilon < y_H - g(d, \boldsymbol{\beta})) \\
&= \Phi\left(\frac{y_H - g(d, \boldsymbol{\beta})}{\sigma}\right) - \Phi\left(\frac{y_L - g(d, \boldsymbol{\beta})}{\sigma}\right) \qquad (7.24)
\end{aligned}
$$

not only depends on $g(d, \boldsymbol{\beta})$ but also also on σ^2. One example of this type of criterion is the success rate of a phase III trial, in which y is the mean response difference. With $y_H = \infty$ and a properly calculated y_L, (7.24) gives approximately the probability of a p-value lower than 0.05, the condition to claim significance in the test for treatment difference.

Although it is relatively simple to eliminate the uncertainty in response $y(d, \xi)$, that in the parameters ξ is much more difficult to deal with. Since parameters ξ are unknown in the frequentist framework, although one can estimate them, it is not possible to evaluate $L(d, \xi)$ at the true value of ξ. This is particularly problematic when the estimates contain considerable variability. An approach is to minimize $L(d, \xi)$ in the worst scenario of ξ (Wald, 1949), which often leads to overconservative or even nonsense decisions. This is a main reason for the use of Bayesian DA (BDA) in order to take advantage of the Bayesian framework. In the Bayesian framework ξ is random, but with prior specified and data H obtained, the posterior distribution can be calculated. Therefore, BDA can minimize the posterior mean loss over the posterior distribution $f(\xi|H)$, i.e.,

$$r(a) = \int L(d, \xi) f(\xi|H) d\xi, \tag{7.25}$$

which is known as Bayesian risk, and the decision that minimizes the risk is the Bayesian decision. The form of (7.25) also suggests that the optimal action d as a function of data only depends on H via $f(\xi|H)$. Specifically, the optimal d based on loss function (7.23) is determined by the equation

$$y_t = E(g(d, \beta)|H) = \int g(d, \beta) f(\beta|H) d\beta. \tag{7.26}$$

The calculation of the right-hand side of (7.26) is very simple. After fitting the models, $f(\beta|H)$ is often represented by its posterior samples $\beta_1, ..., \beta_m$. The right-hand side can be calculated as the mean of $g(d, \beta_i)$s. Then the d satisfying (7.26) can be found by a root finder.

Next we consider the issue of determining the dose to control the exposure level to achieve desired outcomes using joint dose-exposure and exposure–response modeling. One practical situation leading to this type of problem is to select the dose so that the risk of AEs is maintained at an acceptable level. Some drugs control certain biochemistry levels in the body within certain ranges. For example, argatroban can be used to control activated clotting time, as we will use as an example later. A too long clotting time may increase the risk of bleeding, while a too short clotting time may increase another type of risk such as that of deep vein thrombosis, hence all should be avoided. Suppose that after fitting the dose-exposure and exposure–response models

$$\begin{aligned} y_i &= g(h(d_i, \theta), \beta) + \varepsilon_i \\ c_i &= h(d_i, \theta) + e_i \end{aligned} \tag{7.27}$$

to the collection of y_i and c_i data: $H = (\bar{y}, \bar{c})$, we have obtained the posterior distribution $f(\xi|H)$ for $\xi = (\beta, \theta)$. We would like to determine a dose level d so that the response is y_t. However, as ξ are random, there is no dose d_0 such that $g(h(d, \theta), \beta) = y_t$. Instead we may aim at minimizing the difference between the predicted and targeted concentration levels. Accordingly, a mean least square loss function may be defined as

$$L(d, \boldsymbol{\xi}) = (g(h(d, \boldsymbol{\theta}), \boldsymbol{\beta}) - y_t)^2. \tag{7.28}$$

Using it in (7.25) leads to the equation

$$y_t = E(g(h(d_i, \boldsymbol{\theta}), \boldsymbol{\beta}) | H) \tag{7.29}$$

where the posterior mean is taken over parameters in both models. Model (7.27) is easy to use in BDA, as the first one gives the marginal dose–response relationship directly. For other models, such as the classical ME models in Section 4.4, the equation is not easy to calculate, as the dose–response model is not simply $g(h(d_0, \boldsymbol{\theta}), \boldsymbol{\beta})$, due to measurement errors in the exposure. Section 9.1 gives details on the calculation.

There are many other types of loss functions, and the selection of the loss function is a task of multiple disciplines and should involve, e.g., clinicians and pharmacologists, for the example we consider. For example, if the treatment is to control the response within a normal range (L, U), the loss function may be the probability of the response within this range. Outside of this range the distance between the limits and the response may still be penalized. A loss function for this purpose may be written as

$$L(d_0, \boldsymbol{\xi}) = \begin{cases} \int_{\boldsymbol{\xi}} (g(h(d_i, \boldsymbol{\theta}), \boldsymbol{\beta}) - U)^2 dF(\boldsymbol{\xi} | H) & \hat{y}(\boldsymbol{\xi}) > U \\ 0 & L \leq \hat{y}(\boldsymbol{\xi}) \leq U \\ \int_{\boldsymbol{\xi}} (g(h(d_i, \boldsymbol{\theta}), \boldsymbol{\beta}) - L)^2 dF(\boldsymbol{\xi} | H) & \hat{y}(\boldsymbol{\xi}) < L. \end{cases} \tag{7.30}$$

This loss function is a combination of the loss functions (7.28) and (7.24). Alternatively, if the loss depends on the observed y rather than its mean $g(h(d_i, \boldsymbol{\theta}), \boldsymbol{\beta})$, then we should replace it with $y(\boldsymbol{\xi}) = g(h(d_0, \boldsymbol{\theta}), \boldsymbol{\beta}) + \varepsilon$ in (7.30). In this case, the variance of ε should be included in $\boldsymbol{\xi}$.

Although BDA uses a quite different approach from the frequentist one, it is more general, hence the Bayesian decision has some desirable frequentist properties as well. For example, a decision in the frequentist framework is admissible if no other decision is better than it for some $\boldsymbol{\xi}$ values, and not worse than it for any $\boldsymbol{\xi}$ value. Since in the frequentist framework, one cannot find a uniformly optimal decision, an admissible decision is often the best one can find in general. However, it can be shown that every Bayesian decision with a proper prior is admissible, and furthermore, for every admissible decision, there is a prior (which might be improper) that makes the decision a Bayesian decision. This result is a reassurance that Bayesian decisions are not worse and are often better than the frequentist decisions for the same problem.

Consider a simple problem of deciding the dose for a patient to achieve the target exposure c_t when the concentration follows

$$c_i = d_i + u_i. \tag{7.31}$$

Ideally, if we know u_i, dose $d = c_t - u_i$ meets the target exactly. One can use repeated measures $c_{ij} = c_i + e_{ij}$ to estimate u_i, then use $d = c_t - \hat{u}_i$. Without any individual level estimate, one may use $d = c_t - E(u_i)$ if we know the population mean of u_is. One may question what is the value of the information about u_i. This question leads to an important concept in BDA: the value of information (VOI), which is a measure of the benefit of knowing the information about the model, e.g., the parameter values, compared with not knowing it. The most extreme case is when we have prefect information, i.e., the exact value of u_i, and the VOI in this case is called the value of prefect information (VOPI). Suppose that we use criterion

$$r(d) = E(L(d, u_i)) = E((c_t - (d + u_i))^2). \qquad (7.32)$$

With $d = c_t - E(u_i)$, $r = \text{var}(u_i)$, if with repeated measure data u_i has posterior mean m_i and variance σ_i^2 using $d = c_t - m_i$ gives $r(d) = \sigma_i^2$, which in most cases is much lower than $\text{var}(u_i)$. In the extreme case that one knows u_i value, then obviously $r(d) = 0$ is the risk when having prefect information. Therefore, for this scenario, the VOI of knowing m_i is

$$VoI = r(d) - r(d(m_i)) = \text{var}(u_i) - \sigma_i^2 \qquad (7.33)$$

and the VOPI is

$$VoPI = r(d) - r(d(u_i)) = \text{var}(u_i). \qquad (7.34)$$

In most cases, it is impossible have prefect information, but the VOPI is an index for how much benefit is left by gaining all information. A very small VOPI suggests that further information at considerable cost may not be worthwhile to obtain with significant effort. VOI can be used to decide if one should stop a process of sequential decisions and observations, as we will show later.

7.2.2   Example: Dose selection for argatroban

Suppose that the normal range of APTT should be 30 to 40 s and APTT longer than 70 s triggers clinical alert. Here we use the loss function (7.30) and calculate the Bayesian risk (7.25) using the posterior samples generated from the fitted model. These APTT ranges are for illustration of Bayesian decision analysis. The following program extracts the posterior dataset generated by proc MCMC and generates new y_i at grid dose levels, then calculates $r(d)$ for a number of times $t_k.k = 1, ..., K$. Denoting the risk at time t as $r(d, t)$, the overall risk in the time range of 0 to 480 minutes can be written as

$$r(d) \equiv \sum_{k=1}^{K} r(d, t_k) \approx \int_{t=0}^{480} r(d, t) dt. \qquad (7.35)$$

To calculate $r(d, t)$s, we use the posterior samples in a dataset postout from the fitted model with 1:5 thinning in proc MCMC. The calculation repeats that in the proc MCMC in Section 7.1.5.

```
data theta1;  set postout;
keep iteration theta1_1-theta1_37; run;
data theta2;  set postout;
keep iteration theta2_1-theta2_37; run;
data u1;  set postout;
keep iteration u1_1-u1_37; run;

proc transpose data=theta1 out=theta1t(rename=(col1=theta1));
  by iteration; run;

proc transpose data=theta2 out=theta2t(rename=(col1=theta2));
  by iteration; run;

proc transpose data=u1 out=u1t(rename=(col1=u1));
  by iteration; run;

data pred;
  merge postout (keep=iteration emax ec50 e0 var_y) theta1t theta2t u1t;
  by iteration;
cl=exp(theta1);
v=exp(theta2);
do time=10 to 480 by 10;
tinf=240;
t1=1;
if time>tinf then t1=0;
t2=tinf*(1-t1)+t1*time;
do dose=0.1 to 5 by 0.2;
pk=(dose/cl)*(1-exp(-cl*t2/v))*exp(-cl*(1-t1)*(time-tinf)/v);
  emaxi=exp(emax+u1);
  ec50i=exp(ec50);
  pred=E0+(emaxi-E0)*pk/(pk+ec50i);
  newy=pred+rannor(12)*sqrt(var_y);
  if 30< newy and newy<70 then loss=0;
  else loss=max((30-newy)**2,(70-newy)**2);
  output;
end; end; run;
```

Figure 7.2 shows the risk curves for doses up to 5 units, using the loss function (7.28) with $L = 30$, and $U = 70$ and $U = 50$, respectively. It can be seen that the optimal doses are either roughly 2 units or 1 unit, for $U = 70$ and $U = 50$, respectively.

7.2.3 Multistate models

Multistate models play a critical role in BDA, as the utility is often mainly or purely state-driven, i.e., the utility only depends on the state the patient is in. We have introduced time-to-event models for this type of scenario in Chapter 5. However, for decision analysis often a simplified model of discrete time is

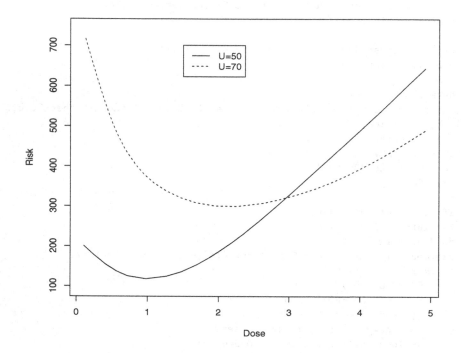

FIGURE 7.2
Risk as a function of argatroban dose based on loss function (7.28) with $L = 30$, and $U = 70$ and $U = 50$.

used to calculate the utility based on health or disease states. For simplicity we assume that time t takes integer values 0, 1, 2,..., and, instead of modeling the exact time of event occurrence, we model the transition from one state to another at each time point. Suppose that there are K possible states and the transition probability from state k to state k' from times $t-1$ to t is $p_{tkk'}$; we can write all the transit probabilities into a matrix $\mathbf{P}_t$ with its kk' element equal to $p_{tkk'}$. As a transition matrix (which is also called a stochastic matrix in matrix theory), the column sums should be 1. With this matrix one can calculate the state occupancy $\mathbf{Q}_T$ at time T, given the initial states $\mathbf{Q}_0$, as

$$\mathbf{Q}_T = \prod_{t=1}^{T} \mathbf{Q}_0 \mathbf{P}_t. \tag{7.36}$$

To use the discrete time multistate model, estimation of the transition matrix is the critical step. It can be estimated by fitting a Markov state model to longitudinal state occupancy data (Mandel, 2010). However, it is often necessary to use the calculation to assess the impact of the exposure in a population for which a discrete time model has already been built, for example, if a clinical trial suggests a 20% reduction in the risk of disease progression via a Cox regression. One may be tempted to simply reduce the transit probability by 20%, i.e., to change the probability by the same proportion as the risk change. This might be a good approximation, but may also be quite wrong. For example, if originally the probability is 0.7, obviously a 50% increase in the risk cannot lead to a 50% increase in the probability. In a complex system, the impact of one risk increase may affect multiple transition probabilities in a complex way. This approximation should be used with caution.

For models with a few (e.g., three) states, the transit probabilities can be derived easily. For the model given in Figure 5.3, the transition matrix from time 0 to t can be written as

$$\mathbf{P}(t) = \begin{pmatrix} e^{-\lambda_s t} & p_{sp} & 1 - e^{-\lambda_s t} - p_{sp} \\ 0 & e^{-h_{sd} t} & 1 - e^{-h_{sd} t} \\ 0 & 0 & 1 \end{pmatrix} \tag{7.37}$$

where

$$p_{sp} = \frac{h_{sp} e^{-h_{pd} t} (1 - e^{-(\lambda_s - h_{pd}) t})}{\lambda_s - h_{pd}}$$

and $\lambda_s = h_{sd} + h_{sp}$. The diagonal elements are in the form of $\exp(-\lambda t)$; then $\exp(-(\lambda + \Delta)t) \approx \exp(-\lambda t)(1 - \Delta t)$ if Δt is small. Therefore, under this condition, the approximation of changing probability by the same proportion as the risk is reasonable. However, for the off-diagonal elements, it is more difficult to judge when this approximation can be used. As this is a time constant system, the transit matrix from time t to t' is $P(t' - t)$. Therefore, one can calculate P_ts for any given time intervals.

The transition probability in a multistate model can also be calculated from a system of ODEs (Kolmogorov equations)

$$\frac{d\mathbf{P}(t)}{dt} = \mathbf{R}\mathbf{P}(t) \tag{7.38}$$

where $\mathbf{R}$ is the rate matrix with element r_{jk} being the transition rate from state j to state k. Setting initial condition $\mathbf{P}(0) = \mathbf{I}$ leads to the solution

$$\mathbf{P}(t) = \exp(\mathbf{R}t) = \sum_{k=1}^{\infty} \mathbf{R}^k t^k / k! \tag{7.39}$$

Note that the elements of $\exp(\mathbf{R}t)$ are not simply $\exp(r_{jk}t)$ unless $\mathbf{R}$ is diagonal. However, if $\mathbf{R}$ has eigenvalues $r_i, ..., r_k$ and their eigen vectors $\mathbf{v}_1, ..., \mathbf{v}_k$, (7.39) can be written as

$$\mathbf{P}(t) = \mathbf{V} diag(\exp(r_k t))\mathbf{V}^{-1}. \tag{7.40}$$

where $\mathbf{V} = (\mathbf{v}_1, ..., \mathbf{v}_k)$. Alternatively, approximations, such as the Pade approximation, can also be used. R library "expm" has a number of options for calculating (7.39).

A special situation worth mentioning is when the transit probabilities are stationary, or tend to stationary, i.e., $\mathbf{P}_t = \mathbf{P}$ for all t when when t is sufficiently large. Since the Markov model describes a dynamic system, for the same reason as given in Chapter 4, we are also interested in, when an action is taken, the outcome at equilibrium . Under this situation, with some conditions on $\mathbf{P}$, the state occupancy tends to its equilibrium $\mathbf{Q}$, which satisfies the relationship

$$\mathbf{Q} = \mathbf{Q}\mathbf{P}, \tag{7.41}$$

hence $\mathbf{Q}$ is the solution to

$$\mathbf{Q}(\mathbf{I} - \mathbf{P}) = 0. \tag{7.42}$$

Note that $\mathbf{P}$ has one eigenvalue equal to 1 and all others have absolute value less than 1 (Horn and Johnson, 1985). Therefore, $\mathbf{Q}$ is the eigenvector corresponding to the eigenvalue equal 1. It can be shown that the kth element in $\mathbf{Q}$ is

$$q_k = a_{kk} / \sum_{l=1}^{K} a_{ll} \tag{7.43}$$

where a_{kk} is the diagonal element of $(\mathbf{I} - \mathbf{P})$. Intuitively, as a_{kk} is the probability of staying in state k, the result shows the state occupancy at equilibrium only depends on the probabilities of staying in individual states.

7.3 Decisions under uncertainty and with multiple objectives

7.3.1 Utility, preference and uncertainty

We introduced BDA using the loss function and Bayesian risk previously, assuming that the outcome of the decision can be measured by the loss function, whose definition is often pragmatic. However, in most cases evaluation of the outcome is a difficult task, as a pragmatic loss function often does not reflect the social benefit and risk. This section gives a brief introduction to the concept of utility and preference of outcomes from practical views. The reader is referred to Parmigiani et al. (2009) for technical details.

The key concept in utility theory is that there exists a utility function $U(y)$, a function of outcome y, whose value reflects the decision maker's (DM's) preference, under some assumptions reflecting rationale preference, so that if DM prefers $y = A$ over $y = B$, then $U(A) > U(B)$. Rationale preference is reflected in the properties of $U(.)$. These include that if $U(A) > U(B)$ and $U(B) > U(C)$, then $U(A) > U(C)$, and if $U(A) \leq U(B)$ and $U(B) \geq U(A)$, then $U(A) = U(B)$, and that $U(y)$ is defined for all possible ys. As the outcome y is often random, $U(y)$ is a random variable. The foundation of classical DA is the expected utility (EU) principle and the Bayesian DA problem becomes finding the optimal decision A to maximize the EU $E(U(y(A)))$. Here we assume that the action affects $U(.)$ only via the outcome y for simplicity.

Sometimes the DM's preference seems obvious. For example, a larger tumor shrinkage is preferred over a smaller one; lower blood pressure (within a certain range) is better than higher blood pressure for hypertension patients. However, one major issue in ranking outcomes and defining $U(.)$ is the uncertainty of outcomes, which may contain uncertainty in their nature or in the model and model parameter. Therefore, the utility should take DM's attitude toward uncertainty into account. One common approach to elicit utility under uncertainty is called a standard gamble . This approach gives the DM a hypothetical scenario to choose between living in the current health state y_0, and taking a gamble with probability p to have $U = 1$ (prefect health) and $1 - p$ to have $U = 0$ (death). The p value that makes the DM indifferent between the two choices is the value of $U(y)$, as it makes the EUs of both choices equal. Taking the gamble, the EU is $1 \times p + 0 \times (1 - p) = p$, which is the same as $u(y_0)$ since getting y_0 is certain. Often the elicited $U(y)$ is a concave function, as normally people are risk averse, i.e., they tend to avoid uncertainty. The concept of risk aversion can be interpreted in monetary terms. Suppose a gamble is set to given choices of taking 10,000 dollars for sure, or taking a 1% chance to win 1,000,000 and otherwise getting nothing. Normally people would take the first choice, although on average the expected reward is the same. One may increase the chance to win to find at which value people would be indifferent among these choices, and the value is the utility of 10,000 dollars relative to unit utility at 1,000,000.

The other issue in ranking outcomes is that in most situations there are multiple, often conflicting, objectives to consider. In the decision to select the dose, often a higher dose may increase not only efficacy, but also the risk of AEs. Therefore, dose selection is balancing two conflicting objects: increasing efficiency and reducing risk of AEs. The most common approach to dealing with multiple objectives is to find a measure that summarizes the multiple objectives; the clinical utility index is an example. For example, for cancer patients, an oncologist gave the utility according to efficacy and safety outcomes (Houede et al., 2010), shown in Table 7.1. Here progressive disease means that the treatment cannot control tumor growth, while stable disease refers to the situation when the tumors are controlled at the current size. The best outcome is that the tumor responds to the treatment and shrinks. The AE grade represents the severity of AEs, with 0 for no AE. But the grade is not cardinal, i.e., a grade 2 AE may not be twice severe as one of grade 1. In the table, apart from the values for the best and worst cases, the expert has to determine seven values. Clinical utility indices have been defined for a

TABLE 7.1
The utility of safety and efficacy outcomes of oncology patients of an oncologist (Houede et al., 2010). The numbers in brackets are approximate utilities using additive weight approximation.

AE grade	Progressive disease	Stable disease	Tumor response
0	40 (46.7)	90 (73.3)	100 (100)
1	10 (23.3)	50 (50)	80 (76.7)
2	0 (0)	40 (26.7)	40 (5.33)

number of situations as the utility of clinical outcomes. Another example, the CUI in Ouellet et al. (2006) for insomnia treatments, takes five components: 1) residual effect; 2) wake after sleep onset; 3) quality of sleep; 4) latency to persistent sleep; 5) sleep architecture, into account. All five terms have quantified criteria to measure if they are achieved. To summarize them into a single index, a weighted mean with relative weights 35, 25, 17, 13 and 10 was suggested. However, we should notice that the these CUIs are disease specific and normally cannot be used to compare outcomes across different indications.

The CUI of Ouellet et al. (2006) is an example of utilities constructed by weighting for multiple outcomes. Weighting is a common way to construct a summary reflecting appropriate tradeoff between them. Suppose that there are K aspects with subutility $U_1, ..., U_k, ..., U_K$, each representing one aspect of the overall outcome. A linear combination of them

$$U = \sum_{k=1}^{K} W_k U_k \qquad (7.44)$$

provides a summary with relative importance of each aspect weighted by W_k. Then the DM may choose a decision to maximize U. The DM is often a panel of experts, hence the weights may reflect views and judgments of multiple persons. The additive utility may not be accurate enough, but may be the only feasible way for a task with, e.g., $K = 5$. A K value as large as 5 is not hypothetical. Drug benefit-risk assessment is a typical example for decision making considering several objectives.

To see how U_k and W_k are set based on utility values for different outcomes, consider the first example where both safety and efficacy endpoints have three categories. So the individual utility should take value 0 for the worst and 1 for the best, leaving the value for the middle category to be determined. After determining the individual U_ks, the only parameter to be determined is W_1, as $W_2 = 1 - W_1$. Therefore, there are in total three values to be determined. We use the additive utility (7.44) to approximate that given by Table 7.1. For this we need to determine the utility of the middle category γ_1 and γ_2 and the relative weight w. We can formally write the approximate utility as

$$U(\gamma, w) = wU_1(\gamma_1) + (1 - w)U_2(\gamma_2) \qquad (7.45)$$

and try to find $\gamma = (\gamma_1, \gamma_2)$ and w to minimize

$$\sum_{i=1}^{n} (u_i - U(\gamma, w))^2 \qquad (7.46)$$

where u_is are the individual utilities for all the categories in Table 7.1. The following codes find the best approximations, as shown in Table 7.1.

```
yv=c(4,9,10,1,5,8,0,4,4)
fun=function(beta){
sdis=rep(c(0,beta[1],1),3)
sae=rep(c(1,beta[2],0),rep(3,3))
sum((yv-10*(beta[3]*sdis+(1-beta[3])*sae))^2)
}

opt=optim(par=c(13/17,6/15,0.5),fn=fun)

$par
[1] 0.7343702 0.3392987 0.5333230
```

Here the marginal mean utilities 13/17 and 6/15 and relative weight 0.5 are used as the initial values for γ_1, γ_2 and w. The optimal ones (given in "par") are not far away from them. However, the approximation for individual utility may be quite poor for some combinations (Table 7.1). This also serves as a reminder of the difficulty of using a simple utility function to represent utility involving multiple aspects.

7.3.2   Cost-benefit, cost-utility and cost-effectiveness analysis

A typical DA problem is how to make a decision to maximize the benefit under budget constraints, which may also be considered an example for decision analyses with two objectives: to maximize the benefit and to minimize the costs, with relative weights reflecting the budget constrain. Cost-benefit (CB) analysis is an approach to explicitly convert the benefit into monetary terms and to compare with the costs incurred. The difference between the two is often referred to as the net benefit (NB) and the DM's task is to maximize the expected NB. The major challenge in CB analyses is eliciting the value of the benefit in monetary terms that really reflect the value of benefit in correct aspects.

In contrast to CB analyses, cost-effectiveness analysis (CEA) deals with situations when the effect of treatment cannot be converted into monetary values. In CEA, effectiveness may be measured in terms of a common measure such as one myocardial infarction (MI) incident avoided. CEA is a typical example of synthesizing multiple objectives yet leaving considerable flexibilities to the DM. However, it cannot be used to compare treatments in different patient populations with different diseases, as effectiveness measures may be very different among them.

Between CBA and CEA, cost-utility analysis (CUA) uses utilities based on, e.g., general quality of life (QoL) measured by some instruments. The major advantage of CUA is that it can be used in medical decision making, such as to determine if drug A or drug B for different diseases should be reimbursed, given the utility measures applied to both disease areas. Most of these QoL instruments are questionnaires with multiple questions covering different health related aspects, weighted by their importance. Therefore, these instruments reflect the relative importance of multiple objectives for health care. Most of them do not directly measure health utility, with a few exceptions of general instruments such as EQ-5D. However, using a general instrument to assess the effect of a specific treatment may not be appropriate, because they may not be sensitive for patients with specific health and disease conditions, and the QoL changes in response to treatments. Mapping methods for some disease specific instruments to a general instrument have been developed to convert disease specific QoL measures to a general utility. The reader is referred to text in this area such as Fayers and Machin (2007). In the following, we assume that the QoL is a general utility, with or without a conversion.

Since we need to assess the benefit within a (often long) period, further summary of QoL measured at different time points is needed. A simple linear time tradeoff is the quality adjusted live year (QALY) calculated as

$$QALY = \int_0^\infty S(t)U(t)dt \tag{7.47}$$

where $S(t)$ is the survival function and $U(t)$ is the utility measured by QoL at time t. QALY can also calculated at an individual level. For example, if someone had a life span of 10 years from $t = 0$, with $U(t) = 0.8$ in the first 5 years and $U(t) = 0.3$ for the rest of the years, then his QALY is $5 \times 0.8 + 5 \times 0.3 = 5.5$ years. Note that QALY has already taken efficacy and safety objectives into account, assuming relevant outcomes can be measured by QoL. The calculation of $U(t)$ often needs modeling activities, in particular multistate models, as described in Section 7.2.3, as $U(t)$ is often state dependent. For example, for cancer patients, those in stable disease may have $U = 0.8$, and those with disease progression may have $U = 0.7$ and a severe AE may further reduce it to 0.4. When patients' health states are modeled by the multi-state model, if the utilities at K states are $\mathbf{U} = (u_1, ..., u_K)$, then the overall expected utility at T is $\mathbf{U}^T \mathbf{Q}_T$ and the cumulative utility until T intervals is

$$QALY = \sum_{t=1}^{T} \mathbf{U}^T \mathbf{Q}_t \Delta t = \sum_{t=1}^{T} \prod_{l=1}^{t} \mathbf{U}^T \mathbf{Q}_0 \mathbf{P}_l \Delta t \qquad (7.48)$$

where Δt is the length of the interval.

The overall cost of a treatment is more straightforward to calculate as a sum of all costs involved, including all relevant costs such as the costs of drug and rescue medication and procedures that occurred during treatment. The next key step to use costs and QALY for decision making is to work out how much a QALY is worth. One may elicit it as willingness to pay, i.e., how much one is willing to pay to buy a QALY. Therefore, to compare a new and expensive treatment t with a standard one s, we calculate the ratio of the QALY gain to the cost increase, the incremental cost-effectiveness ratio (ICER), as

$$ICER = \frac{E_t - E_s}{C_t - C_s}. \qquad (7.49)$$

Then a treatment with ICER lower than the willingness to pay threshold is deemed costeffective and health care providers may reimburse the treatment. This approach gives the health care provider the flexibility to decide which threshold should be used. Note that this may vary substantially among different populations. In the UK, for an approval by the National Institute of Clinical Excellence (NICE) it is often assumed that 30,000 pounds per QALY is an acceptable threshold.

The approach of using ICER to decide reimbursement is a well known decision problem: to allocate resources to multiple projects under an overall budget constraint. It can be shown that the optimal allocation strategy is to order treatments by ICER from low to high (the ordered list is often known as the league table), then to reimburse from the lower end until the budget runs out. This is also equivalent to reimbursing treatments lower than an optimal threshold. Obviously the threshold should be just lower than the ICER of the last reimbursed treatment before the budget runs out.

Sometimes the threshold for ICER can be interpreted willingness to pay (WTP), i.e., the price people are willing to pay to buy a QALY. When WTP is determined, one can calculate the incremental net benefit in monetary terms as

$$INB = WTP(E_t - E_s) - (C_t - C_s) \qquad (7.50)$$

which is a measure for CBA, and a wide range of approaches for CUA are applicable.

7.4   Evidence synthesis and mixed treatment comparison

Decision analysis often needs information from multiple sources such as multiple clinical trials, observational studies, or a combination of them. The approach of synthesizing information from multiple sources is often called meta analysis (Whitehead, 2002). The most fundamental aspect in meta analysis is treatment heterogeneity, i.e., the variability in treatment effects of different studies or sources, and major technical advances in meta analysis deal with this topic. Often a meta analysis uses analysis results rather than raw data to synthesize information obtained from multiple studies or sources, but may also use a mixture of summarized and individual level data. In meta analyses, studies designed for different purposes may compare different treatments. Therefore, to compare specific treatments, one may need to compare them indirectly, if the treatments to be compared are not all from the same sources. The approach using both direct and indirect treatment comparisons is called mixed treatment comparison, in which interest has been growing rapidly in recent years.

7.4.1   Meta analysis

There are a number of excellent textbooks about meta analysis, hence we focus on its key aspects only. Suppose that we are interested in the effect of a treatment, compared with a reference treatment, and have K trials comparing them. The treatment effect is estimated by a comparison between the two treatments in each study as δ_k with SE σ_k. We may assume that

$$\delta_k = \Delta + u_k + \varepsilon_k \qquad (7.51)$$

where Δ is the overall treatment effect, $u_k \sim N(0, \sigma_u^2)$ is treatment heterogeneity, and $\varepsilon_k \sim N(0, \sigma_k^2)$ is the error in δ_k carried over from the estimates with raw data. When σ^2 is the common residual variance for all the K trials, σ_k can be written as $\sigma/\sqrt{n_k}$, and n_k is the sample size of the kth study. Different from linear mixed models introduced in Chapter 3, here we assume that σ_k^2 is known, as σ_k^2 is often estimated in the analysis of individual studies. Therefore, one only needs to estimate Δ and σ_u^2. Fitting such a model is straightforward with common software, except for the need to specify the residual variance σ_k^2. This can be done in proc MIXED by inputting the R matrix from a dataset (SAS, 2011). It can also be fitted using proc NLMIXED, with a variable SEk=σ_k adding to a dataset include variables dk and study, with the following code:

```
mud=Delta+ui;
random uk ~norma(0,sigmau) subject=study;
model dk ~normal(mud,var=SEk**2);
```

proc MCMC uses the same syntax, so this method can also be used in a Bayesian approach. Bayesian approaches are commonly used in meta analysis for their better small sample properties, since the number of studies in a meta analysis is often small. Treatment heterogeneity u_k is the key component in meta analysis, as it determines if treatment effects can be generalized to different studies, which are often considered as representatives of different populations. The ideal situation is when $\sigma_u = 0$, i.e., treatment effects are the same in all trials. It has been a common practice to test heterogeneity, or more specifically $H_0 : \sigma_u = 0$. However, since a meta analysis hardly has sufficient power to detect heterogeneity, it is prudent to keep u_k in the model and take its variability into account when interpreting and using meta analysis results.

The heterogeneity u_k in model (7.51) may depend on some observed covariates, particularly when they are quite different between studies. Meta regression is an approach to model the relationship between the outcome and these factors as well as the treatment. Meta regression adds covariates $\mathbf{X}_k$ into model (7.51):

$$\delta_k = \Delta + \mathbf{X}_k^T \boldsymbol{\beta} + u_k + \varepsilon_k \tag{7.52}$$

where u_k is the residual heterogeneity after adjusting for $\mathbf{X}_k$. This model can be fitted easily in the same way as previously described. Although one may still be mainly interested in Δ, $\boldsymbol{\beta}$ also contains very useful information, as $\mathbf{X}_k^T \boldsymbol{\beta}$ is the predictable part of heterogeneity. For example, if we are interested in the average treatment effect in a mixed population with covariates $\mathbf{X}_{01}, \mathbf{X}_{02}, ..., \mathbf{X}_{0K}$ with proportions $p_1, p_2, ..., p_K$, we can predict it with

$$\hat{\Delta}_m = \hat{\Delta} + \sum_{k=1}^{K} p_k \mathbf{X}_{0k}^T \hat{\boldsymbol{\beta}}. \tag{7.53}$$

Its SE can also be calculated easily from the fitted model.

Sometimes summary data may be given as the mean response $\bar{y}_{k1}$ and $\bar{y}_{k2}$ under treatments 1 and 2, respectively, rather than as treatment difference δ_k. One may work out the treatment differences and their SEs, or may fit the mean responses directly with the model

$$\bar{y}_{k1} = (\Delta + u_k)/2 + v_k + \varepsilon_{k1}$$
$$\bar{y}_{k2} = -(\Delta + u_k)/2 + v_k + \varepsilon_{k2} \tag{7.54}$$

where ε_{kt} is the SE of $\bar{y}_{kt}$, $t = 1, 2$ and v_k is the baseline study effect and is canceled out in δ_k. Note that here we allow for heterogeneity between ε_{kt}s, as the sample sizes of the two treatment arms are often different.

Model (7.54) can be extended to GLMMs. For example, if the response is measured by the number of events y_{kt} out of the total number of patients n_{kt} under treatment t, logistic models with

$$
\begin{aligned}
\text{logit}(P_1) &= (\Delta + u_k)/2 + v_k \\
\text{logit}(P_2) &= -(\Delta + u_k)/2 + v_k,
\end{aligned}
\tag{7.55}
$$

where P_t is the probability of an event under treatment t, can be fitted in either SAS proc GLMMIXED or proc MCMC. When n_{kt} and y_{kt} are large, one may obtain individual treatment effect estimates, then use model (7.51) following the above approach. However, fitting y_{kt} is often even easier, as there is no need to input the SEs. Statistically, this approach is also the preferable one over fitting (7.51), particularly when n_{kt}s are not large.

These meta analysis approaches can be extended to accommodate multiple parameters. For example, the meta regression model for a vector of parameters can be written as

$$
\boldsymbol{\delta}_k = \boldsymbol{\Delta} + \mathbf{X}_k^T \boldsymbol{\beta} + \mathbf{u}_k + \boldsymbol{\varepsilon}_k
\tag{7.56}
$$

where $\mathbf{X}_k$ and $\boldsymbol{\beta}$ are matrices, and $\boldsymbol{\Delta}, \mathbf{u}_k$ and $\boldsymbol{\varepsilon}_k$ are vectors with conformable size with $\boldsymbol{\delta}_k$. To fit the model we may assume $\mathbf{u}_k \sim (0, \boldsymbol{\Sigma}_u)$ and also need to calculate $\text{var}(\boldsymbol{\varepsilon}_k)$ from individual studies. When raw data are available, fitting a mixed model with study-by-treatment effect terms is often an easier approach.

Meta analysis for time-to-event endpoints can be based on fitted survival models such as the Cox model. It is straightforward if the relative risk is our only concern, as we can use (7.51) to fit log-RR as δ_k. The SE of δ_k comes from the fitted Cox model for each study, hence one can input the data and specify it in the way discussed above. This approach is based on good estimates for the individual log-RR and its SE. However, they are difficult to obtain if the events are rare. An alternative when raw data are available is a one step approach fitting raw data with frailty on treatment effect for each study. This approach works better than using log-RR when the number of studies is not small.

In addition to the analyses of a single parameter in a time-to-event model, meta analysis can also be used to synthesize survival and hazard curves. Typical meta data are given as summary statistics by multiple time intervals, e.g., the survival proportion at the end of each interval, or the number of events occurred within each interval. There are multiple ways to model these data. Here we assume data are summarized by treatment and time interval as y_{ijk}: the number of events among n_{ijk} patients at risk in time interval t_j from study i under treatment k. Conditioning on n_{ijk}, y_{ijk} follows a logistic model, $y_{ijk} \sim Bin(n_{ijk}, h_{ijk})$, with the probability of events h_{ijk} (Efron, 1989). Since y_{ijk}s of the same i, may be correlated due to study effects, we may use the following logistic regression model with mixed effects:

$$
\text{logit}(h_{ijk}) = b_{jk} + u_i
\tag{7.57}
$$

where b_{jk}s represent the time profile of $\mathrm{logit}(h_{ijk})$ under treatment k, $u_i \sim N(0, \sigma_u^2)$ is a random baseline for study i. However, often we would like to impose a structure on b_{jk}, e.g., $b_{jk} = \boldsymbol{\beta}_k^T \mathbf{B}(t_j)$, where $\mathbf{B}(t)$ is a set of spline functions, or $\boldsymbol{\beta}_0^T \mathbf{B}(t_j) + \beta_k$. In the latter, treatment effect β_k is separated from the time varying part $\boldsymbol{\beta}_0^T \mathbf{B}(t_j)$, hence the relative risk is proportional to it. The model can be fitted easily with SAS proc GLIMMIX, or approximately with the GEE approach using either function gee(.) in R or proc GENMOD in SAS. From the fitted model we can obtain $\hat{h}_{jk}$, and survival function $S_k(t)$ can be estimated as

$$\hat{S}_k(t) = \prod_{j:t_j \leq t} (1 - \hat{h}_{jk}). \tag{7.58}$$

To obtain a piecewise CI for $S_k(t)$, although one can derive it from asymptotic SEs of $\hat{h}_{jk}$, it is convenient to use a computing intensive approach such as bootstrap, as fitting model (7.57) is easy and efficient. Wang (2012) gives more details and examples of using this approach.

Sometimes the data may be summarized as the survival proportion $\hat{S}_{ijk}$ at the end of each interval. Although one may convert them to y_{ijk}, it is also possible to fit them directly. Recall in Chapter 5, if the risk under different exposures is proportional, one may fit model

$$\log(-\log(S_{ijk})) = \boldsymbol{\beta}_0^T \mathbf{B}(t_j) + \beta_k. \tag{7.59}$$

However, different from fitting model (7.57), here we should take the correlation between $\hat{S}_{ijk}$ and $\hat{S}_{ij'k}$ into account. The correlation has an analytical form (Arends et al., 2008) and can be used in a GLS. Alternatively, one can also use the GEE approach to obtain the parameter estimates and their robust sandwich variance-covariance matrix for statistical inference using a simple working correlation matrix.

7.4.2 Meta analysis for exposure–response relationship

Although commonly a meta analysis is to compare two or more treatments, in some situations one may also need to synthesize dose–response relationship estimates from multiple sources. In fact, technically there is little difference between these two meta analyses. Suppose that the exposure–response relationship in study k is described as

$$y_{jk} = \beta_k d_{jk} + \varepsilon_{jk}, \tag{7.60}$$

with $\beta_k = \beta + u_k$, and the meta analysis is to synthesis evidence about β. We can adapt model (7.51) for the estimate from individual trials as

$$\hat{\beta}_k = \beta + u_k + e_k. \tag{7.61}$$

Then the second stage for inference on β is exactly the same as in the previous subsection. If the outcome (e.g., occurrence of AEs) follows a GLM, then one can also use either a single or a two step approach. In the following we use two examples to illustrate the approach. As the first example, we use the scenario in Zohar et al. (2011) where they reported a meta analysis based on five phase I dose escalation trials for Sorafenib, an anti-cancer agent. The data are read into a SAS dataset and processed for analysis as

```
data p1;
   input study d100 n100 d200 n200 d300 n300
                d400 n400 d600 n600 d800 n800;
cards;
1 0 3 0 3 . . 1 4 1 6 3 3
2  0 4 0 3 1 5 1 10 7 12 1 3
3 0 3 1 6 . .  0 8 3 7 . .
4 1 5 1 6 . .  0 15 4 14 2 7
5  0 3 1 12 . .  0 6 1 6 . .
;


data den (keep=study  n100    n200 n300 n400 n600  n800)
      num (keep=study  d100    d200 d300  d400 d600 d800);
   set p1;
run;
proc transpose data=den out=tden;
   by study;
run;
proc transpose data=num out=tnum;
   by study;
run;


data all;
   merge tden (rename=(col1=num)) tnum (rename=(col1=yi));
   dose=substr(_name_,2,3)+0;
   if num ne .;
run;
```

For the number of dose limiting toxicities (DLT) y_{jk} out of n_{jk} patients in trial k at the jth dose, we assume that $y_{jk} \sim Bin(n_{jk}, p_{jk})$. First a GLMM including a random coefficient for dose

$$\text{logit}(p_{jk}) = \beta_{0k} + (\beta + u_k)d_{jk} \qquad (7.62)$$

was fitted but the variation in u_k is close to zero. Consequently, a logistic model with individual baseline risk for each study is fitted for better small sample properties. We use a simple Bayesian approach with a uniform prior for all parameters and run the following SAS proc GENMOD:

```
proc genmod data=all;
   class study;
   model yi/num=dose study/d=b;
   bayes outpost=pred diag=all nbi=3000 nmc=5000 thinning=5;
run;
```

where the BAYES statement requires 1000 posterior samples generated from the MCMC procedure with a 1:5 sampling ratio under the default uniform prior. From the posterior samples, we can calibrate the dose level corresponding to toxicity probability equal to, e.g., 0.25. As the model allows different baseline risks for individual studies, supposing we are interested in the average probability over these studies, we can take $\beta_0 = \sum_{j=1}^{5} \beta_{0k}$ and calculate the required dose as

$$d = (\log(0.25/0.75) - \beta_0)/\beta. \tag{7.63}$$

From the 1000 posterior samples the median of d is $541(\text{mg/m}^2)$ $(SD = 65$ $\text{mg/m}^2)$. The posterior samples can also be used to study the probability of overdose, e.g., the chance of toxicity probability higher than 35 %. Figure 7.3 gives the mean, median, 75% and 90% percentiles of toxicity probability at a range of dose levels. It can be seen that the target dose is between 500-550 mg/m^2 and, at that dose level, the chance of toxicity probability higher than 35% is less than 10%, as the 90% line is lower than 35% at that dose range. In the analysis we have not considered an issue that the selected dose is in the gap between 400 and 600 mg/m^2 doses, and at a 400 mg/m^2 dose the toxicity probability was even lower than that at a 300 mg/m^2 dose, although only a few patients were given that dose. Therefore, the logistic model may not fit the data well locally.

As a more complex example, we would like to use a meta analysis approach to model the dose-PK and PK-response relationships. Consider a situation when there are K trials, each with multiple dose groups. PK and response measures are taken and only summary statistics are available, as they may come from published papers or databases. We assume the data are given in the following dataset in the SAS data step, which comes from a practical scenario but the numbers are modified for confidentiality reasons.

```
data meta;
   input study dose n y lcon se;
cards;
1  0   55  2 .    .
1  5   53  5 1.67  0.069
1  10  49  8 2.17  0.057
2  0   120 8 . .
2  10  112 22 2.42 0.034
3  0   35  3  . .
3  10  73  12 2.21 0.074
4  0   133 13 . .
4  7.5 127 23 2.22 0.059
;
```

where, for each study and each dose, n and y are the total numbers of patients and those who had a severe AE, and lconc is the log-mean concentration with SE (se) from a power model. There are placebo arms in the data with lconc and se missing. For PK we assume the following model for the mean log-

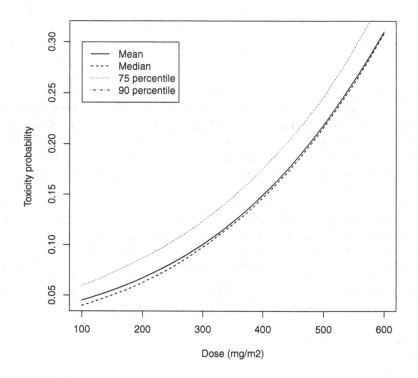

FIGURE 7.3
Mean, median, 75% and 90% percentiles of toxicity probability at different dose levels.

concentration from study k under dose d_{jk},

$$\log(c_{jk}) = \theta_0 + \theta_1 d_{jk} + v_j + e_{jk} \qquad (7.64)$$

where $e_{jk} \sim N(0, \sigma_{jk}^2)$ is the estimation error and σ_{jk} can be calculated from the geometric coefficient of variation (CV), which is often presented in study reports, and here is precalculated and included in the data. This model ensures that the dose-PK relationship passes the origin, hence the missing records in lconc and se are not needed to fit the model. For the AE model we assume that

$$
\begin{aligned}
y_{jk} &\sim Bin(n_{jk}, p_{jk}) \\
logit(p_{jk}) &= \beta_0 + \beta_1 c_{jk} + u_k
\end{aligned}
\qquad (7.65)
$$

where u_k may be correlated with v_k in the PK model.

We can also fit the two models jointly using a Bayesian approach. For this purpose we use the following diffused priors. For simplicity, we reparameterize the correlation between u_i and v_i with a shared latent variable $w_i \sim N(0, \sigma_{uv}^2)$.

$$
\begin{aligned}
\sigma_u^2, \sigma_v, \sigma_{uv}^2 &\sim IG(0.01, 0.01) \\
\theta_0, \theta_1 &\sim N(0, 1000) \\
\beta_0, \beta_1 &\sim N(0, 1000)
\end{aligned}
\qquad (7.66)
$$

The following program fits the two models with these priors.

```
proc mcmc data=meta outpost=postout seed=332786 nbi=10000
              thin=10 nmc=20000;
ods select PostSummaries PostIntervals;
parms beta0 0 beta1 0 theta0 1 theta1 1 s1 1 s2 1 s3 0.01;
prior s: ~ igamma(0.01, s=0.01);
prior beta: ~ n(0,sd=1e3);
prior theta: ~ n(0,sd=1e3);
random ui ~ normal(beta0, var=s1) subject=study;
random vi ~ normal(theta0, var=s2) subject=study;
random wi ~ normal(0, var=s3) subject=study;
lpi = ui+ beta1*conc+wi;
pi = logistic(lpi);
if dose>0 then ldose=log(dose);
model y ~ binomial(n = n, p = pi);
lc=theta1*ldose +vi+wi;
model lconc ~normal(lc,var=se**2);
run;
```

The summary statistics for the posterior samples are

Parameter	Mean	Standard Deviation	Percentiles 25%	50%	75%
beta0	-3.3961	1.2789	-4.2966	-3.5177	-2.4524

beta1	0.1897	0.1333	0.0925	0.2041	0.2833
theta0	0.7005	0.2678	0.5365	0.7078	0.8765
theta1	0.6952	0.1102	0.6181	0.6805	0.7654
s1	0.2141	1.3527	0.0204	0.0518	0.1387
s2	0.0715	0.1853	0.0142	0.0296	0.0675
s3	0.0531	0.1116	0.0121	0.0248	0.0531

where we find a strong dose-exposure relationship, but a weak trend of exposure-risk relationship. One can use the posterior distribution of the parameters in both models in a similar way to that of fitting a joint model to individual level data. For example, for a given dose level d, one may predict the probability of AE by using the posterior samples and models (7.64) and (7.65) to generate concentration $\log(c_k) = \theta_0 + \theta_1 d + v_k + e_k$, then $logit(p_k) = \beta_0 + \beta_1 c_k + u_k$. One can then calculate, e.g., the probability of p_k higher than a given threshold for the purpose of dose escalation with overdose control.

7.4.3   Mixed treatment comparison

We have discussed meta analysis to compare two treatments based on direct comparisons within each study, i.e., we use treatment difference estimates from individual studies, then synthesize them for statistical inference of the population. However, often individual trials may not include both treatments of interest but may still be informative for comparison, for example, if we would like to compare treatments A and B, but some trials compare either A or B and placebo. Under this situation one may bridge the comparison between A and B with the placebo or other common comparators for indirect comparisons . The approach of using both direct and indirect comparisons are referred to as mixed treatment comparison (MTC).

An important early work in the area of MTC is Lumley (2002) who used the following model. Let y_{ij} be the response in study i and under treatment j it follows model

$$y_{ij} = \mu_j + u_i + \eta_{ij} + \varepsilon_{ij} \tag{7.67}$$

where μ_j is the mean of treatment j, and η_{ij} is a term for treatment heterogeneity. He modeled the treatment differences $\hat{d}_{ijk} = y_{ij} - y_{ik}$ between treatments j and k in study i,

$$\hat{d}_{ijk} = \mu_j - \mu_k + \eta_{ij} - \eta_{ik} + \xi_{jk} + e_{ijk}, \tag{7.68}$$

where $\xi_{jk} \sim N(0, \sigma_w^2)$ is a term intended to represent inconsistency or incoherence between direct and indirect comparisons of a treatment pair j and k. The model has been extended to GLMMs, e.g., in Lu and Ades (2006). In addition, they also introduced additional structures including classifying trials by their control groups. They further developed approaches for testing the inconsistency. Dealing with inconsistency remains an active research area for MTC in the last 10 years. If we treat direct comparisons as the gold standard, inconsistency between direct and indirect comparisons may suggest unreliability of the latter.

It is useful to investigate closely the source of the inconsistency and how to deal with it in the analysis. When there is no random variation except in ξ_{jk}, for trial i we should have, for any treatments j, k and k' included, $\hat{d}_{ijk} = \hat{d}_{ijk'} + \hat{d}_{ik'k}$. This leads to a constraint $\xi_{jk'} + \xi_{k'k} = \xi_{jk}$. Note that $\xi_{jk} = -\xi_{kj}$ according to the definition. For a trial with three treatments $j = 1, 2$ and 3 we can write all the constraints as

$$
\begin{aligned}
\xi_{12} + \xi_{23} &= \xi_{13} \\
\xi_{23} + \xi_{31} &= \xi_{21}.
\end{aligned}
\tag{7.69}
$$

Therefore, ξ_{jk}s cannot be independent identical random variables, as previous work assumed. In addition, ξ_{jk} appears not only in indirect comparisons, but also in direct comparisons, since, e.g., $\mathrm{var}(\hat{d}_{i12}) = 2(\sigma_e^2 + \tau^2) + \sigma_w^2$. It is also easy to verify that for an indirect comparison of treatments 1 and 2 via treatment 3, the variance is exactly twice that of the direct comparison. Therefore, ξ_{jk} does not specifically represent inconsistency in the model.

Random study-by-treatment effects only represent random inconsistency. As in any meta analysis, individual studies as outliers, particularly those of high leverage, need special attention. Works such as Lu and Ades (2006) showed examples of how individual indirect comparisons behaved as outliers in MTC. Identifying them is relatively easy, but dealing with them in a meta analysis is more difficult and often requires background information and judgment beyond statistics.

It is interesting to check how the issues in MTC should be dealt with in meta analyses for the exposure–response relationship. The latter is different from the MTC of individual treatments, as there is a quantitative dose–response relationship that involves all doses. In this sense, it seems there is no indirect comparison. However, consider the situation of three trials each with two dose levels (0, 20), (0, 40), and (20, 40), respectively. If we assume the linear model (7.60) is correct, then the meta analysis approach for this model can be applied directly, as the contribution of each trial to the estimation of β is all direct. However, if we treat individual doses as different treatments, then the comparison between 20 and 40 from the third trial is direct and that combining the first two with 0 dose is indirect. Therefore, it seems the central issue of inconsistency still exists here. In fact, in the context of the dose–response relationship, the difference the dose makes is the slope β_k in model (7.60), hence the inconsistency is nothing but the differences between different β_k.

7.5   Comments and bibliographic notes

This chapter attempts to introduce a number of topics about Bayesian approaches and decision analysis without many technical details. There are a number of textbooks the reader can refer to for details and information. For Bayesian analysis Gelman et al. (1995) cover almost all practical aspects, while for Bayesian theory, Robert (2001) is an excellent book. Parmigiani et al (2009) give extensive details on Bayesian decision analysis, including dynamic programming. Parmigiani (2002) focus on applications of Bayesian decision-making with a few worked examples. For meta analysis the reader is referred to Whitehead (2002).

8

Confounding bias and causal inference in exposure–response modeling

8.1 Introduction

In a randomized clinical trial the mean response difference between treatment groups is an unbiased estimate for the causal treatment effect, since potential confounding factors between the treatment and response are eliminated by randomization. However, it is more complex to determine exposure–response relationships since drug exposure, e.g., drug concentration, is generally not controlled even in a randomized clinical trial and is often affected by confounding factors. The Food and Drug Administration (FDA) has issued a technical document for exposure–response analysis (FDA, 2003), with an emphasis on the importance of dealing with confounding bias. The focus of causal effect determination is to eliminate or reduce the bias. Recent statistical developments in causal effect estimation make available several approaches for determining causal effects in exposure–response relationships. See, for example, Hernan and Robins (2015). This chapter introduces relevant current developments and discusses potential applications of classical approaches and recent technical advances in determining causal effects in exposure–response relationships. The scope of this chapter extends to ER modeling in a wide range of situations, in particular, the analysis of observational studies where the concept of exposure is very general and may refer to PK exposure, dose level, or different treatments. An attempt is also made to unify approaches in PKPD modeling and modeling of observational data. Simulated data are also used in examples, since one can assess the bias and performance of different procedures only when one knows the true data generating model. With the general definition of exposure, we will consider causal effect estimation in a general ER model. Our interest may be in parameter estimates in a marginal model that shows the average response for a given exposure level or a model including covariates to describe individual responses with certain covariates. Although in most cases one can reconstruct the former with the latter (Section 9.1), it may be more appropriate to estimate the former directly, especially in the presence of confounders. In this chapter we always assume the stable unit treatment value assumption (Hernan and Robins, 2015), that is, the exposure effect on subject i does not depend on how the exposure is imposed nor on other subjects. For notation purposes, we use doses as examples of categorical exposures and denote it d_i, while for continuous one, we consider it as concentration and denote it c_i.

8.2   Confounding factors and confounding biases

The concept of confounding is the cornerstone of causal inference in statistical modeling. Confounding refers to the situation where the exposure and response are correlated via factors (confounding factors) in the presence or absence of exposure effects. When the exposure is measured by drug concentration, confounding may relate to factors with impact on both the exposure and outcome. As a typical example, elderly people may have lower drug clearance due to reduced hepatic and renal function, hence higher drug exposure than young people. They may also be at higher risk of having adverse events (AE). Therefore, an apparent PKAE correlation may be observed even when the drug does not cause it. Another example is that often a patient at higher risk of AE is likely to get a lower exposure. Assume we have only two dose levels 1 and 2, which produce exposure levels 1 and 2, respectively. We also assume that within the exposure range the exposure-AE relationship is linear, and the AE rate under the high exposure is 10% higher than under the low exposure. Therefore, a 1 unit increase in exposure results in a 10% risk increase. Suppose that there are 100 patients; among them 50 young patients who get the high dose have 20% less baseline risk than the 50 old patients who get the low dose. It is easy to calculate that, on average, the risk difference between the high and low doses would be $(0.4 - 0.3) - 0.2 = -0.1$. Therefore, without any adjustment, the exposure increase apparently reduced the risk, due to the confounding factor of age.

Although the intuition of confounding bias is useful for understanding the concept, it is important to look at the bias from statistical modeling aspects. Consider the simple linear models for the response and exposure

$$
\begin{aligned}
y_i &= \beta c_i + u_i \\
c_i &= \theta d_i + v_i
\end{aligned}
\tag{8.1}
$$

where u_i and v_i are correlated, due to, e.g., age difference in the first example above. The LS estimate for β is inconsistent, since by Slusky theorem (Section A.4)

$$
\begin{aligned}
\hat{\beta}_{LS} &= \sum_{i=1}^{n} y_i c_i / \sum_{i=1}^{n} c_i^2 \\
&\to \beta + \text{cov}(u_i, v_i)/\sigma_c^2
\end{aligned}
\tag{8.2}
$$

where σ_c^2 is the variance of c_i, assuming d_i is also random. Therefore, the bias is proportional to the proportion of the variation in c_i that is correlated to u_i.

From another statistical aspect, recall that the key condition for the LS estimate to be consistent is that the covariate (here c_i) should be independent of the error term, which is clearly violated due to the correlation between u_i and v_i. The existence of biases can also be seen from the fact that β_{LS} is the solution to the estimating equation (EE) $S(\beta) = 0$ with

$$S(\beta) = \sum_{i=1}^{n} c_i(y_i - \beta c_i)/n = \sum_{i=1}^{n} c_i u_i/n. \tag{8.3}$$

In order for $\hat{\beta}$ to be consistent, condition $E(S(\beta)) = 0$ should be satisfied. But with the correlation between v_i and u_i, $E(S(\beta)) = \text{cov}(u_i, c_i) \neq 0$. This condition is crucial in dealing with confounding biases and will frequently be used in this and the next chapters.

As we have discussed in Chapter 4, the true models may not be models (8.1) but rather the classical ME model

$$\begin{aligned} y_i &= \beta c_i^* + u_i \\ c_i^* &= \theta d_i + v_i \end{aligned} \tag{8.4}$$

in which only $c_i = c_i^* + e_i$ is measured. Recall that in the RC approach we regress y_i on $E(c_i^*|c_i)$, or $E(c_i^*|d_i)$ when a model exists between c_i^* and d_i. Following this RC approach, we can write

$$c_i^* = E(c_i^*|d_i) + v_i^* \tag{8.5}$$

where v_i^* includes both v_i and the variation of using d_i to predict c_i. Since (8.5) and (8.4) are in the form of (8.1), we can conclude that it is the correlation between u_i and v_i^* that causes the confounding bias.

The confounding may occur not only on the baseline u_i, but also on treatment heterogeneity. That is, β becomes subject specific in the y-part model $y_i = \beta_i c_i + u_i$, with $\beta_i = \beta + b_i$ and correlated b_i and c_i. For example, it is possible that the exposure effect on the risk of AE in elderly patients may be higher than in young patients. This confounding also results in biased estimation of β. Note that without repeated measures, one cannot fit a model with random coefficients, but confounding biases still exist in the LS estimate when fitting model $y_i = \beta c_i + u_i^*$ with $u_i^* = u_i + c_i b_i$. The following is a simulated numerical example:

```
nsimu=10000
latent=rnorm(nsimu)
ci=2+rnorm(nsimu)+latent
betai=3+latent
yi=betai*ci+rnorm(nsimu)
summary(lm(yi~ci))
```

The simulation gives LS estimate $\hat{\beta}_{LS} = 3.98(SE = 0.017)$, far from the true value 3. We can also check the existence of bias with the estimating equation (EE) approach. In this case,

$$S(\theta) = \sum_{i=1}^{n} c_i((\beta + b_i) + u_i - \beta c_i)/n = \sum_{i=1}^{n} c_i(c_i b_i + u_i)/n. \qquad (8.6)$$

The term $c_i^2 b_i$ causes $E(S(\beta)) \neq 0$ when c_i^2 and b_i are correlated. In this example, the correlation between c_i^2 and b_i is 0.63, hence bias occurs. Note that it is possible that c_i and b_i are correlated, but c_i^2 and b_i are not. For example, with ci generated as

```
ci=rnorm(nsimu)+latent,
```

c_i^2 and b_i are uncorrelated, hence running the simulation, one gets an LS estimate very close to the true value 3. However, this only occurs with a very specific setting and should not be relied on in practice.

For other models with a linear predictor, such as the Cox model, GLM and GLMM, the confounding may be either on the baseline, on β, or on both. But for a nonlinear model, it may occur in multiple places. Taking the Emax model

$$y_i = E_0 + E_{max}/(1 + EC_{50}/c_i) + u_i \qquad (8.7)$$

as an example, confounding may occur on the baseline E_0, as represented by u_i in the current model, but may also be on E_{max}, e.g., if subjects of different age groups have different maximum effects, as well as on EC_{50}, when the drug potency may change with age. The criterion $E(S(\beta)) = 0$ can also be used to check if confounding bias exists, but the magnitude of the bias may not be easy to calculate. In Section 8.6 approaches to assess confounding biases for different types of nonlinear models will be introduced.

8.3   Causal effect and counterfactuals

To introduce the concept of counterfactuals, we start with the following example of only two exposure levels 0 and 1. Let $y_i(d)$ be the response if patient i had exposure d, which may not be the exposure he really received. It may also be denoted as $y_i(do(d_i = d))$, where $do(d_i = d)$ is to force d_i to take value d. The causal exposure effect on subject i can be defined as $E(y_i(0) - y_i(1))$, the mean difference between the responses to $d = 1$ and to $d = 0$. If we assume the same effects among all subjects, $E(y_i(0) - y_i(1))$ also defines the population treatment effect. If not it is the average causal effect over the patient population. However, often subject i only had one exposure d_i, only one among $y_i(0)$ and $y_i(1)$, was realized. $y_i(0)$ and $y_i(1)$ are called counterfactuals. Therefore, one has to estimate $E(y_i(0) - y_i(1))$ by comparing responses of different subjects. However, the naive estimate $\bar{y}_1. - \bar{y}_0.$, where $\bar{y}_d.$ is the mean among all patients with exposure d, is often biased. To see why it is biased, assume that $d_i = 1$ and there is only another patient j who had $d_j = 0$. Then $E(\bar{y}_1. - \bar{y}_0.) = E(y_i(1) - y_j(0)) = E(y_i(1) - y_i(0)) + E(y_i(0) - y_j(0))$, in which the second term is the mean difference in baselines between the subject having exposure 1 and that having exposure 0. Therefore, $\bar{y}.(1) - \bar{y}.(0)$ is an unbiased estimate only if $E(y_i(0) - y_j(0)) = 0$. In general, $E(\bar{y}_1. - \bar{y}_0.) = E(y_i|d_i = 1) - E(y_i|d_i = 0)$, where $E(y_i|d_i = 0)$ is the mean among patients who had exposure 0. They may be different from those who had exposure 1, except, e.g., when d_i is randomized. We can write the observed y_i in terms of $y_i(0)$ and $y_i(1)$ and the actual exposure d_i as $y_i = d_i y_i(1) + (1 - d_i)y_i(0)$.

Consider counterfactuals in a regression model:

$$y_i = d_i\beta + u_i + \varepsilon_i \tag{8.8}$$

where β is the causal effect of exposure d_i and u_i is a confounder, and ε_i is an independent error. The causal effect is determined by β since $E(y_i(1)) - E(y_i(0)) = \beta$. But $E(\bar{y}.(1) - \bar{y}.(0)) = \beta + E(u_i|d_i = 1) - E(u_i|d_i = 0)$, where $E(u_i|d_i = 1) - E(u_i|d_i = 0)$ represents the difference between those with $d_i = 1$ and those with $d_i = 0$. In the previous section, we found the correlation between exposure, denoted as c_i, and u_i is the source of confounding bias, unless $E(u_i|d_i = 1) = E(u_i|d_i = 0)$. For a general exposure d_i, the $E(u_i|d_i = 1) \neq E(u_i|d_i = 0)$ for u_i can be replaced by independence between d_i and u_i.

8.4   Classical adjustment methods

8.4.1   Direct adjustment

Confounding bias can be adjusted in a number of ways if confounding factors are observed. The direct adjustment (DA) includes confounding factors in the dose–response or exposure–response model, given that a correct model can be determined. For example, instead of using models (8.1), we can fit a model including all possible confounders $\mathbf{X}_i$

$$y_i = \beta_c c_i + \mathbf{X}_i^T\beta + u_i^*, \tag{8.9}$$

where u_i^* is the residual error after adjusting for $\mathbf{X}_i$. In order for the DA approach to work, u_i^* has to be independent of c_i. For this not only all confounding factors should be included in $\mathbf{X}_i$, but also the linear structure $\mathbf{X}_i^T\beta$ has to be correct. If confounding occurs because of treatment heterogeneity in $\beta_i = \beta + b_i$, the DA approach can also be applied if, e.g., $\beta_i = \beta_c + \beta^T\mathbf{X}_i + b_i^*$ and c_i is not correlated with b_i^*. In this case, one includes the interaction between $\mathbf{X}_i$ and c_i in the model and fits

$$y_i = \beta_c c_i + c_i\mathbf{X}_i^T\beta + u_i^*. \tag{8.10}$$

Consequently, due to heterogeneity, treatment effects in different patient populations (those with high and low exposures) are also different. Therefore, to assess treatment effects using the fitted model, not only β_c, but also β is important. For models with a linear predictor, e.g., GLM, GLMM and the Cox model, the same approaches for confounding adjustment can easily be adapted by including potential confounding factors and/or their interaction with the exposure in the model.

For nonlinear models, confounding factors may have an impact in multiple places in the model. One may be able to determine a structured model including the confounders. Consider the model

$$y_i = g(c_i, \beta_i) + \varepsilon_i \tag{8.11}$$

with $\beta_i = \beta + \mathbf{b}_i^*$, with $\mathbf{b}_i^*$ being potentially confounded random variations in the parameters, i.e., $\mathbf{b}_i^*$ may be correlated with c_i. If $\mathbf{b}_i^*$ is a function of covariates $\mathbf{X}_i$ including confounders, i.e., $\beta_i = \beta + \mathbf{X}_i\mathbf{B} + \mathbf{b}_i$, where $\mathbf{B}$ is a matrix of parameters, and $\mathbf{b}_i$ is an independent random vector with $E(\mathbf{b}_i) = 0$. then DA can be implemented by including $\mathbf{X}_i$ in the models for the model parameters. Even when $\mathbf{b}_i^*$ does not contains confounders, we can only fit a marginal model

$$y_i = g^*(c_i, \beta) + \varepsilon_i \tag{8.12}$$

with $g^*(c_i, \beta) = E_{b^*}(g(c_i, \beta + \mathbf{b}_i^*))$ as the marginal mean over $\mathbf{b}_i^*$ given c_i. $g^*(c_i, \beta)$ is generally different from $g(c_i, \beta)$, but often the latter is used as an approximation. With DA the marginal model is $g^{**}(c_i, \beta + \mathbf{X}_i\mathbf{B}) = E_b(g(c_i, \beta + \mathbf{X}_i\mathbf{B} + \mathbf{b}_i))$, which can be approximated better by $g(c_i, \beta)$, as the variation in $\mathbf{b}_i$ is smaller than that in $\mathbf{b}_i^*$.

When using DA with an NLMM

$$\mathbf{y}_i = g(\mathbf{c}_i, \beta + \mathbf{X}_i\mathbf{B} + \mathbf{b}_i) + \varepsilon_i, \tag{8.13}$$

fitted to repeated measurement data, the model takes $\mathbf{b}_i$ as random effects on β_i, while without DA, it treats $\mathbf{X}_i\mathbf{B} + \mathbf{b}_i$ as random effects. Therefore, DA may also make the use of a simple model fitting approach possible. For example, when $\mathbf{b}_i$ has a small variation, the first order approximation of Beal and Sheiner (1982) based on $E_b(g(c_i, \beta + \mathbf{X}_i\mathbf{B} + \mathbf{b}_i)) \approx g(c_i, \beta + \mathbf{X}_i\mathbf{B})$, which is often stable and computationally efficient, may be used, while without adjusting for $\mathbf{X}_i$ the Lindstrom–Bates approach or Gaussian quadrature may be needed.

DA is the simplest and, in most cases, the most efficient way to adjust for confounding bias, if the model is correctly specified. However, often model misspecification is very difficult to identify. This is particularly so if the response is a rare event (e.g., in a post-marketing drug safety study, in which, although there might be exposure data from thousands patients, those having rare events are of a small number). Another issue arises when, for the same or similar $\mathbf{X}_i$ values, there is no overlap between exposure levels, e.g., high exposure is mostly among old patients and low exposure among young patients.

Recall that adjusting by covariates relies on comparing the response between different exposure levels at the same or similar covariate levels. No overlap means that the comparison has to be based on extrapolation using the model. The validity of the extrapolation is not testable due to the lack of overlapping data. Furthermore, the lack of overlapping data also causes co-linearity between the exposure and the covariates, which produces poor estimates even when one specifies the right factors in the model.

8.4.2 Stratification and matching

To avoid assumptions about the model structure regarding the confounding factors, if there is one or a few factors, we can stratify subjects according to these factors so that within each stratum the factor values are similar. Therefore, subjects within a stratum can be considered comparable. Consequently, we can determine the exposure effect within strata and pool them to obtain an overall estimate that is (approximately) free of confounding bias. This approach works for almost all models we have covered in this book. For some models, some software such as SAS and R includes options for stratification, while for common models such as the linear model and GLM, one can include a stratum indicator in the model as a categorical covariate. This covariate absorbs the differences between strata such that they do not affect the estimate of exposure effects. Therefore, the response due to the exposure variation between strata cannot be used to estimate the exposure effects. As an example, let strat be a variable taking values 1, 2, ... as an indicator for the stratum the sample belongs to, the following R program fits the first model in: (8.1)

```
lm(yi~ci+as.factor(strat)),
```

in which we estimate the stratum effects explicitly, although they are nuisance parameters for the ER relationship. This approach works for $y_i = \beta c_i + g(u_i)$ with an arbitrary $g(.)$, as long as the $g(u_i)$s are similar within each stratum. Hence it is more robust than the DA approaches, although generally less efficient. In the case of little or no overlapping exposure, there is no or very small exposure variation in some strata, hence these strata have no or only a little contribution to the estimate of the ER relationship.

Matching is a similar approach to stratification, but is only useful for comparing the effects between two or a few exposure levels. For example, one can match one subject having high exposure with one having low exposure, and with similar confounding factor values so that the pair is comparable. When there are multiple factors, matching may be based on a "distance" measure in the space of the factors. Another approach using matching is the case-control study in which we match a subject who experienced an event to one with similar confounding factors but no event, and compare their exposures to determine the exposure effect in the risk of events. Matching provides better control of the confounding factors than stratification, but is more complex to implement. Although, in principle, one can also match a number of subjects so that a model can be fitted between them, this is often infeasible in practice, hence we will not discuss this approach further.

8.4.3   Propensity scores and inverse probability weighting

When there are many potential confounding factors, stratification based on their values becomes infeasible. The propensity score (PS) is a powerful tool for adjustment of multiple confounding factors (Rosenbaum and Rubin, 1983). The classical PS deals with binary exposures in observational studies where each patient either receives the exposure ($d_i = 1$) or does not ($d_i = 0$), and there are confounding factors between the probability of $d_i = 1$ and the outcome of treatment. The PS can be defined as

$$PS(\mathbf{X}_i) \equiv P(d_i = 1|\mathbf{X}_i) \tag{8.14}$$

i.e., the probability of patient i being exposed conditional on covariates $\mathbf{X}_i$, and can be estimated by, e.g., a logistic regression model. $PS(\mathbf{X}_i)$ represents the probability of exposure that can be predicted by confounding factors. For the purpose of confounding control, $\mathbf{X}_i$ should include all potential confounders, under the assumption of no unobserved confounder. Under this assumption the potential response is independent of d_i conditional on $PS(\mathbf{X}_i)$, i.e,

$$y_i(0), y_i(1) \perp d_i|PS(\mathbf{X}_i). \tag{8.15}$$

This condition should be carefully examined, since it cannot be tested based on the data. We also assume that, for any $\mathbf{X}_i$, $PS(\mathbf{X}_i)$ is bounded from 0 or 1:

$$0 < PS(\mathbf{X}_i) < 1 \tag{8.16}$$

which means no complete confounding, e.g., when all males are on a higher dose and all females are on a lower dose. If gender is a confounding factor, propensity of high dose is 1 for male and 0 for female. In this case, dose is completely confounded with gender and its impact cannot be eliminated by any of the approaches below.

There are a number of ways to use PS to eliminate or reduce confounding bias. The most commonly used one is to stratify samples according to their PS values, i.e., the probability of exposure predictable with $\mathbf{X}_i$. To see how this approach works, one can recall ordinary stratification by a potential confounder such as age. When there are many confounders it becomes difficult to stratify. The PS is a single summary measure for how the exposure relates to confounders so that matching by PS is straightforward. A common approach is to stratify samples into, e.g., 10 equal sized strata according to their PS values, which is straightforward to do.

Inverse probability weighting (IPW) is another way to use PS to adjust for confounding biases when the exposure is binary. Let y_i be the response to binary exposure d_i which may be confounded. We are interested in the mean response difference between exposure $d_i = 1$ and $d_i = 0$, i.e., $E(y_i(1) - y_i(0))$. Without assuming any model structure for y_i, one can estimate it by

$$n^{-1} \sum_{i=1}^{n} \frac{d_i y_i}{PS(\mathbf{X}_i)} - n^{-1} \sum_{i=1}^{n} \frac{(1 - d_i)y_i}{1 - PS(\mathbf{X}_i)} \tag{8.17}$$

under the condition of no unobserved confounding factor and that $1 > PS(\mathbf{X}_i) > 0$ for all $\mathbf{X}_i$. This is called an IPW estimate as each y_i under exposure $d_i = d$ is weighted by the inverse probability of under exposure d. The first term is an unbiased estimate of $E(y_i(1))$. This can be shown by

$$
\begin{aligned}
E(n^{-1} \sum_{i=1}^{n} \frac{d_i y_i}{PS(\mathbf{X}_i)}) &= E(\frac{d_i y_i}{PS(\mathbf{X}_i)}) \\
&= E(E(\frac{d_i(y_i(1))}{PS(\mathbf{X}_i)}|\mathbf{X}_i)) \\
&= E(y_i(1)E(\frac{d_i}{PS(\mathbf{X}_i)}|\mathbf{X}_i)) = E(y_i(1)) \quad (8.18)
\end{aligned}
$$

where the second equality is from $y_i = d_i y_i(1) + (1 - d_i)y_i(0)$ and the third follows the fact that conditional on $PS(\mathbf{X}_i)$, d_i is independent of the response (no unobserved confounding) and $E(d_i|\mathbf{X}_i) = PS(\mathbf{X}_i)$.

Estimate (8.17) is not the only IPW estimate. Another one, known as the ratio estimate, is

$$
\sum_{i=1}^{n} \frac{d_i y_i}{PS(\mathbf{X}_i)} / \sum_{i=1}^{n} \frac{d_i}{PS(\mathbf{X}_i)} - \sum_{i=1}^{n} \frac{(1-d_i)y_i}{1-PS(\mathbf{X}_i)} / \sum_{i=1}^{n} \frac{(1-d_i)}{1-PS(\mathbf{X}_i)}. \quad (8.19)
$$

In practice, one needs to calculate $PS(\mathbf{X}_i)$ based on exposure data using, e.g., a logistic regression model. As parameters in this model are estimated, the variability of estimation may be taken into account by, e.g., bootstrapping. However, counterintuitively using estimated propensity reduces the variance of the estimate when the sample size is sufficiently large. Accordingly, using the SE given by commonly used software may lead to a conservative CI or hypothesis test. However, this may not hold when sample sizes are small, hence a bootstrap approach should also be used in this situation.

The use of the standard PS approach in exposure–response modeling is limited since it can only deal with binary exposure. Extensions have been made to extend the standard PS to other types of exposures such as ordered categorical exposures (Imbens, 2000; Wang et al., 2001). A general PS (GPS) approach has been proposed (Imai and van Dyk, 2004) and can be used for a wide range of exposures measured by, e.g., an ordered categorical variable (e.g., exposure at low, medium and high doses) or by a continuous variable (such as drug concentration). In general, for an exposure d_i of arbitrary type, a GPS can be defined as the distribution of exposure as a function of $\mathbf{X}_i$. When it is categorical, the distribution is characterized by the probability of each category. Let $e(\mathbf{X}_i)$ be the GPS, following the notation in (Imai and van Dyk, 2004), under the assumption of no unobserved confounders, we have $y_i(d) \perp d_i|e(\mathbf{X}_i)$ for all ds, i.e., the potential outcome under any treatment is independent of d_i, given the GPS. Often $e(\mathbf{X}_i)$ may only be some characteristics (e.g., the mean and variance) of the full distribution for $y_i(d) \perp d_i|e(\mathbf{X}_i)$ to hold, depending on the model structure. further discussion can be find in Imai and van Dyk (2004), Yang et al (2014) and in Section 8.4.4.

When the exposure level, e.g., dose, is categorical with a few levels, the IPW approach can be used to estimate a dose–response relationship with multiple exposure level data (Feng et al., 2012; Li et al., 2013). Specifically, d_i can take $a_1, ..., a_k$ levels with probability of $P(d_i = a_k) \equiv PS_k(\mathbf{X}_i)$; then $\mu(a_k) = E(y_i(a_k))$ can be estimated by

$$\hat{\mu}(a_k) = n^{-1} \sum_{i=1}^{n} \frac{I(d_i = a_k)y_i}{PS_k(\mathbf{X}_i)}. \tag{8.20}$$

$\hat{\mu}(a_k), k = 1, ..., K$ have asymptotically a joint normal distribution. One may test a linear trend using $\sum_{k=1} B_k \hat{\mu}(a_k)$ with appropriate weights B_ks. The individual estimates may be quite variable. To estimate the exposure–response curve one may take a two stage approach and fit, e.g., a spline function model to $\hat{\mu}(a_k)$ weighted by its inverse variance.

This approach can also be combined with the EE to fit a parametric exposure–response model directly. Let $S_i(d, \boldsymbol{\beta})$ be the contribution of subject i, with possibly counterfactual exposure d, to the EE, and we assume $E(S_i(d, \boldsymbol{\beta})) = 0$ when $y_i = y_i(d)$. We can construct EE with IPW in the same way as for the mean estimate. Writing the observed $S_i(d_i, \boldsymbol{\beta})$ as

$$S_i(d_i, \boldsymbol{\beta}) = \sum_{k=1}^{K} I(d_i = a_k)S_i(a_k, \boldsymbol{\beta}), \tag{8.21}$$

we can construct

$$S(\boldsymbol{\beta}) = \sum_{i=1}^{n} \sum_{k=1}^{K} \frac{I(d_i = a_k)S_i(a_k, \boldsymbol{\beta})}{PS_k(\mathbf{X}_i)} = 0 \tag{8.22}$$

and solve it for β. To check if (8.22) is unbiased, its expectation can be written as

$$
\begin{aligned}
E(S(\boldsymbol{\beta})) &= E(\sum_{i=1}^{n} \sum_{k=1}^{K} \frac{I(d_i = a_k)S_i(a_k, \beta)}{PS_k(\mathbf{X}_i)}) \\
&= \sum_{i=1}^{n} \sum_{k=1}^{K} E(S_i(a_k, \beta)E(\frac{I(d_i = a_k)}{PS_k(\mathbf{X}_i)}|\mathbf{X}_i)) \\
&= \sum_{k=1}^{K} E(S_i(a_k, \beta)) = 0.
\end{aligned} \tag{8.23}
$$

Therefore, the solution to EE (8.22) is consistent, with additional technical conditions.

As a numerical example, we use model $y_i = \beta_0 + \beta d_i + u_i$. Then (8.22) is the EE of a weighted LS. Hence to estimate β, one can use general regression software. The following code shows how R-function lm() can be used for a linear model with two exposure levels. We assume that gender (male=0,1) is a confounding factor and it controls the exposure di via a logistic model. The ER model is fitted by simple LS, IPW with exact and estimated weights, respectively.

```
> nsimu=10000
> male=rep(0:1,nsimu/2)
> pi=1/(1+exp(male))
> di=rbinom(nsimu,1,pi)
> yi=di+male+rnorm(nsimu)
> pdi=predict(glm(di~male))
> pwei=ifelse(di==1,pdi,1-pdi)
> wei=ifelse(di==1,pi,1-pi)
> lm(yi~di)

Coefficients:
(Intercept)          di
    0.5716      0.7959

> lm(yi~di,weight=1/wei)

Coefficients:
(Intercept)          di
    0.4816      1.0383

> lm(yi~di,weight=1/pwei)

Coefficients:
(Intercept)          di
    0.4827      1.0360
```

The following table shows the mean and SD of the three estimates (IPW1 and IPW2 are estimates with exact and estimated propensity, respectively) from 300 simulations with sample sizes 30, 100 and 300.

		Mean			SD	
n	LS	IPW1	IPW2	LS	IPW1	IPW2
30	0.7602	0.9884	1.0166	0.4436	0.4603	0.4385
100	0.7497	0.9898	0.9921	0.2361	0.2449	0.2275
300	0.7549	0.9951	0.9981	0.1263	0.1338	0.1204

One can see that both IPW approaches corrected the confounding bias around -0.25, as shown in the LS estimates. But the one using estimated weights consistently performs better than the one using the exact ones. However, in this situation the weights are generally not close to 1 or 0, hence the results may not hold in scenarios that may make the estimated weights unstable.

To illustrate the EE approach we consider a simple model $y_i = \beta d_i + u_i$ with $d_i = 1$ or $d_i = -1$ and use $S_i(d, \beta) = d(y_i - \beta d)$. In this case one can derive the solution to the EE (8.22) as

$$\hat{\beta} = \frac{\sum_{i:d_i=1} y_i/PS(\mathbf{X}_i) - \sum_{i:d_i=-1} y_i/(1-PS(\mathbf{X}_i))}{\sum_{i:d_i=1} 1/PS(\mathbf{X}_i) + \sum_{i:d_i=-1} 1/(1-PS(\mathbf{X}_i))}, \qquad (8.24)$$

which is different from the nonparametric estimate. But one can check that $E(\hat{\beta}) = \beta$ by taking the mean of $\hat{\beta}$ conditional on $PS(X_i)$. Also note that the denominator has mean 2 and the numerator has mean $E(y(1)) - E(y(-1)) = 2\beta$, hence the ratio is β, which is the limit of $\hat{\beta}$ by Slusky theorem (A.4).

The major issue of the IPW approach is its instability when $PS(\mathbf{X}_i)$ is very close to 0 or 1, hence individual patients may become dominant in the IPW estimate. Therefore, often one has to truncate $PS(\mathbf{X}_i)$ at the two ends for stability. Finally, we note that the properties derived above do not depend on the linear structure in the model. Using the counterfactual framework, the causal effect of d_i in other models can be defined as $E(y_i(1)) - E(y_i(0))$ and can be estimated by $\hat{\beta}_{IPW}$.

8.4.4   Other propensity score–based approaches

A more straightforward way of using PS or GPS is to add them or an appropriate function of them into the y-part model. Consider model

$$y_i = g(d_i, \boldsymbol{\beta}) + u_i + \varepsilon_i \qquad (8.25)$$

with u_i as a potential confounder. If there exists a $e(\mathbf{X}_i)$ such that $u_i = E(u_i|e(\mathbf{X}_i)) + u_i^*$ with independent u_i^*, then it is sufficient to fit model

$$y_i = g(d_i, \boldsymbol{\beta}) + \gamma E(u_i|e(\mathbf{X}_i)) + \varepsilon_i, \qquad (8.26)$$

assuming $E(u_i|e(\mathbf{X}_i))$ can be identified and estimated, as the residual $u_i^* + \varepsilon_i$ is independent of d_i. As the EE of LS estimate for β is

$$S(\boldsymbol{\beta}, \gamma) = \sum_{i=1}^{n} \frac{\partial g(d_i, \boldsymbol{\beta})}{\partial \boldsymbol{\beta}} (y_i - g(d_i, \boldsymbol{\beta}) + E(u_i|e(\mathbf{X}_i))) = 0 \qquad (8.27)$$

assuming $E(u_i|e(\mathbf{X}_i))$ is known. Yang et al (2014) showed that if a set of GPS parameters $e(\mathbf{X}_i)$ can characterize $E(\partial g(d_i, \boldsymbol{\beta})/\partial\boldsymbol{\beta}|e(\mathbf{X}_i))$, then it is sufficient for the solution to EE (8.27) to be consistent. But it may not be the necessary set. For example, if u_i and d_i have joint normal distribution so that u_i can be written as $u_i = a + b(d_i - \mu_d)$, then including $(d_i - \mu_d)$ as a covariate is sufficient, regardless $\partial g(d_i, \boldsymbol{\beta})/\partial\boldsymbol{\beta}$. The major issue with using PS or GPS as a regressor is how to identify $E(u_i|e(\mathbf{X}_i))$ and how to verify if $e(\mathbf{X}_i)$ is sufficient.

Another way of using PS or GPS is also based on separating the exposure into two parts, one predictable and another not predictable by potential confounders, then one only uses the unpredictable part in the y-model. This part is called the intensity function by Brumback et al. (2003), and is simply the residual in the PS model. For example, if we consider drug concentration as exposure, the GPS model is

$$c_i = h(\mathbf{X}_i, \boldsymbol{\theta}) + e_i, \tag{8.28}$$

then $I(\mathbf{X}_i, \boldsymbol{\theta}) = c_i - E(c_i|\boldsymbol{\theta}, \mathbf{X}_i)$ is the intensity function. Note that it is simply the residual e_i in this case, but may also include variation in c_i not related to $\mathbf{X}_i$, e.g., that due to dose changes. Since $I(\mathbf{X}_i, \boldsymbol{\theta})$ is not correlated with u_i, to estimate β, one can replace c_i with $I(\mathbf{X}_i, \hat{\boldsymbol{\theta}})$ in model (8.1). In practice, since $\boldsymbol{\theta}$ is unknown, it has to be replaced by its estimate. Hence we fit model

$$y_i = \beta_0 + \beta I(\mathbf{X}_i, \hat{\boldsymbol{\theta}}) + u_i, \tag{8.29}$$

where $\hat{\boldsymbol{\theta}}$ comes from the fitted PS/GPS model. From the structure of $I(\mathbf{X}_i, \hat{\boldsymbol{\theta}})$ we see that the raw exposure in c_i is still present in the model, but the second term corrects the confounding. However, if $E(E(c_i|\boldsymbol{\theta}, \mathbf{X}_i)) \neq 0$, the second term adds a bias into the baseline estimate. This is normally not a problem, as often we are only interested in the inference of β. Nevertheless it can be adjusted for (Brumback et al., 2003).

8.5 Directional acyclic graphs

When there are multiple outcomes and factors, the causal relationships and confounding between them may be complex, hence a visual tool to represent the relationships is very useful. The development of (DAG) (Pearl, 1995) provides such a tool for investigating potential confounding and finding approaches to deal with it. A DAG is a graph consisting of nodes, each representing a variable (e.g., a measure of exposure, a response, a factor or an intermediate measure), linked by directional edges (indicated by arrows), but not forming any directional cycle, e.g., $x \to y \to z \to x$. For each arrow the starting node is called a parent of the ending node, and the ending node is called a descendant of the starting node. The distribution of the descendant depends on its parents, and the dependence represents the causal effect of the parents on their descendant. The DAG in Figure 8.1 shows the causal relationship between the dose d_i, exposure c_i and response y_i, and also that a confounder, e.g., age, may also have impacts on all three via latent variables s_i, v_i and u_i, hence may confound the causal effects.

A DAG can be used to determine the confounding of the causal effect of one variable on another one. In general, confounding occurs in the relationship between two nodes (e.g., exposure and response) when a factor (e.g., age) affects both, as indicated by paths (connections between nodes regardless of the direction) between them. The DAG can also be used to find a way of

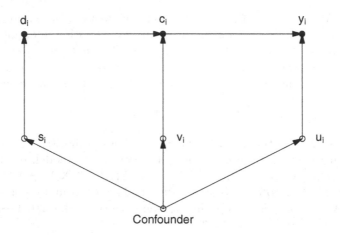

FIGURE 8.1
A directional acyclic graph for the relationship between dose, exposure, response and a confounder.

adjusting for confounding by conditioning on a certain set of variables. An important concept is dependent(d)-separation regarding a DAG. In general, two sets of nodes (C, Y) are d-separated by Z if 1) for all nodes c in C and y in Y, on any paths $c \to z \to y$, z is in Z, i.e., these paths contain one or more nodes in Z, and 2) no z in Z is a collider between any pair of nodes in C and in Y, i.e., there is no path in the form of $c \to z \leftarrow y$. The importance of d-separation is that, conditional on Z, the distribution of Y is independent of C. Therefore, if C contains confounders, conditioning on Z can eliminate their effects. The need of 2) can be seen from the following example. Suppose that c and y are independent but relationship $c \to z \leftarrow y$ holds and the underlying mechanism can be written as a linear model $z = c + y$, then conditional on z, $y = z - c$ becomes correlated with c. Having 2) is to avoid this situation.

Another useful result from DAG theory is the back-door criterion of a set of nodes blocking confounding paths (Pearl, 1995). Conditioning on the set eliminates the confounding bias. Using the DAG in Figure (8.1) as an example, a node set blocks confounding paths between c and y if 1) no node in the set is a descendant of c and 2) the set blocks every path between c and y that has an arrow into c. An important property is that if Z satisfies these criteria with respect to X and Y, then the distribution of Y conditional on a given X (e.g., assigned by an external force) is given by

$$P(Y|do(X = x)) = \int P(Y|X = x, Z)P(Z|X = x)dZ. \qquad (8.30)$$

Hence the causal effect of X on Y can be identified by averaging $P(Y|X, Z)$ over $P(Z|X)$. Furthermore, confounding factors can also be blocked by adjusting elements in Z. Applying the back-door criterion on Figure 8.1, we can see that for the inference of effect of c_i on y_i, the set satisfying the back-door criterion contains either a single node u_i (which blocks the u_i to y_i path), or v_i plus s_i or d_i (which block the two paths to c_i). Therefore, conditioning on u_i (which is the DA approach) can identify the causal effect of c_i on y_i. Note that randomizing dose d_i is not sufficient, as the path linking c_i, v_i, the confounder, and u_i, y_i is not blocked. To block this, one option is to randomize the concentration c_i. A trial using this approach is called a randomized concentration controlled trial.

The DAG is also related to model fitting sequentially following the arrows, as the distribution of a descendant conditional on its parents does not depend on anything else. The joint distribution of all variables in a DAG can be factorized as a product of distributions of descendants conditional on their parents, while for those without any parent, their distributions are unconditional. The approach is known as Markov factorization. For example, the joint likelihood for the model given in Figure (8.1), conditional on the latent variables, can be written as

$$L(d_i, c_i, y_i|s_i, v_i, u_i)) = f(y_i|c_i, u_i)f(c_i|d_i, v_i)f(d_i|s_i) \qquad (8.31)$$

and the marginal likelihood can be obtained by averaging over the latent variables.

8.6 Bias assessment

Often a sensitivity analysis to assess the size of confounding bias under some hypothetical scenarios is useful to determine adjustment strategies and the impacts of confounding factors in the ER analysis. One obvious approach is using simulation to compare parameter estimates with the true values. However, this direct simulation method is often computing intensive. Here we introduce some alternative tools of bias assessment for different models. The focus is on nonlinear models because for these models simulation approaches are often time consuming, and may lead to failure of fitting the model in a considerable number of runs in the simulation. The approach introduced here is easy to use, although it only gives asymptotic bias estimates. This is often sufficient for a sensitivity analysis. Closely related to this topic is the assessment of ignored covariates in regression analysis, but its focus is on nonconfounding covariates.

For illustration, we start from the linear models (8.1), where the confounding bias in the LS estimate has been calculated at the beginning of the chapter. Here we use another approach based on properties of parameter estimates in misspecified models (White, 1981). For a linear model

$$y_i = \beta c_i + u_i \qquad (8.32)$$

with both y_i and c_i centered, the LS estimate $\hat{\beta}$ is the solution to

$$S(\beta) = \sum_{i=1}^{n} c_i(y_i - \beta c_i)/n = 0. \qquad (8.33)$$

Following the approach in White (1981), one can show that $\hat{\beta}$ tends to β^*, which is the solution to EE

$$E(S(\beta^*)) = E(c_i(E(y_i|u_i) - \beta^* c_i)) = 0, \qquad (8.34)$$

where $E(y_i|u_i) = \beta c_i + u_i$ (Appendix A.7). Assuming that c_i is random, the bias is

$$\beta - \beta^* = \frac{\text{cov}(c_i, u_i)}{\text{var}(c_i)} \qquad (8.35)$$

which is equivalent to (8.2). When, e.g., $c_i = d_i + u_i$ and d_i is also random, the variation in d_i goes into $\text{var}(c_i)$ as $\text{var}(c_i) = \text{var}(d_i) + \text{var}(u_i)$, but not in $\text{cov}(c_i, u_i)$, hence the variation in d_i will reduce the bias.

The EE approach can be used for more complex models relatively easily if they have a linear structure. For a GLM or a GLMM with link function $g(.)$, we assume that with confounders u_i

$$E(y_i|u_i) = g(\boldsymbol{\beta}^T \mathbf{x}_i + u_i). \qquad (8.36)$$

Without repeated measurements, we can only fit a marginal model with mean $E_u(g(\boldsymbol{\beta}^T \mathbf{x}_i + u_i))$. With the same approach as for the linear model, we can derive the EE for fitting the marginal model as

$$S(\beta) = \sum_{i=1}^{n} \frac{\partial g_i}{\partial \boldsymbol{\beta}} \mathbf{W}_i(y_i - E_u(g(\boldsymbol{\beta}^T \mathbf{x}_i + u_i)))/n = 0, \qquad (8.37)$$

where $\mathbf{W}_i$ is an appropriate weight matrix and $g_i \equiv g(\boldsymbol{\beta}^T \mathbf{x}_i + u_i)$. With correlation between u_i and c_i, the estimate for $\hat{\beta}$ tends to β^*, the solution to

$$E(S(\beta)) = E_u(\sum_{i=1}^{n} \frac{\partial g_i}{\partial \boldsymbol{\beta}} \mathbf{W}_i(g(\boldsymbol{\beta}^T \mathbf{x}_i + u_i) - E_u(g(\boldsymbol{\beta}^T \mathbf{x}_i + u_i)))/n = 0. \qquad (8.38)$$

To find $\boldsymbol{\beta}^*$, we need to calculate $E(S(\beta))$ and solve (8.38). As noted in Chapter 3, $E_u(g(\boldsymbol{\beta}^T \mathbf{x}_i + u_i))$ may not belong to the same family of $g(.)$, with a few exceptions such as linear, exponential and probit functions, which increases the difficulty of calculating (8.38). But its structure suggests an easy way to estimate $\boldsymbol{\beta}^*$ using simulation and common model fitting software if $E_u(g(\boldsymbol{\beta}^T \mathbf{x}_i + u_i))$ has an explicit parametric form. Instead of simulating y_i we simulate u_i and c_i and generate $g_i = g(\boldsymbol{\beta}^T \mathbf{x}_i + u_i)$. Then we can fit g_i with an appropriate model. For example, for a Poisson model or a binomial model, we could fit a GLM with the log or logit link function, although g_i is not an integer. Most software for Poisson and logistic models accept noninteger responses. Alternatively, one may multiply a large number to g_i and round it to an integer. This approach is similar to the direct simulation approach but it eliminates the variation in y_i, given u_i, hence can use fewer simulation runs to achieve the same accuracy than that needed when simulating y_i. This advantage could be substantial if, e.g., the outcome is a rare AE.

The following program uses simulation to estimate the bias for a Poisson model with a confounding factor when the average event rate is 0.021.

```
nsimu=100000
sigu=1
sige=1
sage=1
dose=rep(1:2,nsimu/2)
lage=rnorm(nsimu)*sage
lc=dose+lage+rnorm(nsimu)*sige
lp=exp(lc+lage-10)
est0=summary(glm(lp~lc,family=poisson))$coef[2,1]
> est0
[1] 1.422632
```

Compared with the true value 1 for β, the bias estimate is 0.423. To compare this approach and the estimate by simulating y_i, we ran 100 simulations with a 5000 sample size, and obtained $SD = 0.076$ in the bias estimate with this approach and $SD = 0.119$ in the y_i simulation estimates. When the event rate is 0.128, the y_i simulation estimates had $SD = 0.082$, close to $SD = 0.075$ from this approach. While in the case of rare events with event rate $= 0.0024$, the SD in the y_i simulation estimate increased to 0.187, more than twice of $SD = 0.078$ with this approach. Note that, if "lc" and "lage" are not correlated, even if the data are generated from a mixed effect Poisson model, fitting a Poisson model without the random effect to either $lp = \exp(lc + lage - 10)$ or to $y_i \sim$ Poisson $(\exp(lc + lage - 10))$ gives consistent estimate for the coefficient of "lc", although not for the intercept.

When repeated measures are available, one can fit a GLMM as a subject-specific model with

$$E(\mathbf{y}_i | u_i) = g(\mathbf{X}_i \boldsymbol{\beta} + u_i), \tag{8.39}$$

where $\mathbf{y}_i$ is a vector of repeated responses and $\mathbf{x}_i$ contains repeated exposure measures $\mathbf{c}_i$. The confounding bias in the fitted subject-specific model is not easy to assess. A practically easy, but computationally intensive, way is the direct simulation approach, as described above. Alternatively, we may use the confounding bias in parameters of the marginal model, i.e., the model fitting simulated $\mathbf{y}_i$ to $E_u(g(\mathbf{X}_i\boldsymbol{\beta} + \mathbf{u}_i))$, as a surrogate, since there is not much extra effort needed to calculate the EE (8.38), apart from generating a vector $g(\mathbf{X}_i\boldsymbol{\beta} + \mathbf{u}_i)$ from vectors $\mathbf{c}_i$ and $\mathbf{u}_i$.

Approximation can also be used to assess the impact of confounder u_i on the estimates. For example, we write

$$E(S(\boldsymbol{\beta})) = \sum_{i=1}^{n}(\frac{\partial g_i}{\partial\boldsymbol{\beta}})^T(g(c_i,\boldsymbol{\beta}_0) + u_i - g(c_i,\boldsymbol{\beta}))$$

$$\approx \sum_{i=1}^{n}(\frac{\partial g_i}{\partial\boldsymbol{\beta}})^T(\frac{\partial g_i}{\partial\boldsymbol{\beta}}(\boldsymbol{\beta} - \boldsymbol{\beta}_0) + u_i) \tag{8.40}$$

which leads to approximation

$$\hat{\boldsymbol{\beta}} - \boldsymbol{\beta}_0 \approx (\sum_{i=1}^{n}(\frac{\partial g_i}{\partial\boldsymbol{\beta}})^T\frac{\partial g_i}{\partial\boldsymbol{\beta}})^{-1}n\mathrm{cov}(u_i, \frac{\partial g_i}{\partial\boldsymbol{\beta}}). \tag{8.41}$$

This simple approximation shows that the bias is approximately proportional to $\mathrm{cov}(u_i, \partial g_i/\partial\boldsymbol{\beta})$. Recall that for a linear model $\partial g_i/\partial\boldsymbol{\beta} = c_i$, the bias is proportional to the correlation between the exposure and the confounding factor u_i, which is the same as in (8.2).

Using this approach for a Poisson model (or a similar multiplicative model) with mean

$$E(y_i|u_i) = \exp(\beta_0 + c_i\beta + u_i), \tag{8.42}$$

the bias can be written as

$$\hat{\boldsymbol{\beta}} - \boldsymbol{\beta}_0 \approx E(P_i(u_i)^2)^{-1}\mathrm{cov}(u_i, P_i(u_i)), \tag{8.43}$$

where $P_i(u_i) = \exp(\beta_0 + c_i\beta + u_i)$.

If the y-part is an NLMM, there are a few model fitting approaches, almost all based on approximations to the marginal likelihood function. When the number of repeated measures is large, the bias can be assessed in a similar way, since, as described in Chapter 3, the model fitting is close to fitting the nonlinear model to individual subject data. When the number of repeated measures is small, one has to calculate the likelihood function directly. In this case, the MS estimates tend to the β that minimizes the difference between the true likelihood with c_i^* in the model and the one using $E(c_i^*|c_i)$. White (1982) gives more details. We can also consider a more general case when the parameters depend on known covariates: $\boldsymbol{\beta}_i = \boldsymbol{\beta}^T\mathbf{X}_i + \mathbf{b}_i$. In this case, $\mathbf{b}_i$ are the residual confounding factors after taking factors in $\mathbf{X}_i$ into account. The simple approach can be used. However, the bias also depends on the distribution of $\mathbf{X}_i$.

8.7   Instrumental variable

8.7.1   Instrumental variable estimates

An instrumental variable (IV) z_i is a variable which correlates to c_i, but does not directly correlate to y_i. The most commonly used IV in ER modeling is the dose, if it is randomized, but some other variables may also serve as an IV. To see why dose satisfies the request for IVs, taking models (8.1) as an example, the correlation between c_i and d_i is due to the c-part model. The second condition holds when d_i is independent of u_i, e.g., when d_i is randomized. In the following we take the dose (assuming randomized) as the IV, i.e., $z_i = d_i$, but the results below hold for any IV. For linear model (8.1) the IV estimate for β is the solution to the EE

$$S_{IV}(\beta) = \sum_{i=1}^{n} d_i(y_i - \beta c_i)/n = \mathbf{d}^T(\mathbf{y} - \beta \mathbf{c})/n = 0 \qquad (8.44)$$

where $\mathbf{y} = (y_1, ..., y_n)^T, \mathbf{c} = (c_1, ..., c_n)^T, \mathbf{d} = (d_1, ..., d_n)^T$. The condition $E(S_{IV}(\beta)) = 0$ is satisfied because of $E(S_{IV}(\beta)) = E(\mathbf{d}^T\mathbf{u}) = 0$ with $\mathbf{u} = (u_1, ..., u_n)$, hence the solution to $S_{IV}(\beta) = 0$ is

$$\hat{\beta}_{IV} = (\mathbf{d}^T\mathbf{c})^{-1}\mathbf{d}^T\mathbf{y} \to \beta. \qquad (8.45)$$

The instrumental variable (IV) approach has advantages over those introduced in the previous sections, as it does not need the assumption of no unobserved confounder. However, although the IV estimate is consistent, its precision may be poor. Its variance $\text{var}(\hat{\beta}_{IV}) = \sigma^2(\mathbf{d}^T\mathbf{c})^{-2}\mathbf{d}^T\mathbf{d}$ is typically higher than the variance of the LS estimate. Therefore, using an IV which is weakly correlated to c_i in (8.44) may lead to a high $\text{var}(\hat{\beta}_{IV})$. Furthermore, in the case of small sample size, $\hat{\beta}_{IV}$ may have high bias as well.

Although $\hat{\beta}_{IV}$ can be estimated by solving the EE (8.44), it can also be obtained by the two stage LS (2SLS) approach. For linear models (8.1), the 2SLS method first fits this linear model to obtain $\hat{c}_i = \hat{\theta}d_i$, where $\hat{\theta}$ is the estimated parameter. The next stage fits a model $y_i = \beta\hat{c}_i + u_i$, and the obtained LS estimate is exactly $\hat{\beta}_{IV}$. To see this, taking $\hat{\mathbf{c}} = \hat{\theta}\mathbf{d}$, and $\hat{\theta} = (\mathbf{d}^T\mathbf{d})^{-1}\mathbf{d}^T\mathbf{c}$ into $\hat{\mathbf{c}}^T\mathbf{y}/(\hat{\mathbf{c}}^T\hat{\mathbf{c}})$, the estimate in the second stage, we find the right hand side in (8.45). This approach not only makes using standard statistical packages for IV methods possible, but also provides insight into the construction of the IV estimate. Since in the second stage, the prediction of c_i by d_i is used to fit the model, only the component of c_i which can be predicted by d_i, and hence does not correlate to u_i, is used in the IV estimate. This eliminates confounding bias, but also leads to a more variable estimate since less variation in c_i is used.

For nonlinear regression models with additive confounding factors, the EE for IV estimates has the form

$$S_{IV}(\boldsymbol{\beta}) = \sum_{i=1}^{n} \mathbf{Z}_i(y_i - g(c_i, \boldsymbol{\beta})) = \sum_{i=1}^{n} \mathbf{Z}_i \delta_i = 0 \qquad (8.46)$$

where $\mathbf{Z}_i$ is a set of IVs, including dose, and $\delta_i \equiv y_i - g(c_i, \boldsymbol{\beta})$. Note that a nonconfounded covariate can be the IV of itself, hence $\mathbf{Z}_i$ can include these covariates. For nonlinear models using dose as the IV is not optimal, but we can derive the optimal IV from the dose. As introduced in Chapter 4, for a nonlinear model $y_i = g(c_i, \boldsymbol{\beta}) + \varepsilon_i$ the optimal IV estimate for $\boldsymbol{\beta}$ is the solution to EE:

$$S_{IV}(\boldsymbol{\beta}) = \sum_{i=1}^{n} E(\frac{\partial g(c_i, \boldsymbol{\beta})}{\partial \boldsymbol{\beta}} | \mathbf{Z}_i)(y_i - g(c_i, \boldsymbol{\beta})) = 0 \qquad (8.47)$$

where $E(\partial g(c_i, \boldsymbol{\beta})/\partial \boldsymbol{\beta} | \mathbf{Z}_i)$ is the optimal IV. As it contains parameters to be estimated, a two stage approach can be used. In the first stage one can fit the y-part model, e.g., ignoring the confounding factor, or using d_i as the IV to obtain a preliminary estimate for $\boldsymbol{\beta}$. In the second stage, the estimated $\boldsymbol{\beta}$ is used to calculate $E(\partial g(c_i, \boldsymbol{\beta})/\partial \boldsymbol{\beta} | \mathbf{Z}_i)$, then we use the optimal EE to obtain a refined estimate for $\boldsymbol{\beta}$. Some software such as proc MODEL implements this approach and offers several options.

We have shown that the IV approach can eliminate confounding bias for ER models with additive confounding terms. For models with multiplicative confounding factors, such as Poisson type models (e.g., negative binomial models), due to a confounding factor on the linear predictor, $S_{IV}(\boldsymbol{\beta})$, (8.47) does not have zero mean. The following EE was proposed:

$$S(\boldsymbol{\beta}) = \sum_{i=1}^{n} \mathbf{Z}_i(y_i g^{-1}(c_i, \boldsymbol{\beta}) - 1) = 0, \qquad (8.48)$$

and it can be shown that $E(S(\boldsymbol{\beta})) = 0$ (Mullahy, 1997). However, the major drawback of using this EE is that general GLS software cannot be used to estimate $\boldsymbol{\beta}$. In fact, it has been shown that even when the EE (8.46) is used, the confounders only make the intercept (the baseline risk in the Poisson model) in $\hat{\boldsymbol{\beta}}_{IV}$ biased. The relative risk for c_i is still consistent (Wang, 2012b). It has also been shown that the IV approach also eliminates confounding biases in treatment effect heterogeneity for simple models such as the linear model (8.1) and multiplicative models such as a Poisson or negative binomial model with a log-link function (Wang, 2012b).

8.7.2   Control function method

The control function (CF) method follows the idea of the DA, but it uses IVs to construct the adjustment. It is also a two stage method very similar to the two stage IV method, but it uses IVs in a slightly different way. Again consider model (8.1). When u_i is available, we can include u_i as a covariate to adjust the confounding bias. But u_i may not be entirely confounded. It is the component correlated to v_i that causes bias and has to be adjusted for. The CF method uses the IV model to predict this component, then includes its prediction to adjust the y-part model. To this end, we need to assume that u_i can be written as

$$u_i = av_i + \varepsilon_i \tag{8.49}$$

where ε_i is a random term independent of v_i. Therefore, u_i has two components; av_i is the source of confounding and the error term ε_i is independent of c_i. The relationship (8.49) holds when u_i and v_i are jointly normally distributed. But it may also be correct empirically without the distributional assumption. Equation (8.49) suggests how to control the confounding component in u_i in the y-part model with a predicted v_i. The following is the CF algorithm as a two stage approach:

- fit the c-part model to obtain $\hat{v}_i$s: a prediction for v_is. $\hat{v}_i$s are the residuals from the fitted model.

- fit the y-part model including c_i and $\hat{v}_i$.

The difference between the IV and CF methods is that, in the y-part model, the former replaces the exposure with the predicted one, and the latter adds a predicted confounding factor into it. As the nature of this approach suggests, it is also known as two stage residual-inclusion (Wooldridge, 1997, Terza et al., 2008). Interestingly, if both the y-part and c-part models are linear, the CF approach is equivalent to the IV approach. However, for multiplicative models, the CF estimate may have significantly lower SE than that of the IV estimate (Wang, 2012b, 2014b).

One significant advantage of the CF approach is that it can be used for a wide range of y-part models, as long as the confounding effects can be summarized by the residual error in the c-part model. However, the simple two stage approach can be justified only for simple models such as the linear model and multiplicative models (e.g., the Poisson or negative binomial model with a log-link function) (Wang, 2012b).

8.8 Joint modeling of exposure and response

This section revisits the joint modeling (JM) approach from the point of view of causal inference. Here we assume that the y-part model cannot include sufficient covariates for DA, as otherwise the JM approach is not needed for bias adjustment. Consider the following models:

$$\begin{aligned} y_i &= \beta c_i + u_i \\ c_i &= h(d_i, \theta) + v_i \end{aligned} \tag{8.50}$$

where u_i and v_i are correlated. Therefore, the LS estimate for β is biased if we fit the first model with c_i. But fitting the two models together gives a valid estimate. The following simulation example illustrates implementation for a simple case of a linear c-part model.

```
%let nsimu=1; %let nsub=2000; %let nrep=6; %let sigv=0.2;
%let sigu=0.2; %let sig=0.4; %let sige=0.2;
data simu;
 do simu=1 to &nsimu;
  do sub=1 to &nsub;
    dose=1+ranbin(12,1,0.5);
    latent=rannor(182)*&sig;
    ui=rannor(12)*&sige+latent;
    vi=rannor(12)*&sigv+latent;
    ci=dose+vi;
    yi=ci+ui;
      output;
   end;
 end; run;

proc mixed data=simu;
  model yi=ci/solution;
run;

proc model data=simu;
EXOGENOUS dose;
parms beta0=0 beta=1 a=0 b=1;
eq.c=a+b*dose-ci;
eq.y=beta0+beta*(a+b*dose)-yi;
fit c y;
run;
```

where in proc model the eq.c and eq.y commands specify the estimating equations and the exogenous statement claims that d_i is not affected by any confounding factor. The LS estimate gives $\hat{\beta} = 1.35, (SE = 0.012)$ with quite large bias, compared with the true value 1. The joint model estimate using proc MODEL is 1.01 $(SE = 0.042)$, almost unbiased but with a much larger SE than that of the LS estimate. The reason for its lack of bias is that in the joint model, we do not condition on c_i. The y-part model in this case has error $\beta v_i + u_i$, and is correlated with the error term in the c-part model. But it is not necessary to specify the correlation structure in proc MODEL. In fact, using dose as an IV, we fit the model

```
proc model data=simu;
instruments dose;
parms beta0=0 beta=1 a=0 b=1;
eq.c=a+b*dose-ci;
eq.y=beta0+beta*ci-yi;
fit c y/ gmm;
run;
```

and the IV estimate for β is 1.01, which is almost identical to the joint model approach, but with a much smaller error ($SE = 0.020$), although still larger than that of the LS estimate.

The approach can be applied to classical ME models with a confounding factor

$$
\begin{aligned}
y_i &= \beta c_i^* + u_i \\
c_i^* &= h(d_i, \boldsymbol{\theta}),
\end{aligned}
\tag{8.51}
$$

where c_i^* is observed via $c_i = c_i^* + v_i$, and v_i may be correlated with u_i. In this case, a joint modeling or RC approach is needed to estimate β even without confounding. However, the ordinary RC approach is also sufficient to eliminate confounding, since after replacing c_i^* with $E(c_i^*|d_i) = h(d_i, \boldsymbol{\theta})$ in the first model, $h(d_i, \boldsymbol{\theta})$ is not correlated with u_i. Although in practice θ has to be replaced with its estimate $\hat{\boldsymbol{\theta}}$, which depends on v_i, the correlation is weak and diminishes to zero when n tends to infinity. Therefore, using $h(d_i, \hat{\boldsymbol{\theta}})$ gives a consistent estimate for β as long as $\hat{\boldsymbol{\theta}}$ is consistent.

8.9 Study designs robust to confounding bias or allowing the use of instrument variables

Although the exposure is often potentially confounded with the response, some study designs may eliminate or reduce the confounding even when an ordinary model fitting approach, such as the LS estimate, is used. The confounding factors are often at the subject level. Therefore, if the model is fitted based on comparing data within subjects, the confounding may be eliminated. Taking study design into consideration helps determine the modeling strategy and judging robustness of the estimated ER relationship.

In the situation that there are repeated exposure and response measures, the ER relationship may be estimated using the information within and between subjects. The impact of confounding depends on the procedure used for the relationship estimation. Recall the two stage procedure (Chapter 3) fits the model to individual data to estimate $\beta + b_i$, then summarizes the individual estimates for population parameter estimation and inference. Here we show that this approach is not affected by individual level confounding factors. Consider the following c-part and y-part models:

$$
\begin{aligned}
\mathbf{c}_i &= h(\mathbf{d}_i, \mathbf{t}_i, \boldsymbol{\theta}_i) + \mathbf{e}_i \\
\mathbf{y}_i &= g(\mathbf{c}_i, \boldsymbol{\beta}_i) + \boldsymbol{\varepsilon}_i
\end{aligned}
\tag{8.52}
$$

where the repeated exposure, response, dose and time are all in vector form. We also assume that $\boldsymbol{\theta}_i = \boldsymbol{\theta} + \mathbf{s}_i$ and $\boldsymbol{\beta}_i = \boldsymbol{\beta} + \mathbf{b}_i$ and confounding factors at the subject level may occur and cause correlation between $\mathbf{s}_i$ and $\mathbf{b}_i$. The two stage approach first obtains individual estimates of $\boldsymbol{\beta}_i$: $\hat{\boldsymbol{\beta}}_i$, which converge to $\boldsymbol{\beta} + \mathbf{b}_i$ when the number of repeated measures is large. To estimate $\boldsymbol{\beta}$, the second stage takes the mean of the individual $\hat{\boldsymbol{\beta}}_i$ and obtains

$$\hat{\boldsymbol{\beta}} = \sum_{i=1}^{n} \hat{\boldsymbol{\beta}}_i / n \tag{8.53}$$

which tends to $\sum_{i=1}^{n}(\boldsymbol{\beta} + E(\mathbf{b}_i))/n = \boldsymbol{\beta}$, since $E(\mathbf{b}_i) = 0$ regardless $\mathbf{s}_i$. Therefore, when the number of repeated measures is large, the two stage method can provide a reference that is robust to confounding factors at the subject level. However, since $\mathrm{var}(\hat{\boldsymbol{\beta}}_i)$ may depend on $\boldsymbol{\theta}_i$, the commonly used weighted mean

$$\hat{\boldsymbol{\beta}} = (\sum_{i=1}^{n} \mathrm{var}(\hat{\boldsymbol{\beta}}_i)^{-1})^{-1} \sum_{i=1}^{n} \mathrm{var}(\hat{\boldsymbol{\beta}}_i)^{-1} \hat{\boldsymbol{\beta}}_i \tag{8.54}$$

is not robust against confounding at the subject level.

From the discussion in Chapter 3, we may expect that when the number of repeated measures is large, even the one stage approach will use mostly the variation within subjects to fit the model. In this case, the impact of subject level confounding factors is likely to be low. But in the case of sparse sampling, the impact may be substantial. The bias assessment approach can be used to assess bias under different plausible scenarios.

As a numerical example, we consider an example with an open one compartmental model with first order absorption as the c-part model

$$c_{ij} = \frac{k_{ai}}{V_i(k_{ai} - k_{ei})}(\exp(-k_{e_i}t) - \exp(-k_{ai}t)) + e_{ij} \tag{8.55}$$

where $k_{ai} = \exp(lk_a + u_{1i})$, $k_{ei} = \exp(lk_e + u_{2i})$ and $V_i = \exp(lv + u_{3i})$ are the random coefficients. For the y-part model we assume a linear mixed model for simplicity:

$$y_{ij} = \beta_0 + \beta c_{ij} + u_i + \varepsilon_{ij}, \tag{8.56}$$

where $u_i = u_{1i} + u_{2i} + u_{3i}$ is the confounding factor. This model can be fitted either as a linear mixed model to use the information between subjects or as a linear model treating u_i as fixed unknown parameters. We run a simple simulation to assess the bias in the estimates for β with the two approaches. The parameters are set as in the following program so that the true value for β is 1.

```
library(nlme)
nsimu=2000
sigu=0.2
sige=0.0
sigep=0.3
ka=0
ke=-1
```

```
v=-1
u1=rnorm(nsimu)*sigu
u2=rnorm(nsimu)*sigu
u3=rnorm(nsimu)
ttime=c(1,2,4,8)
#ttime=c(1,2,4,8,12,24)
nrep=length(ttime)
conf=rep(u1+u2+u3,rep(nrep,nsimu))
sub=rep(1:nsimu,rep(nrep,nsimu))
kai=rep(exp(ka+u1),rep(nrep,nsimu))
kei=rep(exp(ke+u2),rep(nrep,nsimu))
vi=rep(exp(v+u3),rep(nrep,nsimu))
vtime=rep(ttime,nsimu)
cti=(exp(-kei*vtime)-exp(-kai*vtime))*kai/(kai-kei)/vi
                            +rnorm(nrep*nsimu)*sige

yi=cti+conf+rnorm(nsimu*nrep)*sigep
gdata=groupedData(yi~cti|sub,data.frame(yi,cti,sub))
summary(lme(yi~cti,random=~1|sub,data=gdata))

jk=summary(lm(yi~cti+as.factor(sub)))
```

Here we set sige=0, an ideal situation of no error in the PK measure, in order
to concentrate on the confounding bias. The bias is introduced by variable
conf, as a linear combination of random coefficients in the c-part model and
added to the y-part models. Note that the confounding bias is at the subject
level, as it comes from the subject random effects.

The parameter estimates and their SEs are given in Table 8.1. Although
the contribution from the between- and within-subject information is complex
to calculate, a rough measurement of relative proportions is the ratio of the
SEs between the estimates of the two models. We find that the number and
times of the repeated measures play an important role in the sources on which
the mixed model estimate is based. In the first situation with time points 1, 2,
4 and 8, most information for the mixed model estimate comes from within-
subject information, hence there is almost no bias. Under the second scenario
(1, 2, 4), there is a significant bias (-0.11) and also a significant increase of
the information taken from between-subjects. Along with the increase of the
contribution from between subject information, the bias in the mixed model
estimate increases substantially. The fixed effect model estimate is almost
unbiased with all four time patterns, but its SE is consistently higher than
that of the mixed effect model estimate. With repeated time points (1, 2),
the lower than true value estimate of 0.97 is likely due to the large SE (5-
fold of that in the mixed model estimate) rather than confounding bias. The
main message from this example is that confounding bias due to subject level
confounders may occur and can be quite severe with a sparse sampling scheme.
If the y-part model in the mixed model is linear, fitting a fixed effects model
can assess potential bias, so one can decide if an adjustment, as described

in this chapter, should be implemented. If the y-part model is nonlinear or fixed u_i cannot be fitted, it is prudent to apply an adjustment at least as a sensitivity analysis.

TABLE 8.1
Summary of $\hat{\beta}$s and their SEs estimated using either the linear mixed model or a linear model with fixed u_i.

Time pattern	$\hat{\beta}(SE)$ (Mixed model)	$\hat{\beta}(SE)$ (Fixed u_i)
(1,2,4,8)	0.98 (0.0024)	1.00(0.0025)
(1,2,4)	0.89 (0.0044)	1.00(0.0056)
(1,4)	0.85 (0.0052)	1.00(0.0068)
(1,2)	0.78 (0.0052)	0.97(0.0257)

Even with some study designs the naive estimates themselves may not be robust enough; these designs may create IVs ready to use to eliminate the bias. The design with randomized dose levels allowing the use of dose as an IV is a typical example. Designs with repeated measures may also allow for using time or functions of it as IVs. The approach described in Section 4.5 can be used directly for this situation, as IV methods dealing with confounding and measurement error are closely related and often technically the same.

8.10 Doubly robust estimates

We have described approaches which rely on either a correct y-part model (e.g., the DA approach) or a correct c-part model (e.g., the propensity score model). In the case where we do not know which model is correct, it is desirable not to rely on the correctness of one model. Fortunately, there are approaches giving valid estimates for the ER model parameters when any of the models is correct. These approaches are often referred to as doubly robust (DR).

First we note that the IV estimate as the solution to

$$\sum_{i=1}^{n} z_i(y_i - \beta c_i - \mathbf{X}_i^T \boldsymbol{\beta}_x) = 0 \tag{8.57}$$

is DR in the sense that $\hat{\beta}_{IV}$ is consistent if model $y_i = \beta c_i + \mathbf{X}_i^T \boldsymbol{\beta} + \varepsilon_i$ is specified correctly, i.e., $\mathbf{X}_i$ contains all confounding factors. Also note that when the y-model is correctly specified, ε_i is an independent random error, hence any variable that correlates to c_i is entitled to be an IV. This is true for other models such as models leading to valid GEE estimates. The key condition is that with correct covariate inclusion the residual $y_i - \beta c_i - \mathbf{X}_i \boldsymbol{\beta}$ becomes independent to Z_i. Therefore, the IV approach is also robust for nonlinear models satisfying this condition, although for these models, specifying them correctly is more difficult.

The IPW estimates given in Section 8.17 are not DR but can also be made DR. For the exposure relationship estimate with multiple exposure levels, an augmented IPW estimate

$$\hat{\mu}(a_k) = n^{-1} \sum_{i=1}^{n} \frac{I_{ki} y_i}{PS_{ki}} - \frac{I_{ki} - PS_{ki}}{PS_{ki}} m(a_k, \mathbf{X}_i) \qquad (8.58)$$

where $I_{ki} = I(d_i = a_k)$, $PS_{ki} \equiv PS_k(\mathbf{X}_i)$, and $m(a_k, \mathbf{X}_i) = E(y_i | d_i = a_k, \mathbf{X}_i)$ is the prediction for $\mu(a_k)$ based on a model with direct adjustment for $\mathbf{X}_i$. This estimate is DR, as it is consistent either when $PS_k(\mathbf{X}_i)$ or the mean model $m(a_k, \mathbf{X}_i)$ is correct. To verify, we write it as

$$\hat{\mu}(a_k) = n^{-1} \sum_{i=1}^{n} y_i(a_k) + \frac{I_{ki} - PS_{ki}}{PS_{ki}}(y_i - m(a_k, \mathbf{X}_i)) \qquad (8.59)$$

so we only need to check if $E(\frac{I_{ki} - PS_{ki}}{PS_{ki}}(y_i(a_k) - m(a_k, \mathbf{X}_i))) = 0$ in both cases. When $PS_k(\mathbf{X}_i)$ is correct,

$$E(\frac{(I_{ki} - PS_{ki})(y_i - m(a_k, \mathbf{X}_i))}{PS_{ki}}) = E(\frac{(y_i - m(a_k, \mathbf{X}_i))E(I_{ki} - PS_{ki}|\mathbf{X}_i)}{PS_{ki}})$$
$$= 0 \qquad (8.60)$$

using the no unobserved confounder assumption. When $m(a_k, \mathbf{X}_i)$ is correctly specified, i.e., $m(a_k, \mathbf{X}_i) = E(y_i | d_i = a_k, \mathbf{X})$ and $\mathbf{X}_i$ contains all confounders

$$E(\frac{(I_{ki} - PS_{ki})(y_i - m(a_k, \mathbf{X}_i))}{PS_{ki}})$$
$$= E(\frac{E(y_i(a_k) - m(a_k, \mathbf{X}_i)|\mathbf{X}_i)(I_{ki} - PS_{ki})}{PS_{ki}}) = 0 \qquad (8.61)$$

as, conditional on $\mathbf{X}_i$, d_i is independent of $y_i(a_k)$ and $E(y_i(a_k) - m(a_k, \mathbf{X}_i)|\mathbf{X}_i) = 0$.

The estimation is easy to implement. First one should fit a model with conditional mean $m(d_i, \mathbf{X}_i, \boldsymbol{\gamma})$ with parameters $\boldsymbol{\gamma}$ to y_i with direct adjustment for $\mathbf{X}_i$, then plug $m(a_k, \mathbf{X}_i, \hat{\boldsymbol{\gamma}})$ into (8.58). For the SE of the estimate, given the straightforward model fitting and calculation, bootstrapping is a good choice.

8.11 Comments and bibliographic notes

The literature on causal inference and causal effect estimation is huge, so that it is impossible to even list only key papers. Apart from the book by Hernan and Robins (2015), readers can find several landmark papers which also reflect the development in this area starting from the later 1980s, in Robins' website http://www.hsph.harvard.edu/james-robins/bibliography/.

Missing data and the drop out of subjects from a study is a large topic and is often not considered in the context of causal effect analysis. However, as we have seen, the two problems have a lot of similarity and connection that warrants a thorough discussion. Almost all confounding issues can be considered as a missing data problem, using the counterfactual framework. Recall that, in this framework, among all the counterfactuals only one is realized (or observed), hence the others are missing. Therefore, the causal inference could be considered as dealing with the missing counterfactuals. If there is no confounding factor, hence the exposure does not depend on the potential response, then missing counterfactuals are missing completely at random (MCAR). Consequently, we can simply base our analysis on the observed one.

9

Dose–response relationship, dose determination, and adjustment

9.1 Marginal Dose–response relationships

We revisit Dose–response relationships when they are dependent on observed or unobserved factors. These factors may affect the dose–exposure or exposure relationships or both. Often we are interested in the marginal relationship, i.e., the average relationship in a population with a mixture of these factors. For example, to select a dose for a clinical trial to demonstrate a significant treatment effect compared with placebo, the statistical inference is based on the responses from patients with different factors, determined by the marginal Dose–response relationship.

When a model includes covariates $\mathbf{X}_i$,

$$y_i = g(d_i, \mathbf{X}_i, \boldsymbol{\beta}) + \varepsilon_i, \tag{9.1}$$

the Dose–response relationship may be covariate dependent, known as treatment effect heterogeneity. The heterogeneity may be explicitly described by the model, e.g.,

$$y_i = \beta_0 + \beta d_i x_i + \varepsilon_i, \tag{9.2}$$

so that the exposure effect is proportional to covariate x_i. Obviously covariate effects that are additive in a linear model do not cause heterogeneity. However, for nonlinear models, including most generalized linear models, heterogeneity may be introduced due to the coexistence of covariates and nonlinearity. Therefore, the required dose increase to achieve, e.g., a 30% response increase depends on the distribution of $\mathbf{X}_i$ in the target population. Although it is straightforward to find the dose for a particular subject with covariates $\mathbf{X}_0$, in most cases, we would like to find the dose to achieve the target in a population of subjects with different $\mathbf{X}_i$ values. For example, when designing a phase III trial, we would like to find the dose that has satisfactory safety and efficacy profiles for the trial population. Therefore, we need an average, or marginal Dose–response model to determine the dose. For notational simplicity, we consider the $\mathbf{X}_i$ as random variables with distribution $F_x(\mathbf{X})$. Then the marginal mean response of (9.1) can be written as

$$g^*(d, \boldsymbol{\beta}) = \int g(d, \mathbf{X}_i, \boldsymbol{\beta}) dF_x(\mathbf{X}_i). \tag{9.3}$$

Note that $g^*(d,\beta)$ may not be in the same family as $g(d,\mathbf{X}_i,\beta)$, and it may not even have an analytical form. For example, if $g(d,\mathbf{X}_i,\beta)$ is a logistic function given $\mathbf{X}_i$, $g^*(d,\beta)$ may not be. With a fitted model, $g^*(d,\beta)$ can be easily calculated by simulation or analytically. For example, if the only covariate is gender and p_m is the proportion of males, then it can be written as

$$g^*(d,\beta) = p_m g(d,m,\beta) + (1-p_m)g(d,f,\beta), \tag{9.4}$$

where $g(d,m,\beta)$ and $g(d,f,\beta)$ are the responses for males and females, respectively.

Calculating the marginal mean is known as standardization or normalization, which is also a way of eliminating confounding. It is essentially the approach based on the back-door criterion (Section 8.5) for calculating the marginal causal ER relationship. In the example above, suppose that gender is a confounder but we have fitted a gender specific model and obtained $g(d,f,\beta)$ and $g(d,m,\beta)$. Then we can estimate $E(y_i(d))$ by $g^*(d,\beta)$, which is unbiased as the average response of this population. But the group d mean $\bar{y}_d$. is an estimate of $E(y_i|d_i=d)$, depending on the proportion of males or females receiving d, is biased. s the back-door criterion (Section 8.5) for calculating the marginal causal ER relationship.

If the Dose–response model has random effects, e.g.,

$$y_{ij} = g(d_{ij},\beta_i) + \varepsilon_{ij} \tag{9.5}$$

with $\beta_i = \mathbf{X}_i\beta + \mathbf{u}_i$ and $\mathbf{u}_i$ follows distribution $F_u(\mathbf{u})$, specification of the distribution of $\mathbf{u}_i$ is needed to determine the marginal Dose–response model, although we can find the dose of a desired response of a typical patient with covariate $\mathbf{X}_0$ and $\mathbf{u}_i = 0$. A typical situation is dose selection for a phase III trial, as we just discussed for patients with different covariate values. The only difference here is that we don't know the individual $\mathbf{u}_i$, but we know their variability. In order to calculate the marginal Dose–response relationship, we have to integrate the Dose–response relationship over both $\mathbf{X}_i$ and $\mathbf{u}_i$. The marginal Dose–response relationship can be written as

$$g^*(d,\beta^*) = \int\int g(d,\mathbf{X}_i,\beta_i)dF_x(\mathbf{X}_i)dF_u(\mathbf{u}). \tag{9.6}$$

It can also be calculated easily by simulation with $\mathbf{u}_i$ sampled from $F_u(\mathbf{u})$. These models are known as structure mean models for Dose–response relationships.

As an example of marginal Dose–response models, we consider the Emax model

$$y_i = E_{max}/(1 + (EC_{50i}/d_i)^{\rho_i}) \tag{9.7}$$

where EC_{50i} and the Hill parameter ρ_i are random variables, and we have omitted variability in E_{max} and no measurement error in y_i, since they do not affect the shape of the marginal model. We assume that $\log(EC_{50i}) \sim$

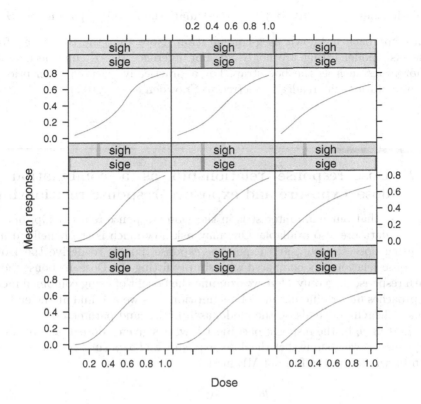

FIGURE 9.1
Marginal Dose–response relationships generated from an Emax model with
low to high heterogeneity in EC_{50i} and ρ_i (low (0.01), medium (0.3) and high
(1) σ_e and σ_h).

$N(1, \sigma_e^2)$ and $\log(\rho_i) \sim N(1, \sigma_h^2)$. Figure 9.1 gives marginal Dose–response
relationships with low, medium and high (0.1, 0.3 and 1, respectively) σ_e and
σ_h, respectively. The plots show that random effects on EC_{50} cause substantial
effect attenuation, and that on the Hill parameter leads to slight shape change.

In practice, often we assume that $g^*(d, \boldsymbol{\beta}^*)$ can be calculated easily, e.g,
it belongs to the same family as $g(d, \mathbf{X}_i, \boldsymbol{\beta}_i)$ exactly or approximately. This
assumption allows an easy calculation of (9.6) with standard model fitting
tools. The approach consists of the following steps:

- Fit model (9.1) with all covariates.

- Generate $\mathbf{X}_i \sim dF_x(\mathbf{X}_i)$ and $\mathbf{u}_i \sim dF_u(\mathbf{u})$ from the target population.

- Generate y_i from model (9.1) with the generated $\mathbf{X}_i$ and $\mathbf{u}_i$ and a wide
  range of d_i.

- Fit model $y_i = g^*(d_i, \boldsymbol{\beta}^*) + \varepsilon_i$ to estimate the marginal parameters $\boldsymbol{\beta}^*$.

This approach is known as g-computation. Note that the SEs given from the last model fitting are not valid. For inference on $\boldsymbol{\beta}^*$, one needs other approaches such as the bootstrap. For a practically oriented description of g-computation, the reader is referred to Snowden et al. (2011).

9.2   Dose–response relationship as a combination of dose–exposure and exposure–response relationships

Suppose that our main interest is in the Dose–response relationship, but exposure data are also available. One may ask how much is the benefit, if any, of using dose–exposure and exposure–response models to derive the Dose–response relationship, compared with simply fitting the Dose–response model with response data only. Here we examine the benefit of using combined model approaches under different model assumptions. As we will find at the end, the benefit heavily depends on the model assumption and parameters.

Letting y_i be the response of subject i, who is given a dose d_i and results in exposure c_i, we assume that the dose–exposure and exposure–response models can be written in the form of ME models

$$
\begin{aligned}
y_i &= g(c_i^*, \boldsymbol{\beta}) + \varepsilon_i \\
c_i^* &= h(d_i, \boldsymbol{\theta}) \\
c_i &= c_i^* + e_i
\end{aligned}
\tag{9.8}
$$

where $g(c_i^*, \boldsymbol{\beta})$ and $h(d_i, \boldsymbol{\theta})$ are known regression models with unknown parameters $\boldsymbol{\beta}$ and $\boldsymbol{\theta}$, respectively, and c_i^* is not observed. For these models we can easily obtain the model between d_i and y_i,

$$
y_i = g(h(d_i, \boldsymbol{\theta}), \boldsymbol{\beta}) + \varepsilon_i.
\tag{9.9}
$$

Note that often not all parameters in the two models in (9.8) are not estimable by fitting this model, since some parameters have been absorbed into other parameters. For a given target response level y_0, the corresponding dose level can be obtained from

$$
d_0 = h^{-1}(g^{-1}(y_0, \boldsymbol{\beta}), \boldsymbol{\theta})
\tag{9.10}
$$

with the parameters replaced with their estimates.

Now we consider whether using both models for parameter estimation and putting them into model (9.9) is more efficient (giving a more accurate Dose–response relationship) than fitting model (9.9) directly. To fit the first model in (9.8), one has to use an appropriate approach to deal with the measurement error. In this case, the best approach would be the regression calibration (RC), which is to replace c_i^* with its estimate based on the second model. Therefore, this approach leads to model (9.9). In particular, when both models in (9.8) are linear, e.g., $y_i = \beta_0 + \beta c_i^* + \varepsilon_i$ and $c_i^* = \theta_0 + \theta d_i$, the two approaches lead to the same $g(h(d_i, \boldsymbol{\theta}), \boldsymbol{\beta})$, or equivalently the same (9.10). The following program gives empirical evidence for the equivalence.

```
> d=rnorm(100)
> x=1+d+rnorm(100)
> y=2+d+rnorm(100)
> jk1=lm(y~d)
> jk1$coef
(Intercept)          d
  2.119824    0.991632
> jk2=lm(x~d)
> jk3=lm(y~predict(jk2))
> jk2$coef[2]*jk3$coef[2]
         d
  0.991632
```

which shows the β estimates by both methods are the same. Therefore, under this situation, the combined model approach has no advantage for dose finding. This result is expected from the point of view of using dose as an instrumental variable (IV), as the RC approach is equivalent to using d_i as the IV. The IV theory says that only the exposure change projected to the IV can be used to estimate the exposure–response relationship, hence the marginal Dose–response model uses all information.

Model $c_i^* = h(d_i, \boldsymbol{\theta})$ assumes that the exposure driving the response for any individual only depends on the given dose, which is hardly true. In fact, it is more likely that the c-part model is

$$c_i^* = h(\boldsymbol{\theta}, d_i, \mathbf{u}_i) \tag{9.11}$$

and $\mathbf{u}_i$ are random effects, e.g., in random coefficients $\boldsymbol{\theta}_i$. Here we assume that c_i^* can be observed or estimated. For example, if repeated measures $c_{ij}, j = 1, ..., n_i$s are available, it can be estimated as $\hat{c}_i^* = \sum_{j=1}^{n_i} c_{ij}/n_i$, when n_i is sufficiently large.

Furthermore, we assume that the marginal model of the Dose–response relationship can be written as

$$y_i = g^*(d_i, \boldsymbol{\beta}^*) + \varepsilon_i^*. \tag{9.12}$$

Suppose that we can estimate parameters $\boldsymbol{\eta} = (\boldsymbol{\beta}^T, \boldsymbol{\theta}^T)^T$ by fitting the dose–exposure and exposure–response models and $\boldsymbol{\beta}^*$ by fitting model (9.12). If $\boldsymbol{\beta}^*$ can be written as a function of $\boldsymbol{\eta}$ such as $\boldsymbol{\beta}^* = M(\boldsymbol{\eta})$, one can compare the variance of $\hat{\boldsymbol{\beta}}^*$ from fitting model (9.12) with that of $\tilde{\boldsymbol{\beta}}^* = M(\hat{\boldsymbol{\eta}})$,

$$\mathrm{var}(\tilde{\boldsymbol{\beta}}^*) = \mathbf{G}\mathrm{var}(\hat{\boldsymbol{\eta}})\mathbf{G}^T, \tag{9.13}$$

where $\mathbf{G} = \partial M(\boldsymbol{\eta})/\partial \boldsymbol{\eta}$ and $\mathrm{var}(\hat{\boldsymbol{\eta}})$ can be calculated from the exposure and response models.

For linear models

$$
\begin{aligned}
y_i &= \beta_0 + \beta c_i + \varepsilon_i \\
c_i &= \theta_0 + \theta d_i + e_i,
\end{aligned}
\tag{9.14}
$$

some interesting analytical results can be derived. For the Dose–response relationship, conditional on d_i we can write the marginal model as

$$
\begin{aligned}
y_i &= \beta_0 + \beta(\theta_0 + \theta d_i) + \beta e_i + \varepsilon_i \\
&= \beta_0^* + \beta^* d_i + \varepsilon_i^*
\end{aligned}
\tag{9.15}
$$

where $\beta_0^* \equiv \beta_0 + \beta\theta_0$, $\beta^* = \beta\theta$, and $\varepsilon_i^* \sim N(0, \beta^2\sigma_e^2 + \sigma_\varepsilon^2)$. Here the key parameter for the dose–exposure relationship, and also for dose finding, is β^*. Hence we examine the accuracy of its estimates either based on fitting the dose–exposure data to model (9.15) or the joint modeling approach fitting both models in (9.14) as $\hat{\beta}\hat{\theta}$. In this simple case, we can calculate $\mathrm{var}(\hat{\beta}\hat{\theta})$ directly without using the general formula (9.13).

To further simplify the problem, we assume $\beta_0 = 0$, $\theta_0 = 0$, $\sum_{i=1}^n d_i = 0$. This assumption has no material impact on the generality of the derived results, as for the estimation of the slope, one can center the data to eliminate the intercept. When fitting model (9.15), the LS estimate for β^* has variance

$$
\mathrm{var}(\hat{\beta}^*) = \frac{\beta^2\sigma_e^2 + \sigma_\varepsilon^2}{n\sigma_d^2},
\tag{9.16}
$$

where $\sigma_d^2 = \sum_{i=1}^n d_i^2/n$. The approach uses both model estimates β^* by $\tilde{\beta}^* = \hat{\theta}\hat{\beta}$, where $\hat{\theta}$ and $\hat{\beta}$ are the LS estimates from fitting the two models in (9.14). To calculate $\mathrm{var}(\hat{\theta}\hat{\beta})$ with an approximation based on the large sample properties of $\hat{\beta}$ and $\hat{\theta}$, first we calculate

$$
\mathrm{cov}(\hat{\theta}, \hat{\beta}) = E(\mathrm{cov}(\hat{\theta}, \hat{\beta}|\mathbf{e})) + \mathrm{cov}(E(\hat{\theta}|\mathbf{e}), E(\hat{\beta}|\mathbf{e}))
\tag{9.17}
$$

where $\mathbf{e} = (e_1, ..., e_n)$. Since conditional on e_i, $\hat{\theta}$ is fixed, hence $\mathrm{cov}(\hat{\theta}, \hat{\beta}|\mathbf{e}) = 0$. The second term is also 0 since $E(\hat{\beta}|\mathbf{e}) = \beta$, hence $\mathrm{cov}(\hat{\theta}, \hat{\beta}) = 0$. This allows the use of the well known formula

$$
\mathrm{var}(uv) = \mathrm{var}(u)\mathrm{var}(v) + \mathrm{var}(u)E(v)^2 + \mathrm{var}(v)E(u)^2
\tag{9.18}
$$

when $\mathrm{cov}(u, v) = 0$. Since $\mathrm{var}(\hat{\theta}) = O(n^{-1})$ and $\mathrm{var}(\hat{\beta}) = O(n^{-1})$, $\mathrm{var}(\hat{\theta})\mathrm{var}(\hat{\beta}) = O(n^{-2})$ can be ignored. This leads to

$$
\mathrm{var}(\hat{\theta}\hat{\beta}) \approx (\beta^2\sigma_e^2/\sigma_d^2 + \theta^2\sigma_\varepsilon^2/\sigma_c^2)/n
\tag{9.19}
$$

where $\sigma_c^2 = \theta^2\sigma_d^2 + \sigma_e^2$.

To assess the relative efficiency of the two approaches, we calculate the variance ratio

$$\frac{\text{var}(\hat{\theta}\hat{\beta})}{\text{var}(\hat{\beta}^*)} = \frac{\beta^2 \sigma_e^2/\sigma_d^2 + \theta^2 \sigma_\varepsilon^2/\sigma_c^2}{(\beta^2 \sigma_e^2 + \sigma_\varepsilon^2)\sigma_d^{-2}}$$

$$= 1 - \frac{\sigma_\varepsilon^2 \sigma_e^2}{(\beta^2 \sigma_e^2 + \sigma_\varepsilon^2)(\sigma_d^2 \theta^2 + \sigma_e^2)}. \tag{9.20}$$

Since the second term is positive, this result suggests that joint modeling gives a better estimate than fitting the marginal dose–exposure model. From this we can also find when the maximum efficiency gain occurs. To this end, we use the fact that the maximum of function $f(x) = x/((a + x)(b + x))$ occurs when $x = \sqrt{ab}$, which can be easily verified. Applying this to the second term in (9.20) with $a = \sigma_\varepsilon^2/\beta^2$ and $b = \sigma_d^2\theta^2$, we find that the maximum gain occurs when

$$\sigma_e^2 = \sqrt{ab} = \sqrt{\sigma_\varepsilon^2 \sigma_d^2 \theta^2/\beta^2}. \tag{9.21}$$

As a numerical example, if $\beta = 1, \theta = 1, \sigma_\varepsilon^2 = 1$ and $\sigma_d^2 = 1$, then the maximum gain occurs when $\sigma_e^2 = 1$, and the maximum gain in terms of the variance ratio is $\frac{\text{var}(\hat{\theta}\hat{\beta})}{\text{var}(\hat{\beta}^*)} = 3/4$. This is equivalent to a reduction of 15% in the SE. Whether the $1/4$ variance reduction is worthwhile for the effort of joint modeling is, however, sometimes a difficult practical question to answer.

The two approaches can also be compared in terms of the accuracy of predicting the response at a given dose level d_0. For the direct approach, it is $\hat{y}(d_0) = g^*(d_0, \boldsymbol{\eta})$. For the combination approach, one can use $\hat{y}(d_0) = g^*(d_0, M(\hat{\boldsymbol{\theta}}, \hat{\boldsymbol{\beta}}))$ and their variance can be calculated using the delta approach (Section A.4). When the marginal model does not have a closed form, the variance calculation becomes difficult and a direct simulation using the construction of the marginal model may be a better approach.

9.3 Dose determination: Dose–response or dose–exposure–response modeling approaches?

This section considers a similar topic to that in the last section, but from the point of view of dose determination. Dose determination is a key task during drug development to select the right dose for further development. In a phase II dose finding trial patients are randomized to take the drug at multiple dose levels. In this case, modeling the Dose–response relationship (which we also consider as a type of ER relationship in places) can be fitted and the appropriate dose level determined from the fitted model. There are many papers and monographs dealing with this type of approach, hence we refer the reader to these sources (e.g., Ting, 2006). Although dose determination is a complex task requiring consideration of multiple objectives such as the balance between efficacy and safety profiles, we first focus on the task concerning only one outcome and show even under this scenario there are still a number of issues in need of careful consideration.

First we review the routine procedure of dose–determination based on Dose–response model

$$y_i = g(d_i, \boldsymbol{\beta}) + \varepsilon_i. \tag{9.22}$$

Our goal is to find a dose level d so that $E(y(d)) = y_0$, the target response level. With this simple model, the required dose can be calculated as $d = g^{-1}(y_0, \boldsymbol{\beta})$, where $g^{-1}(y_0, \boldsymbol{\beta})$ may have to be calculated numerically. In most clinical settings the target response is relative to a placebo or a baseline response and we would like to have $E(y(d) - y(0)) = y_0$. For the linear model $y_i = \beta_0 + \beta d_i + \varepsilon_i$, $E(y(d) - y(0)) = \beta d_i$ so the response change is proportional to the dose. This simple relationship is very useful to determine the amount of dose increase needed to achieve, e.g., a 30% response increase. However, when treatment heterogeneity exists, one needs the marginal model, as introduced in Section 9.1, to determine the dose level for a specific population. In the following we will use marginal models.

Suppose that the marginal Dose–response model is

$$y_i = g^*(d_i, \boldsymbol{\eta}) + \varepsilon_i^* \tag{9.23}$$

and we are interested in finding the dose for a given response (e.g., the maximum tolerated dose) we need to find $g^{*-1}(y, \boldsymbol{\eta})$ so that we can calculate, for a given response y_0, the corresponding dose d_0 as

$$\hat{d}_0 = g^{*-1}(y_0, \hat{\boldsymbol{\eta}}). \tag{9.24}$$

Then the accuracy of $\hat{d}_0$ can be assessed by

$$\text{var}(\hat{d}_0) = \mathbf{H}\text{var}(\hat{\boldsymbol{\eta}})\mathbf{H}^T \tag{9.25}$$

where $\mathbf{H} = (\partial g^{*-1}(y_0)/\partial \eta_1, ..., \partial g^{*-1}(y_0)/\partial \eta_K)$, evaluated at d_0, and η_k is the kth element in $\boldsymbol{\eta}$. However, there is no need to calculate $\mathbf{H}$ from $g^{*-1}(y)$, due to the well known fact that $\partial g^{*-1}(y)/\partial \eta_k = (\partial g^*(d)/\partial \eta_k)^{-1}$. This property has a significant advantage when $g^{*-1}(y)$ has no analytical form, since it allows the calculation of $\mathbf{H}$ from $g(d, \boldsymbol{\eta})^*$ directly.

Specifically, for an Emax model with $g^*(\alpha, \beta, \gamma) = \alpha/(1 + \beta/d_i^\gamma)$, we have

$$
\begin{aligned}
\frac{\partial g^*}{\partial \alpha} &= (1 + \beta/d_i^\gamma)^{-1} \\
\frac{\partial g^*}{\partial \beta} &= \frac{-\alpha}{(1 + \beta/d_i^\gamma)^2 d_i^\gamma} \\
\frac{\partial g^*}{\partial \gamma} &= \frac{-\alpha\beta}{(1 + \beta/d_i^\gamma)^2 d_i^\gamma \log(d_i)}
\end{aligned} \tag{9.26}
$$

and $\mathbf{F}$ can be calculated based on them. From here we can find easily that, for example,

$$\frac{\partial g^{*-1}}{\partial \alpha} = (\frac{\partial g^*}{\partial \alpha})^{-1} = 1 + \beta/d_i^\gamma, \tag{9.27}$$

hence **H** can also be calculated easily.

Although it is possible to calculate the information matrix manually, for applied statisticians, it may be more convenient to use software with the option of automatic derivative generation. We will consider the probit model with random effects as an example of using such software, because its marginal model is also a probit model, hence it is easy to compare the two approaches by simulations. In practice, the logistic and probit models behave very similarly, hence the results here also apply to logistic regression models.

Specifically, let y_i be the indicator of an event and the models for c_i and y_i are

$$
\begin{aligned}
c_i &= \theta_0 + \theta d_i + e_i, \\
y_i &\sim Bin(P_i) \\
P_i &= \Phi(\beta_0 + \beta c_i)
\end{aligned}
$$
(9.28)

where $Bin(P)$ is a binary distribution with probability P and $\Phi(.)$ is the distribution function of the standard normal distribution. The marginal model for the dose-event relationship is also a probit model with

$$
\begin{aligned}
E(P_i|d_i) &= E(\Phi(\beta_0 + \beta(\theta_0 + \theta d_i + e_i))) \\
&= \Phi((\beta_0 + \beta\theta_0 + \theta\beta d_i)\gamma^{-1}) \\
&= \Phi(\beta_0^* + \beta^* d_i))
\end{aligned}
$$
(9.29)

where $\gamma = \sqrt{1 + \beta^2\sigma_e^2}$, $\beta_0^* = (\beta_0 + \beta\theta_0)\gamma^{-1}$, and $\beta^* = \beta\theta\gamma^{-1}$. To estimate β^*, one can directly fit the marginal model for y_i and d_i data with $E(P_i|d_i)$ given in (9.29), or fit both models in (9.28) and estimate β^* as $\hat{\beta}\hat{\theta}\hat{\gamma}^{-1}$. With the estimated parameters, one can also calibrate the dose level for a given risk level p_t as

$$
d_t = (\Phi^{-1}(p_t) - \hat{\beta}_0^*)/\hat{\beta}^*.
$$
(9.30)

The following R-codes, which call function deriv3(.) to generate the derivatives and to calculate **H**, and to use simulation to estimate var($\hat{\eta}$), are used to calculate asymptotic variances of the estimated dose by the two approaches and to generate the results in Table 9.1, with $d_i \sim N(0, \sigma_d^2)$.

```
#information matrix for probit model with and without the PK model
library(Matrix)
calc=function(beta=1, theta0=0, theta=1, sig=1,
sigd=0.3, nsimu=10000,p0=0.1){
gamma=1+beta^2*sig
dose=rnorm(nsimu)*sigd
conc=theta0+theta*dose+rnorm(nsimu)*sig
beta0=qnorm(p0)
lp=(beta0+beta*conc)
yi=1*(rnorm(nsimu)+lp>0)
fitc=summary(lm(conc~dose))
fity=summary(glm(yi~conc,family=binomial(link = "probit")))
```

```
fityd=summary(glm(yi~dose,family=binomial(link = "probit")))

#PKPD
Vall=bdiag(fitc$cov*fitc$sigma^2,fity$cov.un,2*sig^4/nsimu)*1000

gd2=deriv3(expression((sqrt(1+beta^2*sig)*lp0-beta0-beta*theta0)
                                                    /beta/theta),
    c("beta0","beta","theta0","theta","sig"),hessian=F,func=T)

jk=as.matrix(attr(gd2(beta0,beta,theta0,theta,sig),"gradient"))
vpkpd=jk%*%Vall%*%t(jk)

#DosePD
Vall=fityd$cov.un*1000

gd2=deriv3(expression((lp0-beta0)/beta),c("beta0","beta"),hessian=F,
                                                    func=T)
jk=as.matrix(attr(gd2(beta0,beta),"gradient"))
vpd=jk%*%Vall%*%t(jk)
return(c(p0,sigd,as.vector(vpkpd),vpd))
}

lp0=qnorm(0.3)
Out=NULL
Out=rbind(Out,calc(sigd=0.1,p0=0.1))
Out=rbind(Out,calc(sigd=0.2,p0=0.1))
...
```

In general, the joint modeling approach performs well with large σ_ds. However, the value of $P(d = 0)$ also plays an important role, especially when σ_d is small. The same approach can also be used to calculate the asymptotic variance of the estimated dose for nonlinear models such as the Emax model. One only needs to change the formula in

```
expression((sqrt(1+beta^2*sig)*lp0-beta0-beta*theta0)/beta/theta)
```

to that of the Emax model. However, it is not easy to compare the two approaches, since the marginal model is no longer an Emax model and does not have an analytical form.

We may be interested in their small sample properties under these scenarios. Hence the following simulation was conducted. The SAS code for this simulation is simple and straightforward, hence is omitted here. Table 9.2 gives the 5th, 25th, 50th, 75th, and 95th percentiles of the estimated dose level for two different sample sizes and $P(d = 0)$ based on 1000 simulations. We set $\sigma_d = 0.3$, which is between the two nearest σ_d values, one in favor of (var$_{der}$) and another in favor of (var$_{dr}$) in Table 9.1. The true doses for target probability 0.3 with $P(d = 0) = 0.1$ and $P(d = 0) = 0.5$ are 0.540 and -0.742, respectively. In general, the quartiles show that the combination approach is better, particularly when $n = 500$. However, the 5th, and 95th percentiles are

TABLE 9.1
Asymptotic variances of the estimated dose for target probability equal to 0.3 using Dose–response model (var_{dr}) and dose–exposure–response models (var_{der}). $P(d = 0)$ is the baseline probability.

$P(d = 0)$	σ_d	var_{dr}	var_{der}
0.1	0.1	8.840	12.388
0.1	0.2	2.385	3.148
0.1	0.5	0.745	0.611
0.1	1.0	0.533	0.280
0.2	0.1	2.552	1.896
0.2	0.2	0.937	0.594
0.2	0.5	0.497	0.233
0.2	1.0	0.456	0.204
0.5	0.1	1.917	4.578
0.5	0.2	0.858	1.231
0.5	0.5	0.562	0.360
0.5	1.0	0.560	0.256

wider than the dose–exposure model based estimates, suggesting instability with small sample sizes.

TABLE 9.2
Quantiles of estimated dose levels for target probability equal to 0.3 using Dose–response and dose–exposure–response models, with probabilities 0.1 and 0.5 at zero log dose based on 1000 simulations. The true doses for $P(d = 0) = 0.1$ and $P(d = 0) = 0.5$ are 0.540 and -0.742, respectively.

	Quantile(%)	$P(d = 0)$	0.1		0.5	
		n	500	100	500	100
Dose–	95		1.0265	1.8620	-0.3660	-0.1637
exposure	75		0.7908	0.9688	-0.5196	-0.4225
model	50		0.6492	0.6725	-0.6421	-0.6374
	25		0.5278	0.4458	-0.7786	-0.9241
	5		0.3607	0.1794	-0.9765	-1.6494
Dose–	95		1.0722	2.7402	-0.4805	-0.1668
exposure–	75		0.6840	0.7659	-0.6218	-0.4575
response	50		0.5312	0.4630	-0.7508	-0.6840
model	25		0.4322	0.2935	-0.9293	-1.0613
	5		0.3143	-2.0594	-1.3607	-3.1524

9.4   Dose adjustment

From this section onward, we turn our focus to dose adjustment and its interaction with ER modeling. Here dose is used as a general term in order to cover a wide range of treatments adjustment. For example, we may use dose = 1 or 0 to represent treatment on and off. Dose adjustment also refers to either pre-planned or spontaneous treatment changes. Dose adjustments are common in both clinical practice and clinical trial settings. They are often triggered by exposure or response–dependent events. For example, dose reduction or interruption may be triggered by adverse events, or by exposure reaching the therapeutic target. In the latter situation, dose adjustment may be used to control the exposure level.

To see how dose adjustments may affect the modeling of dose–exposure relationships, consider therapeutic dose monitoring (TDM), which is common when drug exposure is highly variable and we wish to ensure most patients have exposure within a target range. Consider the power model for repeated exposure measures:

$$c_{ij} = \theta d_{ij} + v_i + e_{ij}. \tag{9.31}$$

To control the concentration for subject i within a given range R, i.e., we would like to find dose d so that $\theta d + v_i \in R$. Suppose that R is very narrow and the adjustment can finally put each patient's exposure into the range. Then at the end of the adjustment, everyone has a similar exposure, but under different doses. Therefore, simply looking at the dose and exposure at this point, we cannot find a relationship between them, although the exposure depends on dose according to the model. In the next two subsections we will discuss a number of issues relating to dose adjustments and modeling under dose adjustments.

Closely related to dose adjustment is dynamic treatment regimens (DTR) (Murphy, 2003), which refers to response–dependent treatment changes in a general sense. The development has been mainly focused on simple treatment switching between a few alternatives, since dealing with continuous changes is not only more complex in theory, but also more difficult in practice. A key assumption for determining causal effects is that treatment changes can be considered as the consequence of a sequential randomization at any time point given observed history, or equivalently that there is no unobserved confounding factor affecting treatment changes. We will show that the same principle also applies to treatments as a continuous variable (e.g., dose).

9.4.1   Dose adjustment mechanisms

There are many types of treatment adjustment; some are not planned and may be difficult to describe exactly. Nevertheless it is important to understand and to model, if necessary, the mechanism for causal inference. Two common types of dose adjustment are adjustments based on drug exposure and adjustments based on drug response. The former is often planned and serves purposes such as TDM, while the latter is often spontaneous. But both share the same type of mechanism: the adjustment depends on the outcome of dosing.

9.4.1.1 Exposure-dependent dose adjustment

Often the purpose of dose adjustment is to keep individual exposure levels within a target exposure range (L, U), for example, in a trial with therapeutic dose monitoring. For this purpose a simple adjustment rule is to escalate the dose one level higher when $c_{ij} < L$ and to de-escalate one level lower when $c_{ij} > U$. This adjustment is a trial-and-error approach and needs no dose–exposure model. Formally, this mechanism can be written as

$$d_{ij+1} = \begin{cases} d_{ij} + \Delta & c_{ij} < L \\ d_{ij} & L \leq c_{ij} \leq U \\ d_{ij} - \Delta & c_{ij} > U \end{cases} \tag{9.32}$$

where Δ is the dosing step, which may also be variable. In practice, often the dose adjustment stops when the exposure becomes stable within the target range. This adjustment is widely used for TDM because of its simplicity.

Knowing the dose–exposure model is often beneficial, as dose adjustment can be made more efficient and safer by using this model. For example, for repeated exposure measures following the power model

$$c_{ij} = \theta d_{ij} + v_i + e_{ij}, \tag{9.33}$$

the required dose to achieve exposure level c_0 for subject i after the jth visit can be found as

$$\hat{d} = (\log(c_0) - \hat{v}_i)/\hat{\theta} \tag{9.34}$$

where $\hat{\theta}, \hat{v}_i$ are estimated based on dosing and exposure data observed until visit j (Daiz et al., 2007). Another model based approach using an empirical model was proposed by O'Quigley (2010). All these authors examined asymptotic properties of the dose adjustments and showed consistency of sequential estimation of θ under some technical conditions. Since often there are only a few available doses, e.g., when the drug formulation is a tablet or capsule, the simple algorithm (9.32) is the most commonly used. We will concentrate on these simple adjustments.

In practice, the adjustment described by (9.32) may not be followed exactly, since some factors that are often unknown or not recorded may also have an impact on dose adjustment. To model this mechanism, one has to consider them as random factors. A straightforward approach is to use two binary (e.g., logistic) models. The probability of a dose increase at step j may be expressed as

$$P(d_{ij+1} = d_{ij} + \Delta) = 1/(1 + \exp(-\eta(L - c_{ij}) - \eta_0 + s_i)) \tag{9.35}$$

where η and η_0 are parameters. In addition to the part $\eta(L-c_{ij})$ depending on the exposure, s_i is a factor representing the impact of subject level characteristics on dose escalation. For example, a subject may have a lower probability of a dose increase if he has a higher risk of an AE than someone with the same c_{ij} but having a lower risk. One can set up the model for dose reduction in the same way. Alternative models treating dose levels as an ordered categorical variable may also be used if appropriate. Since there are unknown parameters in the model, if the dose adjustment mechanism is to be considered in causal effect estimation, one needs to fit the dose adjustment model as well.

9.4.1.2   Response-dependent dose adjustment

A common dose response–dependent adjustment in clinical practice is dose reduction due to AEs. Here we introduce the following model assuming the probability of an AE occurrence depending on the exposure level, hence on the dose indirectly. Let $y_{ij} = 1$ if an AE occurs between visits j and $j-1$ and $y_{ij} = 0$ otherwise. A dose reduction is triggered when $y_{ij} = 1$, i.e., $d_{ij+1} = d_{ij} - \Delta$ if $y_{ij} = 1$. We assume that the risk of the AE relates to the exposure via a logistic model with

$$P(y_{ij} = 1|c_{ij-1}) = 1/(1 + \exp(-\beta c_{ij-1} - \beta_0 + u_i)) \qquad (9.36)$$

where $u_i \sim N(0, \sigma_u^2)$ is a subject level effect. With this model we link drug exposure to dose adjustment, hence the exposure–response model forms a part of the dose adjustment mechanism.

A similar dose adjustment to (9.32), but depending on efficacy measurements, may also be used. For example, when y_{ij} is the blood pressure level, then a dose increase of an anti-hypertension drug may be granted when y_{ij} is higher than a certain level. Again we find the exposure–response model in the dose adjustment mechanism.

A typical example for response–dependent dose adjustment can be found in phase I dose escalation trials. The objective of dose adjustment in this context is to find the dose that achieves the target probability of toxicity (known as dose limiting toxicity (DLT)), which is often $1/3$ in cancer trials. Commonly used and simple to implement is the "3+3" escalation rule, using cohorts of 3 patients each. To start we need to pre-specify available dose levels. Often the trial starts from the lowest dose and escalates using the following escalation rule. After testing a cohort at the current dose, if there is no DLT, the dose is escalated to the next higher level. If there is 1 DLT, treat the next cohort at the same dose. If the are 2 or 3 DLTs, the dose is reduced to the next lower level. If two consecutive cohorts on the same dose all result in 1 DLT, then this dose is recommended as the target dose and the trial ends. Otherwise, the trial continues, but may often end when a large number of cohorts have been tested but still the target dose cannot be found. The most remarkable difference from the previous examples is that the adjusted dose applies to a different group from the patients from whom the response is measured. To distinguish the two classes, we call this adjustment "between patient" adjustment and those described previously "within patient" adjustment.

9.5   Dose adjustment and causal effect estimation

9.5.1   Dose adjustment and sequential randomization

We have introduced a number of dose adjustment mechanisms that are either exposure or response–dependent. Intuitively, the mechanisms may introduce confounding bias in the dose–exposure or Dose–response relationships if the adjustment mechanisms are not taken into account in the analysis. However, under some situations, modeling the dose adjustment mechanism may not be necessary. Recall that for the analysis of missing data, missing mechanisms can be classified as missing completely at random (MCAR), missing at random (MAR) and nonignorable missing. For the first two classes, a valid analysis may not necessarily take the missing mechanism into account. We follow a similar route to find conditions under which the modeling of exposure and response data may not necessarily include the dose adjustment model. In particular, we are interested in a class of dose adjustments that only depends on the observed previous dosing, exposure and response history and known factors. This class is very similar to the missing data mechanism of MAR. It is also very similar to sequential randomization in DTR, in which treatment assignment at visit j may be considered as randomized, conditional on the history.

Formally the distribution of the next dose at visit j follows

$$d_{ij+1} \sim g(d; \bar{F}_{ij}, s_i) \qquad (9.37)$$

where $g(d; \bar{F}_{ij})$ is an arbitrary density function, conditional on $\bar{F}_{ij}$ containing the history of dosing, response and exposure information until visit j, and known constant or time varying covariates in $\mathbf{X}_{ij}$ and s_i is a random variable. For convenience, we call a dose adjustment mechanism a dose adjustment at random (DAR), named after MAR, when it satisfies condition (9.37) with independent or fixed s_i. See Wang (2014b) for more details. The key feature of DAR is that d_{ij} can be considered as sequentially randomized given previous history, which links to the concept of sequential randomization in dynamic treatment regimens (Murphy, 2003).

The key characteristic of DAR is that it allows fitting the Dose–response model separately from the dose adjustment model, given proper adjustment for factors in $\bar{F}_{ij}$ is made in the model fitting. Sometimes dose adjustment may depend on models sequentially estimated. These adjustments may also be a DAR. For example, when a dose–exposure model is fitted to data up to visit $j - 1$, hence only depends on $\bar{F}_{ij-1}$, an adjustment of (9.32) or (9.51) based on a predicted y_{ij} rather than the observed one is also DAR.

For similar motivations, we will consider the distribution of c_{ij} as

$$c_{ij+1} \sim h(c; \bar{G}_{ij}, v_i) \qquad (9.38)$$

where $\bar{G}_{ij}$ is defined as F_{ij} but may also include d_{ij}. Similarly, it is sufficient to adjust factors in G_{ij} for fitting exposure–response models.

9.5.2   Directional acyclic graphs and the decomposition of the likelihood function

The mechanism of dose adjustment can also be presented in a DAG such as Figure 9.2. Rules applied to DAGs can also be applied here. For example, one advantage of using ER modeling is that it is not impacted by response–dependent dose adjustment. Indeed, one can find in Figure 9.2 that c_{ij} blocks the path from d_{ij} to y_{ij}, hence, if there is no confounding between u_i and v_i, the ER relationship can be determined independently. In longitudinal settings it has a clearer meaning as the correlation of random effects at the patient level (s_i, u_i and v_i in Figure 9.2). This discussion also brings another concept: DAR conditional on random effects, that is, (9.37) holds when v_i is fixed or is included in F_{ij}. Under this condition, often causal inference can be based on within patient comparisons.

DAGs can also be used to write the likelihood function for the triple (y_{ij}, c_{ij}, d_{ij}) in a sequential manner using the decomposition demonstrated in Chapter 6. The joint distribution of all variables in a DAG can be factorized as a product of distributions of descendants conditional on their parents. When a node has no parent, its distribution does not depend on any other node. The joint distribution of y_{ij}, c_{ij} and d_{ij}, conditioning on subject level factors, can be factorized into

$$\begin{aligned} & f(y_{ij}, c_{ij}, d_{ij} | u_i, v_i, s_i, \bar{F}_{ij-1}, \bar{G}_{ij-1}) \\ = \; & l(y_{ij} | c_{ij}, u_i) h(c_{ij} | \bar{G}_{ij-1}, v_i) g(d_{ij} | \bar{F}_{ij-1}, s_i), \end{aligned} \qquad (9.39)$$

where s_i, u_i and v_i are subject level random variables as defined in the exposure–response, dose exposure and dose adjustment models, respectively, and $\bar{F}_{ij-1}$ and $\bar{G}_{ij-1}$ are history data for c_{ij} and d_{ij}. From the decomposition, given c_{ij}, y_{ij} is independent of other variables, the exposure–response model does not need to consider the dose adjustment mechanism. This observation also shows that a mixed model approach based on the marginal likelihood

$$L(\Omega) = \prod_{i=1}^{n} \int_{(s_i, u_i, v_i)} \prod_{j=1}^{J} f(y_{ij}, c_{ij}, d_{ij} | s_i, u_i, v_i, \bar{F}_{ij-1}, \bar{G}_{ij-1}) dH(s_i, u_i, v_i),$$

$$(9.40)$$

where Ω contains all parameters, is also valid, as long as the distribution of random effects $H(u_i, v_i)$ is correctly specified. If $\bar{F}_{ij-1}, \bar{G}_{ij-1}$ are sufficient so

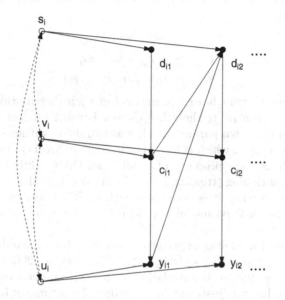

FIGURE 9.2
A DAG for sequential dose adjustment with potential exposure and response–dependence (showing only first 2 periods). Dashed bi-direction arcs represents potential confounding factors between patient level random effects u_i, v_i and s_i.

that conditional on them, $dH(s_i, u_i, v_i) = H_u(u_i)H_{vs}(s_i, v_i)$, then

$$L(\mathbf{\Omega}) = \prod_{i=1}^{n} \int_{u_i} \prod_{j=1}^{J} l(y_{ij}|c_{ij}, u_i)dH_u(u_i)$$

$$\int_{(s_i, v_i)} \prod_{j=1}^{J} h(c_{ij}|\bar{G}_{ij-1}, v_i)g(d_{ij}|\bar{F}_{ij-1}, s_i)dH_{vs}(s_i, v_i) \quad (9.41)$$

then the exposure–response models may be fitted separately. The ML approach can be used to on these.

DAR can also be used for non-ML approaches. For example, (9.2) describes the relationships

$$\begin{aligned}
c_{ij} &= \theta d_{ij} + v_i + e_{ij} \\
y_{ij} &= \beta c_{ij} + u_i + \varepsilon_{ij} \\
d_{ij} &= g(c_{ij-1}, y_{ij-1}, s_i)
\end{aligned} \quad (9.42)$$

where the random terms have zero means but their distributions are unspecified. An easy alternative to the ML approach is using models, including s_i, u_i and v_i as fixed unknown parameters. It leads to using within subject comparisons for estimating β, which is free from individual level confounding factors. This approach uses the property of conditional DAR, while the path of u_i is blocked by conditioning (treating it as a fixed effect), so that it does not cause confounding even when it is correlated with v_i. Therefore, with sufficient repeated measures, a fixed model approach is rather robust to most common confounders.

Nonlikelihood modeling approaches may also be adjusted by the control function (CF) approach (Wooldridge, 2007, Terza et al., 2008). Since the confounding in u_i is due to its correlation with v_i, if one can estimate v_i, then one may use a direct adjustment by including the estimates in the exposure–response model. A key assumption for this approach is that the conditional mean of u_i given v_i can be written as $E(u_i|v_i) = av_i$, where a is a constant. It is satisfied when u_i and v_i are jointly normally distributed, or $u_i = av_i + w_i$ so that v_i is a shared latent variable between the the dose–exposure and exposure–response models. The efficiency of this approach depends on the prediction of v_i, while repeated measurements provide data for the prediction. The approach consists of two stages of simple model fitting:

- Fit the mixed effect dose–exposure model to repeated exposure data and obtain prediction $\hat{v}_i$ for each subject, using common approaches for mixed models such as the best linear unbiased prediction (BLUP).

- Fit the exposure–response model adding $\hat{v}_i$ from the first step as a covariate.

Again the u_i path is blocked with direct adjustment. This approach has been widely used in combination with IV in social science. For an application in dose–exposure modeling, see Wang (2012b, 2014).

9.5.3   Exposure–response relationship with dynamic treatment changes

After introducing dose adjustment mechanisms, we revisit the marginal dose–exposure relationship under dose adjustment. We consider the linear dynamic model in Chapter 6

$$y_t = \rho y_{t-1} + \beta d_t + \gamma d_t y_{t-1} + \varepsilon_t \tag{9.43}$$

where we add a term $\gamma d_t y_{t-1}$ to make treatment effects depend on the previous state, and d_i is a binary treatment indicator. y_t may be the primary outcome in a study, and the primary endpoint is the outcome measured at a given time. But there are often repeated measures before that. From the model we find

$$E(y_t(d_t)|y_{t-1}) = \rho y_{t-1} + \beta d_t + \gamma d_t y_{t-1} \tag{9.44}$$

hence it does not depend on the previous treatment conditional on y_{t-1}, but y_{t-1} carries the effect of d_{t-1}. The treatment effect of d_t is

$$E(y_t(1)|y_{t-1}) - E(y_t(0)|y_{t-1}) = \beta + \gamma y_{t-1}. \tag{9.45}$$

This type of model is called structure nested mean models (SNMM) (Hernan and Robins, 2015). A SNMM can be specified by the mean difference between two treatment sequences with exposure different at one period only, e.g., between $d* = (0,0,1,0,0)$ and $d = (0,0,0,0,0)$, the change from 0 to 1 in d^* at a time is called a "blip", so the mean difference represents the effect of the blip. From the dynamic model (9.43) we can find, for any $T > t$,

$$E(y_T(\bar{d}_{t-1}, 1, \bar{0}_{T-t}))|y_{t-1}) - E(y_T((\bar{d}_{t-1}, 0, \bar{0}_{T-t}))|y_{t-1}) = \rho^{T-t}(\beta + \gamma y_{t-1}) \tag{9.46}$$

where $\bar{d}_{t-1}$ are exposures before t and $\bar{0}_{T-t}$ are $T - t$ zeros.

    As an example of exploring how treatment effects in the dynamic model are represented in a SNMM, consider the situation of a trial with two periods initially randomized, but with exposure change for period 2, potentially response related. Our interest is in the evaluation of the response at period 2 $y_2(d)$ for a given sequence of exposure $d = (d_1, d_2)$, for example, $d = (1, 1)$, and to compare different treatment sequences. Based on the dynamic model(9.43), we can calculate the mean difference in $y_2(d)$ for pairs of sequences with one and only one difference. The calculation is straightforward.

$$
\begin{aligned}
E(y_2(1,0)) - E(y_2(0,0)) &= \gamma \rho y_0 + \beta \rho \\
E(y_2(0,1)) - E(y_2(0,0)) &= \gamma \rho y_0 + \beta \\
E(y_2(1,1)) - E(y_2(1,0)) &= \gamma(\rho y_0 + \beta) + \beta \\
E(y_2(1,1)) - E(y_2(0,1)) &= \rho^*(\gamma y_0 + \beta),
\end{aligned}
\tag{9.47}
$$

where $\rho^* = \rho + \gamma$. From these we find that, compared with no treatment

$d = (0,0)$, the effect of both $d = (1,0)$ and $d = (0,1)$ have a component depending on y_0. The effect difference between $d = (1,1)$ and $d = (1,0)$ has a mediated effect part $\gamma(\rho y_0 + \beta)$, as well as the direct effect β. The effect difference between $d = (1,1)$ and $d = (0,1)$ is only the ρ^* percent of effect difference from time 1 carried to y_2. Also of interest is $E(y_2(0,1)) - E(y_2(1,0)) = (1-\rho)\beta$, which shows the difference with the same exposure but at different times. Without assuming the underline linear model the saturated mean model has 4 parameters as the 4 different means for $d = (0,0), (0,1), (1,0)$ and $(1,1)$. Therefore, the linear dynamic model has one parameter less. When number of periods increase, the number of parameters in the saturated model increases exponentially. One needs to enforce a model structure in the SNMM. The SNMM can still be more flexible than the 3-parameter linear dynamic model. However, the linear dynamic model can also be made flexible by, e.g., adding time-varying coefficients (See Chapter 6), time-varying covariates, or treatment-covariate interactions.

Next we consider a simple treatment adaptation rule $d_2(y_1) = I(y_1 > y^0)$ so that d_2 is response dependent. Note that this rule means that d_2 does not depend on any other factors, hence there is no unobserved confounder. We would like to compare $d = (1, d_2(y_1(1)))$ with $d = (0, d_2(y_1(0)))$. To this end, we first calculate

$$E(y_2(d_1, d_2)|y_1(d_1)) = \rho y_1(d_1) + (\beta + \gamma y_1(d_1))d_2. \qquad (9.48)$$

From this

$$
\begin{aligned}
E(y_2(1, d_2(y_1(1)))) &= E_{y_1(1)}[E(y_2(1, d_2)|y_1(1))] \\
&= \rho(\rho^* y_0 + \beta) + E(I(y_1(1) > y^0)(\beta + \gamma y_1(1))) \\
E(y_2(0, d_2(y_1(0)))) &= E_{y_1(0)}[E(y_2(0, d_2)|y_1(0))] \\
&= \rho^2 y_0 + E(I(y_1(0) > y^0)(\beta + \gamma y_1(0))), \qquad (9.49)
\end{aligned}
$$

where $I(A) = 1$ if A is true and $I(A) = 0$ otherwise. Combining them, we get the effect of exposure at period 1, with that at period 2 response adapted, as

$$
\begin{aligned}
&E(y_2(1, d_2(y_1(1)))) - E(y_2(0, d_2(y_1(0)))) \\
&= \rho(\beta + \gamma y_0) + E(I(y_1(1) > y^0)(\beta + \gamma y_1(1))) \\
&\quad - E(I(y_1(0) > y^0)(\beta + \gamma y_1(0))). \qquad (9.50)
\end{aligned}
$$

To calculate $E(I(y_1(0) > y^0)(\beta + \gamma y_1(d)))$, we have to specify the distribution of ε_t. The same approach applies when $d_2(y_1) = I(y_1 > y^0)$ is replaced with a randomization rule with $P(d_2 = 1)$ depending only on y_1 or other observed factors at time 1. Therefore, d_2 can be considered as sequentially randomized.

The calculation above closely relates to using g-formula for SNMM (Hernan and Robins, 2015). These models are powerful tools for modeling effects of treatment sequences, with possible treatment change depending on intermediate outcomes i.e. DTR. Our example shows how a dynamic model with DTR behaves when put into the context of SNMM.

9.5.4   Dose adjustment for causal effect determination: RCC trials

Although dose adjustment may introduce bias, it can also be used to reduce or eliminate the bias in exposure–response relationships, by controlling individual patient's exposure levels. It is often not possible to control individual concentrations, however, a partial control can be achieved by randomized concentration controlled (RCC) trials (Sanathanan et al., 1991; FDA, 2003) in which patients are randomized into two or three exposure ranges and the PK exposure is measured repeatedly and the dose adjusted, if necessary, until the individual exposure level is within the range the patient is randomized to. Design and analysis issues in RCC trials are discussed in Karlsson et al. (2007). This approach does not control the exposure level within a range, hence the exposure still depends on confounding factors. Therefore, routine analysis procedures may still lead to confounding bias in the estimated drug effect.

Suppose that in an RCC trial patients are randomized into $k = 1, ..., K$ groups with exposure ranges (L_k, U_k), and often with $L_k = U_{k-1}$. Then the dose for a patient is adjusted so that the exposure falls into the range corresponding to the group that the patient is randomized to. For RCC the simple adjustment (9.32) can be written as

$$d_{ij+1} = \begin{cases} d_{ij} + \Delta & c_{ij} < L_k \\ d_{ij} & U_k \geq c_{ij} \geq L_k \\ d_{ij} - \Delta & c_{ij} > U_k \end{cases} \qquad (9.51)$$

where subject i is randomized to the kth group. Δ is the step size for dose adjustment, which may not necessarily be constant.

For analysis of the RCC trial, we assume that the exposure is measured repeatedly but the response is measured only until the target exposure has been achieved, e.g., at the jth visit. This scenario is typical for RCC trials, as they are designed to adjust most individual exposures into randomized ranges after a ceratin period, then the response is measured afterwards. Let y_i and c_i be the response and exposure of patient i at visit j and we suppress the subscript j. We assume the following exposure–response model:

$$y_i = \boldsymbol{\beta}_x^T \mathbf{X}_i + \beta c_i + u_i + \varepsilon_i \qquad (9.52)$$

where $\mathbf{X}_i$ is a set of covariates including the intercept, $\boldsymbol{\beta}_x$ are the corresponding parameters. Here $u_i + \varepsilon_i$ is treated as a single error term, but potential confounding comes from u_i only. To see why fitting an ER model to RCC data without adjusting for confounding factors will lead to a biased estimate for β, recall Chapter 8, where we classified exposure into two parts, one depending on confounding factors and the other part not. For a highly variable drug, the first part may still be quite large even in an RCC trial, since due to feasibility most common RCC designs use only two ranges. Although patients are randomized into these ranges and a part of the variation in c_i is indeed randomized, there is often still a significant variation potentially confounded among those randomized into the same range. Therefore, a least squares estimate for β is still biased.

An obvious approach is to use methods robust to confounding factors, such as joint mixed models with an ML method. Karlsson et al. (2007) showed a considerable advantage of the joint modeling approach over fitting a simple ER model. The joint modeling approach, in fact, does not use the fact that the range is randomized. Its validity still depends on the conditions we discussed in Chapter 8. However, the approach is more robust in RCC trials in the sense that the bias is less than without RCC, if the model is not fully correctly specified.

The IV approach is an alternative to joint modeling but its validity does not depend on the model structure, hence it is more robust. In RCC trials randomization is a natural IV, since randomization is independent of the response, and dose adjustment to achieve a certain range of exposure makes it strongly related to the exposure. For linear and a few special nonlinear models, the following two stage IV (2SIV) approach is very easy to implement (Wang, 2014). For example, to estimate β in model

$$y_i = \beta_0 + \beta s(c_i) + u_i \tag{9.53}$$

with a function of the exposure $s(c_i)$ to represent the exposure, the following two steps approach can be used.

- Fit the randomization exposure model

$$s(c_i) = \boldsymbol{\alpha}^T \mathbf{R}_i + v_i + e_i, \tag{9.54}$$

where $\mathbf{R}_i = (R_{i1}, ..., R_{iK})$ with $R_{ik} = 1$ if subject i is randomized to group k and $R_{ik} = 0$ otherwise, to the exposure data and obtain the predicted mean exposure for each group.

- Fit the exposure–response model with $s(c_i)$ replaced by the predicted mean exposure of the group subject i is randomized to. The IV estimate $\hat{\beta}_{IV}$ is the coefficient for the predicted exposure.

Note that, although $\hat{\beta}_{IV}$ does not have confounding bias, its small sample bias depends mainly on how closely the dose is correlated to the exposure. It is important that the exact exposure term in the exposure–response model is modeled in (9.54). It is incorrect to model, e.g., c_i to obtain a mean estimate $\hat{c}_i$, then replace $s(c_i)$ with $s(\hat{c}_i)$.

Compared with the IV approach for randomized dose trials, a difference is that at the first stage we estimate the mean $s(c_i)$ for each randomized range, rather than that of a randomized dose, and in the second stage we replace $s(c_i)$ with the mean of the randomized range the subject is randomized to. However, the IV approach for RCC design is less robust than in dose randomized trials, as it cannot eliminate the bias due to confounding factors in treatment heterogeneity, e.g., u_{bi} in the model

$$y_i = \beta_x^T \mathbf{X}_i + (\beta + u_{bi})c_i + u_i + \varepsilon_i, \qquad (9.55)$$

although in general it helps to reduce the bias. For more details and the impacts of confounding in u_{bi}, see Wang (2014).

9.5.5 Treatment switch over and IPW adjustment

Treatment switchover refers to changing from one treatment to another, often triggered by patients' disease state changes. A typical one is to offer patients in a control arm the active or test treatment upon disease progression in an randomized trial. Hence we can consider the switching as changing the active treatment exposure from zero to an appropriate level. A common approach for analysis of randomized trials follows the intend-to-treat (ITT) principle that censors those who switch over at the switch over time and analyzes the data as they are randomized. This approach is to control the type I error of the hypothesis of no treatment difference. However, if there is a difference, this approach is generally biased, as those switched over may have a better or worse prognosis than those not switched. Therefore, those not censored to to switching is not representative for all those who had a chance to switch, e.g., those who experienced a disease progression. As we have shown in Chapter 8, if the propensity of switch over is known, one may adjust the analysis by weighting the data based on the probability of switching, hence censored. We will use the Cox model as an illustration, but the approach can be used for other models as well as nonparametric estimation.

Recall that the parameters in a Cox model can be estimated by solving the following EE for failure times $t_1, ..., t_n$:

$$\sum_{i=1}^{n} \delta_i \left(\mathbf{X}_i(t_i) - \frac{\sum_{j \in R_i} \mathbf{X}_j(t_i) \exp(\mathbf{X}_j^T \beta)}{\sum_{j \in R_i} \exp(\mathbf{X}_j^T(t_i)\beta)} \right) = 0. \qquad (9.56)$$

Here we assume that the only reason for censoring is switch over, hence $\delta = 1$ is an indicator of not switching, and $\mathbf{X}_i$ only contains the treatment indicator. Switching is typically a decision based on the current disease state and other factors. Let $G(t|\mathbf{Z}_i(t))$ be the probability that subject i has not switched up to time t, where $\mathbf{Z}_i(t)$ contains all potential confounders including time varying ones such as intermediate outcome of the treatment, then the IPW adjusted estimate is the solution to

$$\sum_{i=1}^{n} W_i \delta_i \left(\mathbf{X}_i(t_i) - \frac{\sum_{j \in R_i} W_i \mathbf{X}_j \exp(\mathbf{X}_j^T(t_i)\beta)}{\sum_{j \in R_i} W_i \exp(\mathbf{X}_j^T(t_i)\beta)} \right) = 0, \qquad (9.57)$$

where $W_i = G(t_i|\mathbf{Z}_i(t_i))^{-1}$ (Robins and Finkelstein, 2000). From EE (9.57) we find that W_i is needed for each failure time t_i for all subjects who are still at risk. One can always fit a survival model to the switching data, with factors in $\mathbf{Z}_i(t_i)$ as time varying covariates. For example, after fitting a Cox model, an estimate for $G(t|\bar{F}_i(t))$ can be required from most survival analysis software.

If switchover can only happen at disease progression time s_i, then $G(t|\mathbf{Z}_i(t)) = 0$ for $t < s_i$ and $G(t|\mathbf{Z}_i(t)) = G(s_i|\mathbf{Z}_i(s_i)), t > s_i$. If only those in the control group can switch to the active treatment, then $G(t|\mathbf{Z}_i(t)) = 1$ for the whole active arm. Therefore, one only needs to estimate those in the control group who had a disease progression. This can be done with a logistic regression including patient characteristics and disease state at or before time s_i. Input from clinical experts is invaluable for building a good model, as one should not decide which factor to exclude based on its statistical significance. To avoid instability of the weighted EE, the stable weights as described in Chapter 8 is generally a better choice. Note that if there are missing values in $\mathbf{Z}_i(t)$, even if they are MCAR, ignoring subjects with missing $G(t|\mathbf{Z}_i(t))$ is generally incorrect. Instead, they should be considered as censored and counted in the calculation of W_i. For example, if there are $K\%$ subjects with missing $G(t|\mathbf{Z}_i(t))$, then then the weight for subject i with non-missing $G(t|\mathbf{Z}_i(t))$ should be increased to $W_i = (G(t|\mathbf{Z}_i(t))(1 - K))^{-1}$.

Another practical situation where simple approaches can also be applied is when switching can only occur at a number of fixed time points, such as predetermined visit times in a clinical trial. In this case, one may fit a survival model for discrete event times using the data structure above. Alternatively, one may also fit a logistic regression at each visit time and calculate the probability of switching for each subject. Letting the visit times be $V_1, ..., V_K$, for a failure time t_i with $V_k \leq t_i < V_{k+1}$,

$$G(t_i|\mathbf{Z}_i(t)) = \prod_{i=1}^{k} P(\delta_i = 0|\mathbf{Z}_i(V_k)). \tag{9.58}$$

This approach in fact allows us to use different models at different visits, hence is more flexible than a model with constant coefficients. However, it may also be more unstable.

To use the IPCW approach, as the weights are time varying, we may use the counting process data structure and cut the survival times for those with positive $P(\delta_i)$ into two pieces. Suppose that subject 1 had an event at time 2 without disease progression, and subject 1 had disease progression at time 1 and is not switched with wi=1.5 then had an event at time 2, while subject 3 had a progression at time 3 and switched to the active treatment; then the model can be fitted to the following data:

id	start	stop	event	wi	treat
1	0	2	1	1	0
2	0	1	0	1	0
2	1	2	1	1.5	0
3	0	3	0	1	0

. . .

The following codes generate data under the scenario of switch over at progression, fit the propensity model and calculate weights.

```
nsimu=5000
t0=rexp(nsimu) # time to progression
beta=0.5  # log-RR
sig1=1
eta=1
x1=rnorm(nsimu)*sig1  # confounding factor
u0=rexp(nsimu)*exp(x1)  #control group event time
stt=rep(0,nsimu)
stp=pmin(u0,t0)
event=ifelse(u0<t0,1,0) # event of the first piece
ps=1/(1+exp(eta*x1))  #prob of switch
xo=rbinom(nsimu,1,ps)
stt=c(stt,t0)
stp=c(stp,ifelse(xo==0,t0+rexp(nsimu)*exp(x1),0)) #second piece
zv=rexp(nsimu)*exp(beta+x1)  #active event time
stt=c(stt,rep(0,nsimu))
stp=c(stp,zv)
event=c(event,rep(1,2*nsimu))

pps=predict(glm(xo~x1,family=binomial), type = "response" )

wei0=1/(1-pps)
wei=c(rep(1,nsimu),wei0,rep(1,nsimu))

treat=rep(c(0,1),c(2*nsimu,nsimu))

simud=data.frame(stt=stt,stp=stp,treat=treat,wei=wei,
                                    event=event)[stp>0,]
```

where some variable names are slightly different from the data example. Then one can use the coxph call

```
coxph(Surv(stt,stp,event)~treat,data=simud)
coxph(Surv(stt,stp,event)~treat,weights=wei,data=simud)
```

which gives unweighted log-RR estimate 0.34, and IPCW log-RR estimate 0.49, respectively. The latter is almost the same as the true value 0.5.

9.6   Sequential decision analysis

We move to another topic to explore how ER modeling helps with decision making in a number of practical scenarios. Multiple decisions are often made sequentially, with later decisions depending on the outcome of earlier decisions. Sequential decision analysis (DA) aims to find a series of actions to be taken sequentially to maximize the overall utility over a (often long) period. The framework of sequential decision analysis falls into the general decision analysis topic in Chapter 7, but is much more complex due to its dynamic nature. This topic also brings another aspect from preference and utility theory: time trade-off between benefits and costs in the present and those in the future. Here we will not consider this direction further, except when using a discount factor to discount the value of the reward as well as costs in the future.

9.6.1   Decision tree

A decision tree is a graphic tool for sequential decision making of finite steps and actions widely used in project management, and recently in medical decision making. A decision tree consists of nodes and directional paths to illustrate the relationship between series of decisions and outcomes. In a decision tree a square denotes a decision node with outward paths representing different decisions. A circle is a chance node from which outward paths lead to random outcomes with specific probabilities. On each node, there may be rewards as well as costs. A project starts at the root of a decision tree and ends at one of the multiple leaves, with a certain reward or loss. Which leaf the project ends at is random, depending on both decisions and random outcomes. Our goal is to take the optimal decision to maximize the expected value (EV), the expected cumulative reward less costs. Figure 9.3 shows such a tree representing a typical project in drug development, in which the nodes are numbered from 1 to 8. Node 2 is denoted by both a circle and a square, i.e, it may be considered as a decision or a chance node, depending on whether a decision rule is prespecified and depends on random outcomes. At the root (far left) node, suppose that after initial safety and efficacy assessments, e.g., a proof-of-concept study, the safety and efficacy effects and dose–exposure relationship can be summarized by the prior distribution of the model parameters. The decision maker (DM) has three choices: 1) to start a phase III trial directly; 2) to conduct a phase II trial to assess the efficacy and safety profile and to select an appropriate dose, then decide if a phase III trial should be carried out; 3) to stop the project immediately. These are represented by the three paths from the root, one leading to node 2 then to 3, one to node 4 and the last going directly to stop.

To decide which of the three actions to choose, the DM needs to calculate and to compare their EVs. Let the reward of a successful phase III trial be R and the costs of the phase II and III trials be C^2 and C^3, respectively. For the chance nodes probability $p_{kk'}$ is the probability moving to node k' from k, where we code stop as 0. In particular, p_{35} and p_{47} are the probability of success in the phase III trial with and without the phase II one. The reason they are different will become clear later. Assume that we always continue with the phase III trial after a successful phase II one, which is often the case in practice, node 2 becomes a chance node and p_{23} is the success rate of the

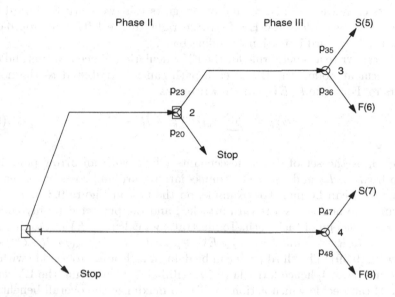

FIGURE 9.3
A decision tree for making sequential decisions in drug development after preliminary efficacy and safety evaluation. The rewards for success (S) and failure (F) are 100 and 0, respectively. The cost of a phase II trial is $C^2 = 2$ and that of a phase III trial is $C^3 = 10$. Based on the numerical example, the probabilities are $p_{47} = 0.5824, p_{23} = 0.8737, p_{35} = 0.7498$.

phase II trial. In general, a decision node is converted to a chance node if a decision rule is prespecified. For node 2 we will discuss how success should be defined later. We may also assume that p_{23} and p_{20} are known, e.g., from the historical success rate of relevant phase II trials. With these, we have all the necessary elements for the EV calculation. Starting from the top path, we need to work from the leaves and to calculate the EV of phase III trial after the phase II one, which is the reward of success less the costs $p_{35}R - C^3$. As node 2 is considered as a chance node, the EV of the top path from the root is $p_{23}(p_{35}R - C^3) - C^2$. The calculation of EV for the other two paths is much easier. The middle path has EV= $p_{47}R - C^3$ and the last one is 0, as there are no costs or rewards. This calculation suggests that we work backward from the leaves to calculate the EV from the right to the left. This approach is known as backward induction or rolling back.

We can write a general rule for the EV calculation. Since rewards and costs only occur at nodes, the EV of each path can be attributed to the node it points to. For node k, EV_k can be written as

$$EV_k = \sum_{k' \in S_k} p_{kk'} EV_{k'} + R_k - C_k \tag{9.59}$$

where S_k is the set of direct descendants (those with an arrow pointing to them from k). R_k and C_k are the immediate reward and costs when reaching k, but both can be null. For example, on the tree in Figure 9.3, none of the intermediate nodes has a reward attached and the project can only collect a reward at the end of each path. To illustrate, we calculate $EV_3 = p_{35}R + p_{36} \times 0 - C^3 = p_{35}R - C^3$ and $EV_2 = p_{23}EV_3 + p_{36} \times 0 - C^2 = p_{23}(p_{35}R - C^3) - C^2$, following (9.59). The third path can be defined as leading to a node with zero costs and reward, hence formula (9.59) still holds. Comparing the EV values, the DM can decide which action to take to maximize the overall benefit.

Using a decision tree to make optimal decisions seems easy. But we have simplified the example to concentrate on the basic concept by assuming that the probabilities are known. In most cases, one needs to estimate them. Also we assumed p_{35} is fixed, e.g., as the average success rate after a successful phase II trial of similar compounds. However, after the phase II trial, new information allows a better estimate for p_{35}, hence the DM can make a better decision than one based on the average rate. Let us consider a numerical example, which is an extraction from a real scenario but has been substantially simplified, to illustrate how parameters on the decision tree are calculated. Suppose that efficacy is measured by the response given by

$$y_i = \beta d_i + \varepsilon_i \tag{9.60}$$

and safety is measured by the logit of the probability of having an AE, p_i, by model

$$\text{logit}(p_i) = \theta d_i + e_i, \tag{9.61}$$

in which, for simplicity, we assume that $e_i \sim N(0, \sigma_t^2)$ approximately. We also assume that d_i is on a log-scale with range $-2, 2$. To maximize the chance of phase III success, we have to select a dose d^* to maximize the probability of both the efficacy and safety profiles meeting the target.

$$P(\beta d^* > y_0 \cap \theta d^* < z_0) \tag{9.62}$$

where y_0 and z_0 are given thresholds. As for the prior information, we assume that β and θ are independent and follow distributions

$$\begin{aligned} \beta &\sim N(\beta_0, \sigma_b^2) \\ \theta &\sim N(\theta_0, \sigma_t^2). \end{aligned} \tag{9.63}$$

The independence is not practically restrictive, since if they are correlated, one can always use a linear transformation to obtain a pair (θ^*, β^*) which is independent. Without the dosing finding study we can calculate

$$P(\beta d^* > y_0 \cap \theta d^* < z_0) = \Phi(\frac{z_0 - \theta_0}{d^* \sigma_t})(1 - \Phi(\frac{y_0 - \beta_0 d^*}{d^* \sigma_b})). \tag{9.64}$$

With the parameter values given below, this probability under the optimal dose $= 0.97$ is 0.5824. Therefore, we have calculated $p_{47} = 0.5824$ in Figure 9.3. The calculation uses the following R code.

```
sigb=1
sigt=0.5
b0=1.5
bt0=0.5
y0=1
z0=1
obj=function(dos,mub,sb,mut,st){
  (1-pnorm((y0-dos*mub)/abs(dos)/sb))* pnorm((z0-dos*mut)/abs(dos)/st)
}

op=optimize(f=obj,maximum=T, interval=range(dose),mub=b0,mut=bt0,
                                        sb=sigb,st=sigt).
```

Suppose that in the dose finding study, we use five dose levels $-2, -1, 0, 1, 2$ on the log-scale and y_i and $\text{logit}(p_i)$ are generated from (9.60) and (9.61). For simplicity we assume that the variances are known parameters so that the posterior distribution of β is

$$\begin{aligned} \beta &\sim N(m_b, v_b), \\ v_b &= 1/(\sigma_b^{-2} + \sum d_i^2), \\ m_b &= (\beta_0 \sigma_b^{-2} + \hat{\beta} \sum d_i^2) v_b. \end{aligned} \tag{9.65}$$

Here we have used the fact that $\sum d_i = 0$, and $\hat{\beta}$ is the LS estimate for β.

The posterior distribution of θ has the same formulas. Now we can calculate, for a set of y_i and logit(p_i), the conditional probability

$$P_a(\beta d^* > y_0 \cap \theta d^* < z_0) = \Phi(\frac{z_0 - m_t d^*}{d^* \sqrt{v_t}})(1 - \Phi(\frac{y_0 - m_b d^*}{d^* \sqrt{v_b}})) \qquad (9.66)$$

where m_t and v_t are the posterior mean and variance for θ, respectively. With the following code and parameter settings

```
sigp=1
sigp2=1
dose=(1:5)-3
vdose=sum(dose^2)
ldose=length(dose)
calc=function(nsimu){
Out=NULL
for (i in 1:nsimu){
beta=rnorm(1)*sigb+b0
theta=rnorm(1)*sigt+bt0
yi=beta*dose+rnorm(ldose)*sigp
zi=theta*dose+rnorm(ldose)*sigp2
bhat=lm(yi~dose-1)$coef
that=lm(zi~dose-1)$coef
vb=1/(1/sigb^2+vdose)
mb=(b0/sigb^2+bhat*vdose)*vb

vs=1/(1/sigt^2+vdose)
ms=(b0/sigt^2+that*vdose)*vs

op=optimize(f=obj,maximum=T, interval=range(dose),mub=mb,mut=ms,
            sb=sqrt(vb),st=sqrt(vs))
Out=rbind(Out,c(op$max,op$obj,mb,ms))
}
return(Out)
}
```

we simulated y_i and logit(p_i) 10,000 times to obtain posterior samples for $p_{35} = P_a(\beta d^* > y_0 \cap \theta d^* < z_0)$, m_b and m_t. Note that at node 2, we have the option to decide whether to go ahead with the phase III trial or to stop. To maximize the EV, the optimal decision is to stop if $Rp_{35} < C^3$. With this predetermined rule, we can now calculate EV_2 as

```
EV2=mean(pmax(reward*res[,2]-C3,0)-C2)
```

which estimates

$$\int_{\xi} \max(Rp_{35}(\xi) - C^3)dF(\xi) - C^2 \qquad (9.67)$$

where $dF(\xi)$ is the posterior distribution for parameters in the models above. From this calculation, we get $EV_3 = 54.77$, much higher than $EV_4 = 100 \times$

$0.5824 - 10 = 48.24$. Therefore, the optimal strategy is to take the top path and to use a phase II trial first. Based on the results, we can also calculate p_{23} as

```
> mean(reward*res[,2]-C3>0)
[1] 0.8737
```

and p_{35} as

```
> mean(res[reward*res[,2]-C3>0,2])
[1] 0.7497676
```

as in Figure 9.3.

For simplicity we have assumed that the reward of success is the same with or without the phase II trial. In practice, the time delay due to running the phase II trial makes the reward with the phase II trial much lower than those without it. A simple approach is to use a discount rate (e.g., 3% each year) to adjust for the delay. Although this approach is commonly used for public health policy making, e.g., reimbursement, making a investment decision is more complex. Using investment theory on decision making for projects like this one has been developed into a system of approaches known as real options, which is far beyond our scope. See Burman and Senn (2003) for an introduction to potential applications in drug development.

9.6.2   Dynamic programming

Decision trees seem to be a universal approach to solving sequential decision problems. However, they have a number of limitations. First, the decision has to be discrete, i.e, one can only select one from multiple paths at a node. Second, the nodes can only represent discrete events rather than a continuous measure, so the outcome has to be categorical or have discrete values. For example, suppose that from a chance node the outcome is the development of hypertension. A decision tree can only represent a binary outcome or multiple category outcome, e.g., no, mild and severe hypertension based on the blood pressure values. Third, the time scale has to be discrete and finite. Dynamic programming (DP) is a general approach to finding optimal sequential decisions without these limitations. However, complexity increases with reducing limitations. We therefore will not consider the most complex situation: DP with continuous time scales.

Let $j = 1, ..., J$ be the index of stages at which the decision maker can take an action a_j, leading to the system state moving from y_j to y_{j+1} at the next stage via, e.g., a Markov dynamic model:

$$y_j = g(y_{j-1}, a_j) + \varepsilon_j, \tag{9.68}$$

with initial state y_0. For example, this may be a model for tumor size dynamic and the action is to choose a treatment or a dose. The state may also be the posterior distribution of parameters in the model, e.g., β and θ in the previous example, and the action is to decide if further information collection is beneficial in terms of the reward and costs. Specifically for the example, the posterior means and variances of β given by (9.65) can be considered as states, as they, together with those for θ, determine the EV of the phase III trial. For example, with the option of using more than one phase II trial to evaluate drug profiles before making the decision of a phase III trial, the posterior mean and variance of β: m_b^k, v_b^k after the kth phase II trial can be written as

$$
\begin{aligned}
v_{bk} &= 1/(v_{bk-1}^{-2} + \sum d_i^2) \\
m_{bk} &= (m_{bk-1}v_{bk-1}^{-2} + \hat{\beta}_k \sum d_i^2)v_{bk},
\end{aligned}
\tag{9.69}
$$

if the action is to do the kth phase II trial. In this case, the state model represents the dynamic of information update described in Chapter 7 and the value of information regarding the model determines if the cost of further information collection is justifiable.

Let the instant value under states y_j and a_j be $v(y_j, a_j)$, which is a summary of the associated utility and costs. In particular, it is often the net benefit: $v(y_j, a_j) = R(y_j) - C(a_j)$, where $R(y_j)$ and $C(a_j)$ are the rewards for state y_j and the costs of taking action a_j. Note that $v(y_j, a_j)$ implicitly depends on y_{j-1}, and in fact the entire history of the states and actions, via model (9.68). The overall expected value with starting state y_0 can be written as

$$
EV(\mathbf{a}, y_0) = \sum_{j=1}^{J} \rho^{j-1} E(v(y_j, a_j))
\tag{9.70}
$$

where $1 - \rho$ is the discount rate, $\mathbf{a} = (a_1, ..., a_J)$ are the actions taken over all the J stages. This setting reflects many common situations in which the action controls the outcome of the next stage. Taking the expectation of $v(y_j, a_j)$ is necessary since y_j is a random variable.

Our task is to find the optimal $\mathbf{a}^*$ so that $EV^*(y_0) \equiv EV(\mathbf{a}^*, y_0) = \max EV(\mathbf{a}, y_0)$. In principle, the backward induction approach for decision trees can apply here for finite J. To do so first we try to write (9.70) in a consecutive manner. Let $\mathbf{a}_j = (a_j, ..., a_J)$ be the actions taken from stage j onwards. We can write

$$
EV(\mathbf{a}, y_0) = \max_{a_1} v(y_1, a_1) + \rho \sum_{j=2}^{J} \rho^{j-2} E(v(y_j, a_j))
\tag{9.71}
$$

where $\sum_{j=2}^{J} \rho^{j-2} E(v(y_j, a_j))$ is the EV after stage 1. Letting $EV_j(\mathbf{a}_j, y_{j-1}) = \sum_{k=j}^{J} \rho^{k-j-1} E(v(y_k, a_k))$ with starting state y_{j-1}, the optimal EV at stage j is $EV_j^*(y_{j-1}) = \max_{\mathbf{a}_j} EV_j(\mathbf{a}_j, y_{j-1})$ with optimal $\mathbf{a}_j^*$, which can be written as

$$
\begin{aligned}
EV_j^*(y_{j-1}) &= \max_{a_j} v(y_j, a_j) + \rho \sum_{k=j+1}^{J} \rho^{k-j-1} E(v(y_k, a_k^*)) \\
&= \max_{a_j} v(y_j, a_j) + \rho EV_{j+1}^*(y_j). \quad (9.72)
\end{aligned}
$$

The relationship between $EV_{j+1}^*(y_j)$ and $EV_j^*(y_{j-1})$ in (9.72) is known as the Bellman equation.

To use the backward induction we start from stage J, so given state y_{J-1}, a_J^* can be found easily as it maximizes $v(y_J, a_J)$, which depends on y_{J-1} implicitly via model (9.68). The optimal action a_{J-1}^* is much more difficult to find since its outcome y_{J-1} affects a_J^*. Specifically, a_{J-1}^* is the action that $\max_a [v(y_{J-1}, a) + \rho EV_J^*(y_{J-1})]$ and the maximization has to search through all possible states of y_{J-1}. Therefore, although DP provides a general framework to deal with general sequential decision problems, the computational complexity is still a major issue to overcome. The complexity of finding the optimal decision lies in two factors. To find the optimal action for a given EV function, one needs to go backward (backward induction) from the last stage, while to calculate the states in the system, one needs to simulate the system forward from stage 1 using model (9.68). A combination of the two tasks is sometimes referred to as forward simulation, backward induction (Muller et al., 2007). The complexity increases exponentially with the number of stages. Often an approximation is needed for $J > 3$. But for problems on a similar scale as (9.6.1), the optimal solution can be found by a grid search. Note that for decision tree problems such as that of Figure 9.3, the forward simulation part is presented in the transit probabilities, hence is often not shown explicitly.

A specific DP problem is to determine a stopping rule with $a_j = 0$ to stop and $a_j = 1$ to continue. In this case, at stage j, one decides to stop to claim $E(v(y_{j-1}, 0))$, or to continue for $\rho EV_{j+1}^*(y_j)$ in the future. If $a_j = 0$, then $EV_{j+1}^*(y_j) = 0$. Therefore, the optimal stopping rule is to stop if $EV_{j+1}^*(y_j) \leq 0$. Consider the DT example, where the phase II trial is to collect information on the drug safety-efficacy profile, which we consider as the state, and the instant reward is either $Rp_{31} - C^3$, if we start the phase III trial, and 0 if we stop the project. Intuitively, we should only conduct the phase II trial if its costs are less than the reward of using the information to make the phase III decision, i.e., the value of information (VoI, see Chapter 7) of the phase II trial is higher than its costs. Recall the state model (9.69) that represents information update with multiple phase II trials. The stopping rule is to stop when there is sufficient information to make a decision on the phase III trial. In general, we can extend the VoI concept to the optimal stopping rule, which is to stop when the cost of continuing is no less than the reward of going ahead. However, since $EV(\mathbf{a}_j^*, y_j)$ needs backward induction to calculate, even for finding a stopping rule, computational complexity is still a difficult technical issue.

The extension from a finite J to $J \to \infty$ is fundamentally difficult, as without the last J, the backward algorithm has no place to start. However, the Bellman equation still holds as long as EV_j^* is defined. The only problem is in the calculation of its value. There are several approximations that make the problem solvable. One, particularly with discounted future values with factor $\rho < 1$, is to assume that the process will stop after k more steps from the current stage. This assumption makes it possible to use the backward algorithm. Often, even for a small k, this approach provides a good approximation to the true Bayesian decision. Nevertheless, in practical medical decision problems, there is often a finite k which is sufficiently large to be considered as an approximation to the scenario of infinite J. Another way is to start with an arbitrary action series $a_1, ..., a_j, ...$, then find an approximate solution to (9.72) with an iterative algorithm. This approach is known as the policy (an action is often referred to as a policy in DP) iteration approach.

9.6.3   Optimal stopping for therapeutic dose monitoring

Following the thread of the last subsection, we consider a rather simplified situation where the optimal stopping rule can be easily calculated via simulation. Assume the power model on log-scale

$$c_{ij} = \theta d_{ij} + u_i + e_{ij} \qquad (9.73)$$

for repeated measures at visits $j = 1, ...,$. The goal is to find a dose d so that $c_d = \theta d + u_i$ is within a given range $R = (-L, L)$. To concentrate on the key aspects in dynamic programming, we further assume that $\theta = 1$, i.e., the exposure is proportional to the dose. Suppose that the reward for $c_d \in R$ in one period is just the probability $P(c_d \in R)$, and we have options to do an assay at each visit and adjust the dose accordingly. An interesting question is when there is sufficient information about u_i so that the costs and inconvenience associated with the assay and dose adjustment offset the benefit of further adjustment. Here we take a Bayesian view but simply assume a uniform prior for u_i and var$(e_{ij}) = 1$ as known. We can write the posterior distribution of u_i as $u_i \sim N(\mu_i, \sigma_j^2)$. It is easy to see that taking dose $d_i = -\mu_i$ places $c_d = \theta d_i + u_i$ in the center of R. Consider the situation of period t at which we already have t measures taken. Then the posterior distribution of the exposure under dose $d_i = -\mu_i$ is

$$c(d) = d + u_i \sim N(0, 1/t). \qquad (9.74)$$

Denoting by $\bar{F}_t$ the data until visit t, we can derive

$$P(c_d \in R | \bar{F}_t) = 1 - 2\Phi(-L\sqrt{t}). \qquad (9.75)$$

Suppose we are dosing patients a total of T periods and the reward of correct dosing for each period is $R_t = P(c_d \in R | \bar{F}_t)$. The assay costs C to do. Our question is when the cost of doing further assays offsets the benefit, hence we should stop. This is a simple case of the optimal stopping problem. Suppose that at stage t one can stop and claim reward R_t, or continue with costs C_t. The optimal value at t can be written as

$$V_t = \max(R_t, V_{t+1} - C), \tag{9.76}$$

that is, we should stop when $R_t > V_{t+1} - C$. For a general optimal stopping problem, we have the same problem as in general dynamic programming, since 1) we need to decide the optimal stopping time backward from T and 2) the calculation of V_i may need to start from $t = 1$.

For our specific problem, there is no forward simulation needed to calculate the posterior probability, so it is relatively easy. Assuming we always need the first assay, since we have assumed an noninformative prior, we can write out the total net benefit (the total reward less costs) under the optimal stopping rule consecutively to V_1 as

$$
\begin{aligned}
V_T &= \max[R_T - C, R_{T-1}] \\
V_{T-1} &= \max[V_T + R_{T-1} - C, 2R_{T-2}] \\
&\cdots \qquad \cdots \\
V_t &= \max[V_{t+1} + R_t - C, (T - t + 1)R_{t-1}] \\
&\cdots \qquad \cdots \\
V_2 &= \max[V_3 + R_2 - C, (T - 1)R_1], \\
V_1 &= R_1 T - C
\end{aligned}
$$

where at each t, taking the maximum between the value of "go": $V_t(go) \equiv V_{t+1} + R_t - C$ and that of "stop": $V_t(stop) \equiv (T - t + 1)R_{t-1}$ reflects the optimal stopping rule. In fact for this example, we can calculate them from $t = T, T - 1, \ldots$ to $V(go) < V(stop)$, which is the time to stop. To illustrate, for each t we calculate $V_t(go) \equiv V_{t+1} + R_t - C, V_t(stop) \equiv (T - t + 1)R_{t-1}$ along with V_t and R_t under a hypothetical parameter setting, and plot them in Figure 9.4. The following program and parameters are used for the calculation.

```
nrep=10
C=0.1
L=0.25
opstop=function(C, L){
Rt=function(t) 1-2*pnorm(-L*sqrt(t))
Vt=max(Rt(nrep)-C,Rt(nrep-1))
Vp=Vt
Rv=Rt(nrep-1)
Ind=1*(Rt(nrep)-C<Rt(nrep-1))
Stop=Rt(nrep-1)
Go=Rt(nrep)-C
for (i in 1:(nrep-2)){
  Rtl=Rt(nrep-i-1)
  Rt1=Rt(nrep-i)
  print(Rt1)
```

```
  Stop=c(Stop,(i+1)*Rtl)
  Go=c(Go,Vt-C+Rt1)
  Vt=max(Vt-C+Rt(nrep-i),(i+1)*Rtl)
  Vp=c(Vp,Vt)
  Ind=c(Ind,1*(Vt-C+Rt1>(i+1)*Rtl))
  Rv=c(Rv,Rtl)
 }
return(cbind(Stop,Go,Vp,Rv))
}


outv=opstop(C=0.2,L=0.25)
out=outv[9:1,]
plot(1:9,out[,3],ylim=c(0,max(out)),ylab="Value",lty=1,
                                  type="l",xlab="Period")


lines(1:9,out[,1],lty=2)
lines(1:9,out[,2],lty=3)
lines(1:9,out[,4],lty=4)
legend(7,2.8,lty=1:4,legend=c("Vt","Vgo","Vstop","Reward"))
```

The figure shows that before the 5th period, V_t is the same as V_{go} and after that V_t is the same as V_{stop}. Therefore, the optimal stopping rule is to stop at period 5. The curve of R_t shows a slow increase in the reward of cumulative information. The optimal stopping time is when the costs of assay offset the benefit of information accumulation.

One may question that if T is very large, we will not be able to compare V_{go} and V_{stop} for all ts. In fact, there is no need to do so for this problem. Suppose that starting from $t = 1$ we find the first t such that it is better to stop at t than to stop at $t + 1$ since $(R_{t+1} - R_t)(T - t) < C$. Note that $R_{t+1} - R_t$ decreases with t, due the the property of normal distribution, as does $(R_{t+1} - R_t)(T - t)$. Therefore, at $t = T - 1$, $(R_{t+1} - R_t)(T - t) < C$ means $R_T - C < R_{T-1}$, so we should stop at $T - 1$, had we not stopped at $T - 2$. But at $T - 2$, we find we should have already stopped there since $2(R_{T-1} - R_{T-2}) < C$. Step by step, we can go back to time t and show that we should stop at the first t with $(R_{t+1} - R_t)(T - t) < C$. There is a large class of problems known as the monotone Markov decision problem. Informally, a key result for solving this problem is that if the set of stopping (the result covers stopping as a random event) does not reduce over time, one can look only one step ahead and stop at the first t when going one step further is worse than stopping right now. One can also view this problem wit the concept of VoI, as $(R_{t+1} - R_t)(T - t)$ is the value of information from going one step further. As it is an decreasing function of t, once it is lower than the costs, it will always be.

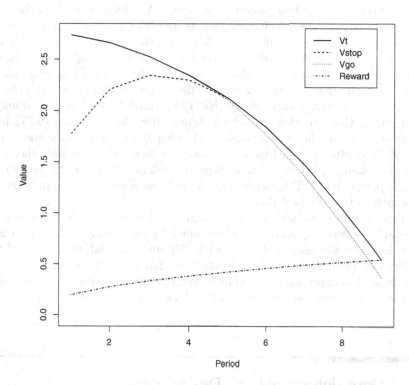

FIGURE 9.4
The net benefit of continuously monitoring drug exposure: an illustrative example. $V_t, V_t(go), V_t(stop)$ and R_t are labeled Vt, Vgo, Vstop and Reward, respectively.

9.6.4   Dose adjustment in view of dynamic programming

Since the aim of dose adjustment is to achieve a desired response or exposure, searching for the optimal dose adjustment may be considered a problem of DP. An empirical DA mechanism can be considered as a DP approach to find an approximate solution. For example, the mechanism (9.32) can be written as

$$d_{j+1} = d_j + K(c_j(d_j))   \tag{9.77}$$

where $K(c_j(d_j))$ may take values Δ, 0, and $-\Delta$, which depend on the value of $c_j(d_j)$, under dose d_j. This is, in fact, a simple policy iteration approach in which we modify the current dose d_j based on the exposure it has produced. Since if d^* satisfies $K(c(d^*)) = 0$ (when $c(d^*)$ is in the right range, in this case), then $d_j = d^*$ will be the dose at steady state. In the (9.32) example, d^* may not be unique, but this does not affect using the approach in practice. If, e.g., c_0 is the target exposure, $K(c(d^*)) = \Delta(c(d^*) - c_0)^2 = 0$ has a unique solution. In this situation, it can be shown that the algorithm (9.77) leads to convergence to the optimal dose level, with Δ satisfying some conditions. In DP, Δ is often referred to as the learning factor that controls the speed of action changes. When the dose–exposure relationship is static, e.g, driven by the power model, DA can also be considered a stochastic approximation approach to find the best dose.

This simple approach can also be extended to cope with a dynamic relationship between c_j and d_j, without actually estimating it. The idea is to learn how the the reward changes with different dose adjustments and find empirically the best one. The approach is known as Q-learning, a popular approximation approach to dynamic programming problems. One application on dose adjustment can be found in Henderson et al. (2010).

9.7   Dose determination: Design issues

A large area this book does not cover is study designs for exposure–response modeling (Ting, 2006). The main design issue in this context is how to determine the exposure levels to obtain the maximum information for, e.g., finding a dose that achieves the target response and to the best estimate the ER relationship. With ER models, the latter is often simplified to best estimation of model parameters. Therefore, the latter task is to find optimal exposure levels to maximize the determinant of the information matrix. Specifically, for ER model

$$y_i = g(d_i, \boldsymbol{\beta}) + \varepsilon_i   \tag{9.78}$$

we would like to find d_is that maximize the information matrix $|I(\boldsymbol{\beta})|$, or for non-ML approaches $|\text{var}(\hat{\boldsymbol{\beta}})^{-1}|$. A design that maximizes the determinant of the information matrix is called the D-optimal design. This design typically leads to a few d_i levels with optimal sample proportions. A number of issues we have discussed are relevant to study designs. For example, to find the dose level d_0 to achieve response y_0 based on a dose finding study, we need $\hat{d}_0 = g^{*-1}(y_0, \hat{\boldsymbol{\eta}})$, which has variance $\text{var}(\hat{d}_0) = \mathbf{H}\text{var}(\hat{\boldsymbol{\eta}})\mathbf{H}^T$, where $\text{var}(\hat{\boldsymbol{\eta}})$ depends on the dose levels used in the study. Therefore, when designing the study, one may want to minimize $\text{var}(\hat{d}_0)$. This type of design is known as a c-optimal design. Although special algorithms have been developed to find D- and c-optimal designs, practically one may use universal optimization algorithms to find the optimal dose levels and corresponding sample proportions.

Using optimal design in practice, one needs to adapt the approach to meet multiple practical objectives. One practical as well as conceptual issue in using optimal design is that the optimal design depends on the true parameter values, which are typically unknown at the design stage. The Bayesian framework offers a solution to the conceptual issue. If the prior distribution for $\boldsymbol{\beta}$, one can calculate $E_\beta(|I(\boldsymbol{\beta})|)$: the average $|I(\boldsymbol{\beta})|$ and maximizes it. However, if the prior is vague, the design may be far from optimal. The issue of lack of information can only be solved by making the design adaptive to information update during the study. Some designs have robust properties+ under some common conditions. But, in general, a two stage or adaptive design is needed. The idea is to run the first part of the study with a design that is not necessarily optimal, then adapt the dose levels to optimize the study. Draglin et al. (2007) have used this approach to find designs for Emax models. The approach in general can be used on other models. Another extension from the standard optimal design approach is to deal with multiple responses such as safety and efficacy endpoints simultaneously (Fedorov et al., 2012). An example of optimal design for a bivariate probit model for binary efficacy and safety endpoints can be found in Dragalin et al. (2008). Fedorov and Leonov (2013) gives a systemic account of using the classical design theory on nonlinear exposure–response models.

In contrast to dose–finding approaches based on parametric models, some nonmodel based or semiparametric model based approaches are more commonly used. A non-model based approach is the $3 + 3$ design to find the MTD in Section 9.4.1.2, which is based on a systematic trial-and-error method. In a phase II dose finding trial, flexible models such as a spline function model may be used to determine the phase III dose. Optimal design for determining PKPD relationship is not commonly used. One additional issue is one can only control the PK exposure indirectly by, e.g., dose levels. For a given dose level, the PK exposure follows a distribution. One approach to find optimal dose design so that the PK exposure distribution is optimal for PKPD relationships can be found in Wang (2006).

9.8   Comments and bibliographic notes

In this chapter we have considered the role and impacts of dose adjustments, particularly individual exposure-dependent and response-dependent dose adjustments in the analysis of dose exposure and exposure response relationships. Exposure–response modeling is a key part of modeling and simulation during drug development with increasing application in industry as well as in academic research. Recent research in causal effect determination in econometrics and medical statistics has led to useful techniques, and some are closely related to causal effect estimation in studies with dose adjustments. This chapter has shown a number of possible combinations of approaches from both areas and potential applications in drug development.

A number of interesting topics cannot be covered in this chapter due to the space limit. Exposure may be considered as a mediator of dose, and the exposure–response relationship as representing indirect effects of dose changes. We have assumed that there is not a direct effect of the dose on the response in any of the models. This is a reasonable assumption in common cases since a drug is normally absorbed into the central system (blood or plasma) then transported to target organs where the drug effects take place. Therefore, PK samples taken from the central system are a good surrogate for the change in exposure at target organs due to dose change in general. However, it is possible a drug may bypass the central system. A well known example is the first pass scenario in clinical pharmacology. Another interesting topic is the heterogeneity in the exposure–response relationship among patients with different observed characteristics. When the characteristics are observed and included in the models, the g-computation (Snowden et al., 2011) is a powerful tool to calculate causal effects in the population, e.g., the average diastolic blood pressure reduction by 1 unit concentration increase in a mixed population of different characteristic values. In practice, the g-computation may be implemented by simulation.

We have concentrated on individual response–dependent and exposure–dependent dose adjustment. The approaches can be adapted for some other scenarios. For example, in a typical phase I dose escalation trial with a 3 + 3 design, a number of doses are tested with cohorts of three subjects. At the beginning, a dose is given to one cohort. If no AE occurs, the next higher dose will be tested on another cohort; if there is one AE, the same dose will be tested, and if there is more than one AE the next lower dose will be tested. In this case the cohort can be considered as an individual and the exposure-AE model (9.36) can be used to model the number of subjects with AEs. Then the model can be used to describe the dose adjustment mechanism.

One key feature of a dynamic treatment regimen is optimal treatment selection and adjustment, which we have not considered in this chapter. The omission is partially due to the technical complexity of the method and partially the feasibility of implementing complex dosing formulae in practice. Nevertheless, some quasi-optimal approaches might be feasible. In some special situations optimal dose adjustments using complex algorithms can also find applications. Some general principles discussed here, e.g., when the dose adjustment process can be ignored also apply. Some approaches, such as the control function approach, that needs fitting in the dose adjustment model, may depend on specific situations.

10

Implementation using software

We have discussed a large number of approaches for fitting dose–exposure and exposure–response models. The models and model fitting algorithms are given as mathematical formulas, together with some program codes and output. The codes are to clarify which model and model options are implemented, rather than to teach how to use the software. For those without much experience of using the aforementioned software, a more systematic guidance on model implementation may be helpful. That is the purpose of this chapter.

Although different software may have different syntaxes to specify a model, a number of key elements are common in the three types of software, i.e., SAS, R and NONMEM. Therefore, one should be able to switch from one to another without major problems on model specification after reading the chapter. Obviously, learning the programming environment for each type of software is still needed. The reader will need to read specific books and manuals and the practice with the software.

10.1   Two key elements: Model and data

For any software for model based analyses there are two key elements: the model and the data. All three types of software provide an environment to extract data and to feed the data into a specified model. NONMEM reads data directly from files using

`$DATA   path/datafile`

However, the datafile should follow a specific structure. See NONMEM manuals and sample datasets for details. It is much easier to call a dataset within SAS and R since they allow data operations and storage. Model specification in all the three are similar. To fit a linear model

$$Cmin_i = \theta_1 + dose_i\theta_2 + age_i\theta_3 + \varepsilon_i \qquad (10.1)$$

to data "cmin", in which there are variables cmin, dose and age, the model specification in SAS proc glm is

`proc glm data=cmin;`

```
model cmin=dose age;
...
```

and in R using function lm() it is

```
lm(cmin~dose+age,data=cmin).
```

In both the intercept θ_1 is added automatically, as is the error term ε_i. Note that the coefficients θ_1, θ_2 and θ_3 are omitted in the model specification and are named after the variable names, e.g., θ_3 is named "age" in the fitted model. However, in NONMEM one has to specify the coefficients and the error term manually:

```
$PK
IPRED=theta(1)+dose*theta(2) +age*theta(3)
$Error
Y=IPRED + eps(1)
```

where line "IPRED=..." specifies $\theta_1 + dose_i\theta_2 + age_i\theta_3$ and ε_i is added in line Y=IPRED + eps(1).

For nonlinear models, both R and and SAS need exact model specifications including all parameters. For example, fitting the Emax model

$$y_i = Emax/(1 + EC_{50}/Conc_i) + \varepsilon_{+}i \qquad (10.2)$$

to data "emax" can be implemented with function nls(.) in R as

```
nls(yi~Emax/(1+EC50/Conc),start=c(Emax=1, EC50=5),data=emax).
```

Since there we need starting values for the parameters, these are specified in "start". Implementation using SAS procedure NLIN needs the same specification split into multiple statements:

```
 proc nlin data=emax;
   parms Emax=1 EC50=5;
   model yi=Emax/(1+EC50/Conc);
```

Note that one can also fit the model with proc NLMIXED, a procedure for fitting linear mixed models.

The model specification in NONMEM is similar to the linear one, since the model in the $PK block has to be explicit for any model. Hence for the Emax model it is

```
$PK
IPRED=theta(1)/(1+theta(2)/Conc)
```

and the starting values need to be specified in the $THETA block

```
$THETA
(0, 1, 1000) ; Emax
(0.01,5,1000); EC50
```

where the middle number in each bracket gives the starting value and the values before and after it are the lower and upper bounds. In SAS NLIN the bounds are specified in the "bounds" statement, while in R function nls(.) the lower and upper bounds are specified in arguments "lower" and "upper", respectively.

10.2 Linear mixed and generalized linear mixed models

The most difficult part to specify in a mixed model is the random effects. In SAS MIXED and GLIMMIX this part is specified jointly by the covariance of the random effects and the design matrix $\mathbf{Z}$. In nlme() it is specified by the random function using a class of positive definite matrices known as the pdMat class. General mixed model specification can be found in books such as Pinheiro and Bates (2000) and Littell et al. (2006). Assuming there are n subjects each having r repeated measures, both SAS and Splus use the standard formulation

$$\mathbf{y}_i = \mathbf{X}_i \boldsymbol{\beta} + \mathbf{Z}_i \mathbf{u}_i + \boldsymbol{\varepsilon}_i \qquad (10.3)$$

where $\mathbf{X}_i$ and $\mathbf{Z}_i$ are called design matrices for the fixed and random parts, respectively, and $\boldsymbol{\varepsilon}_i$ is a random vector of measurement errors. It is easy to see that

$$\text{var}(\mathbf{y}_i) = \mathbf{Z}_i \mathbf{G} \mathbf{Z}_i^T + \mathbf{R} \qquad (10.4)$$

where $\mathbf{G} = \text{var}(\mathbf{u}_i)$ and $\mathbf{R} = \text{var}(\boldsymbol{\varepsilon}_i)$. Here we use notations $\mathbf{G}$ and $\mathbf{R}$ since in the SAS documents they are controlled by two statements "Random" and "Repeated" and are often referred to as the "G-side" and "R-side" in model specification. Therefore, one can change $\text{var}(\mathbf{y}_i)$ in three ways: via $\mathbf{Z}_i$, $\mathbf{G}$ or $\mathbf{R}$. In NONMEM, $\mathbf{G}$ and $\mathbf{R}$ are directly specified, but with SAS and R, they are specified by the model to be fitted with the aid of several options.

We will concentrate on random effect specification for mixed models with continuous exposure effect, e.g., a model allowing the exposure–response relationship varying between subjects:

$$\mathbf{y}_i = \beta_0 + \beta_{ci} \mathbf{c}_i + \varepsilon_{ij} \qquad (10.5)$$

where $\mathbf{c}_i = (c_{i1}, ..., c_{ir})$, and β_{ci} is our main focus. Supposing that $\beta_{ci} = \beta_c + u_{ci}$ with $u_{ci} \sim N(0, \sigma)$, we can write $\beta_{ci} \mathbf{c}_i = \beta_c \mathbf{c}_i + u_{ci} \mathbf{c}_i$, hence $\mathbf{c}_i$ are in both $\mathbf{X}_i$ and $\mathbf{Z}_i$. In SAS proc MIXED, this model is specified as

```
model y=conc;
random conc/subject=i;
```

in which the "model" statement specifies the $\beta_c \mathbf{c}_i$ part and the "random" statement is for $u_{ci} \mathbf{c}_i$. The second one needs the subject identifier "i", since although c_i is in both $\mathbf{X}_i$ and $\mathbf{Z}_i$, in $\mathbf{X}_i$ all c_is are stacked as one vector, but in $\mathbf{Z}_i$ they form n columns, one for each u_{ci}. The identifiers determine into which column each c_i goes. One may also introduce random subject effects on the intercept, assuming each subject has own baseline response when there is no exposure. The model becomes

$$\mathbf{y}_i = \mathbf{X}_i\boldsymbol{\beta}_i + \beta_{ci}c_i + u_{0i} + \varepsilon_{ij} \tag{10.6}$$

where u_{0i} is the individual baseline response. To specify this model, we may use

```
random intercept conc/type=UN subject=i;
```

where we have also specified the correlation between u_{ci} and u_{0i} as unstructured, although in the case of two random variables other structures lead to the same $\mathbf{G}$. With this model, we have the variances and covariance for u_{ci} and u_{0i} to be estimated.

To specify $\mathbf{R}$, we need the repeated statement. A commonly used one is

```
repeated /type=SP(POW)(day) subject=i;
```

which leads to a correlation between two repeated measures from the same subject i at days j and k given by a power function $\rho^{|j-k|}$, but measures in different subjects remain independent.

10.3   Nonlinear mixed models

Fitting an NLMM is among the most difficult tasks in ER modeling due to not only the complexity of model fitting algorithms, but also that of model specification. In this section we focus on using SAS proc NLMIXED, as it is the most commonly used tool for statisticians. The next section gives a quick introduction to NONMEM, the tool for pharmacometricians, which has considerable similarity to proc NLMIXED. Proc NLMIXED uses the ML approaches to fit NLMMs. Therefore, it contains two components: one for specifying the likelihood function based on an NLMM, the other for the maximization of the likelihood function. The latter consists of a set of optimization algorithms, which we will discuss briefly from practical aspects.

There are a few options for likelihood calculation specified by option "method": "method=FIRO", based on the simple approximation of Beal and Sheiner (1982), and "method=GAUSS" using Gaussian quadratures are two common choices. The first one is the most robust method and is often used for the first attempt to fit a model to a dataset. The default is the Gaussian quadrature with adaptive selection of quadrature points to ensure sufficient accuracy. However, the selected number of quadrature points may be very large so that it may take a long time for SAS to find that convergence cannot be achieved finally. Therefore, it is useful to restrict the number to, e.g., 10 by option "qpoints=10" before a reasonably good model fitting is achieved.

Proc NLMIXED offers a number of optimization algorithms and also user's control to key parameters in these algorithms. The algorithms range from conjugate-gradient, a very efficient algorithm for large problems, to Nelder–Mead simplex algorithm using function values only. The default, and also the most common one, is the quasi-Newton algorithm, which uses the function value and its derivatives, and is often very efficient. Two other useful options are to select the method and precision of line search, a very important component in optimization algorithms. "LINESEARCH=i", where "i" is the code of ith line search method (see SAS, 2012, for details), allows choices among a number of methods, while "LSPRECISION=r" specifies r as the search precision. Sometimes increasing the precision may help with algorithm convergence. The selection of methods, however, needs experience in using a wide range of optimization algorithms.

Although the document for proc NLMIXED is of large volume, the following statements are the most important ones and often sufficient for model tasks.

- PARMS and BOUNDS: PARMS specifies parameters and their initial values in models to be estimated and BOUNDS specifies the range for them. Different from linear mixed models where we only specifies effects, here the model has to be explicitly parameterized, including the part of categorical covariates and variance components. For example,

  ```
  PARMS beta=1 sigma=1;
  BOUNDS 0 < sigma 1;
  ```

  specifies value 1 as the initial value for both parameters beta and sigma and sigma only takes positive values. The two statements should be consistent, i.e., the initial values should be within the specified boundary.

- MODEL: This statement in fact specifies the distribution of the response variable conditional on the random variables. Apart from the normal distribution commonly used for NLMMs, other distributions such as binomial, gamma, Poisson and negative binomial can also be specified. For example,

  ```
  pi=1/(1+exp(beta0+ui-Emax/((EC50/conci+1))));
  yi~binomial(ni,pi);
  ```

  specifies a nonlinear logistic regression model with random effects ui and an Emax model that connect the exposure conci with the probability of event pi.

  In addition, it can also specify that the likelihood is user specified.

```
yi ~general(ll);
```

specifies the log-likelihood function is given by variable ll and programmed by the user. Note yi here is symbolic and has no impact on the model or on the calculation.

- PREDICT: This statement specifies what the user wants to predict based on the fitted model. For example,

```
PREDICT pi out=pred;
PREDICT 1/(1+exp(beta0-Emax/((EC50/conci+1))) out=pred;
PREDICT 1/(1+exp(beta0-Emax/((EC50/100+1))) out=pred;
PREDICT 1/(1+exp(beta0+ui-Emax/((EC50/100+1))) out=pred;
```

The first one simply takes the pi as in the previous MODEL statement and outputs it to dataset pred, together with variables in the input dataset. The second one predicts the probability without ui, hence it only depends on conci. The third one is a prediction at the conci=100 level, so in dataset pred, every record has the same predicted value. The last one predicts individual's probabilities at the conci=100 level and their mean is the marginal probability at this exposure level. One can also require a 90% CI for the prediction. Without this statement, although in principle one can calculate the prediction and its CI from fitted models and predicted random effects, the calculation can be very difficult and requires intensive programming.

- RANDOM: This is the counterpart of the random statement in proc MIXED.

```
random beta1 beta2 ~ normal([0,0],[sig11,sig21,sig22]) subject=id;
```

specifies a zero mean (specified by [0,0]) bi-normal distribution for parameters beta1 and beta2 with the lower covariance matrix [sig11,sig21,sig22]. In SAS V9.3 or older, only normal distributions are allowed. However, one may generate other distributions including copulas based on them.

All variances in a model should be nonnegative and a lower bound should be set for them. It is also common that an estimate may be at the zero boundary. In this case, one should remove the corresponding random effect and rerun the model.

In addition to the above statements, one also needs to use programming statements the same as those for data step. Among them, the ARRAY statement is very important, as proc NLMIXED can only access variables of the current record, except using the lag(.) function to access the previous ones, which is not formally documented to use in proc NLMIXED. Using arrays one can use a wide range of intermediate variables, e.g., for the calculation of cumulative exposures. Often after specifying the model, proc NLMIXED cannot even start due to problems of calculating the likelihood function. This might be due to problems in calculation commands for model specification. A major cause of the problems is division by a number very close to zero. As the total log-likelihood for the whole dataset needs to be calculated, the problem might occur not immediately at the first subject, but only when the issue accumulates and beyond the limit SAS can handle. Debugging a proc NLMIXED program is not easy, although a number of options for debugging are available, in particular, the TRACE, and a number of LISTxx options. These options list intermediate results while calculating the likelihood function, and produce a large volume of listings that are hard to read even for experienced users. Often a user may start with their alternatives to start debugging. For readers who are interested in using this option, the first step is to find out what is in the produced list. For this one may try a toy dataset and a very simple mixed model and look at the produced list, in combination with some of the following approaches. When there are small to moderate numbers of parameters in the model and subjects in the data, one may use

```
PUT _PDV_;
```

to display all variables in a formatted manner. But often it is better to specify a list of key variables and parameters in the PUT statement. Another debugging approach is to extract the programming statements and to run them in a data step. This approach is particularly useful when the likelihood function is user defined. Values of all parameters have to be set, including random effects. In this way one can examine the dataset the data step created, which is much easier than reading the list of parameter values in SAS logs.

The likelihood function for NLMMs may have a complex shape with potentially multiple local maximums. This is particularly so when the sample size is not large. It is quite common that even if the model is correctly specified, and there is no numerical instability problem, the model fitting still does not converge. The most common problem is inappropriate initial parameter values. SAS provides a few options to help with solving these issues. SAS can do a grid search before starting the optimization algorithm and allows the user to specify the grid. For example,

```
PRAMS  beta=-1 to 1 by 0.2 sigma=0.1 to 1 by 0.1
```

asks proc NLMIXED to search the beta and sigma combination that gives the maximum likelihood, then start the optimization from there. A grid search of a few dimensions is inexpensive and one can specify a rather dense grid. However, the computing burden increases exponentially with the number of parameters in the grid search. The user should identify parameters that are sensitive to initial values and not include all parameters in a grid search. Have I found the true (global) maximum? This is a valid question even when the fitted model seems reasonable and the algorithm converges properly. However, in principle one cannot verify it without an exhaustive search for all parameters. One may assess the fitted model in the aspect of goodness-of-fit, so at least one may be satisfied even if the global maximum was not reached, that the current model fitting is not far from the global optimal for some specific purposes, e.g., prediction. Proc NLMIXED does provide an option OPTCHECK to check if there is a point near the final estimates that has a higher likelihood function value.

Proc NLMIXED provides the covariance matrix for all estimated parameters based on the information matrix. In some situations the matrix may not be positive-definite, hence the full covariance matrix cannot be calculated. There are a number of reasons that make the fitted model have no full covariance for the parameters. In most cases, the parameters are not converged at the peak of the likelihood function. Sometimes this problem may be resolved by changing the initial values or optimization algorithms, but often it is difficult to overcome.

10.4   A very quick guide to NONMEM

NONMEM is the most commonly used model fitting tool among pharmacometricians. It was first written specifically for NLMMs in PK and PKPD modeling. But it also allows the user to write the likelihood function, hence can fit most NLMMs. It has been further developed in parallel with proc NLMIXED and offers some options proc NLMIXED does not have, for example, the use of ODE solver and MCMC approaches for Bayesian modeling. NONMEM still lacks key features of a model software, e.g., there is no graphic interface, but some packages based on R, e.g., XPOSE, can be used in combination easily.

A NONMEM program consists of a number of "blocks" starting with a dollar. The following is a summary of what each block does to specify the data, model, parameter and outputs.

```
$PROBLEM PHENOBARB   ; Project identifier
$INPUT ID TIME AMT WT APGR DV ; variables in dataset
$DATA PHENO          ; dataset
$SUBROUTINE ADVAN1   ; ADVAN1 is a one-compartment model
$PK                  ; Specify parameters in the model
TVCL=THETA(1)        ; LHS: parameters the model required
CL=TVCL+ETA(1)       ; RHS: model of the parameters
...                  ; THETA(x): xth fixed effect
...                  ; ETA(x): xth random effect
$ERROR               ; Specify error model
Y=F+ERR(1)           ; F is the mean and ERR(1) is the error
$THETA (0,.0047) (0,.99) ; Ranges of fixed effects (as in $PK)
$OMEGA .0000055, .04 ; Initial values of Var(ETA)
$SIGMA 25            ; Initial values of Var(ERR(1))
$ESTIMATION PRINT=5  ; Require model estimation
$COVARIANCE          ; Require estimated covariance
$SCATTER PRED VS DV  ; Require scatter plot of predicted vs DV
```

In the PK block programming steps are allowed so that the parameters may vary, e.g., according to different groups.

```
$PK
IF (SEX EQ. 1) CL=THETA(1)
ELSE CL=THETA(1)+THETA(2)
```

specifies THETA(1) as the clearance for males and THETA(2) as the difference between males and females. The same approach can be used to specify error terms in the error block.

A NONMEM dataset should have a special structure with specific variables:

- DV: dependent variable, could be PK or PD

- ID: Subject identifier

- MDV: 0 if a value in DV is present and 1 if it is missing

- TIME: needed for specific libraries

- AMT: amount of dose

- EVID: 0 if the record is an observation, 1 if the record is a dose of AMT amount, and 2 if neither a dose nor an observation. This option may be used to specify a condition for prediction, e.g., a predicted exposure at a given time where there is no observation.

- CMT: Compartment identifier for either the observation or the dose records.

One needs to use a command line such as "nonmem commandfile outputfile" to run NONMEM. But, in general, NONMEM runs quickly due to its relatively simple structure and the lack of the burden of interfaces.

A

Appendix

A.1 Basic statistical concepts

This section describes basic statistical concepts used in this book, except for Bayesian analysis. A statistical analysis starts from data and the model from which the data are generated. In a parametric model there are unknown parameters to be estimated. Often not all parameters are of interest and some are nuisance parameters. We use the data to estimate the parameters, hence parameter estimates are functions of the data. A function of data independent of any unknown parameter is called a *statistic*. A parameter estimate $\hat{\theta}$ that on average equals the true value, i.e., $E(\hat{\theta}) = \theta$, is an *unbiased* estimate. We often consider the asymptotic properties of parameter estimates when sample size is large. An important property called consistency refers to the fact that the estimate based on n samples $\hat{\boldsymbol{\theta}}_n$ tends to $\boldsymbol{\theta}$ when $n \to \infty$, namely the difference between them can be made as small as we like by increasing the sample size n, and this is denoted by $\hat{\boldsymbol{\theta}}_n \to \boldsymbol{\theta}$. Note that a consistent estimate may not be unbiased with finite sample size. An estimate is call asymptotic unbiased if the bias diminishes to zero with increasing sample sizes. It may not be a consistent estimate since its variation may not necessarily diminish, but counterexamples are rare, if any, in common situations in this book.

One important aspect in statistical inference is hypothesis testing. For example, in a model

$$y_i = \beta_0 + \beta c_i + \varepsilon_i \tag{A.1}$$

one key question to answer is if $\beta = 0$, i.e., c_i has no effect on y_i. The question can be answered by testing a hypothesis $H_0 : \beta = 0$ based on the data observed. Put another way, we examine if there is sufficient evidence in the observed data against H_0. Based on test statistics, we may reject H_0 if there is sufficient evidence. When testing a hypothesis, we may commit two types of mistakes; one is to reject H_0 when it is true, while the other is not to reject H_0 when it is not true. The two types of mistakes are called type I and type II errors, respectively. The amount of information against H_0 is measured by the probability of making a type I error; a common level in practice is 5%. Therefore, hypothesis testing is a task to control the type I error at a given level, and to maximize the probability of not committing a type II error. The probability is known as the power of the test. However, the power depends on

303

the true value of the parameter. Obviously, the power of a test is higher when $\beta = 1$ than when $\beta = 0.5$. A test controlling its type I error at α (referred to as a test of level α) is a uniformly most powerful test (UMP) if its power is not less than other tests of the same level α for any β value. As an example, suppose that $y_1, ..., y_n$ follows normal distribution with $y_i \sim N(\mu, \sigma^2)$ and we would like to test $H_0 : \mu \leq 0$; then test statistic

$$t = \bar{y}. / \sqrt{S^2/n}, \tag{A.2}$$

where S^2 is the sample variance, follows the t distribution with $n - 1$ degree of freedom. Therefore, if we would like to test $H_0 : \mu \leq 0$ at the 5% level, we reject it when $t > t_{n-1,0.95}$, where $t_{n-1,0.95}$ is the 95% percentile of the t distribution. However, to test $H_0 : \beta = 0$ we need a two-sided 5% test which rejects H_0 when $t > t_{n-1,0.975}$ or $t < t_{n-1,0.025}$. A common test for a model parameter $H_0 : \theta = \theta_0$ is based on (asymptotic) normal distribution $\hat{\theta} \sim N(\theta, se^2)$, where se is the standard error of $\hat{\theta}$. Then $(\hat{\theta} - \theta_0)/se \sim N(0, 1)$ under H_0. Percentiles of t and standard normal distributions can be find using R or SAS.

Relevant to both parameter estimation and hypothesis testing is the confidence interval (CI) of a parameter. A $1 - \alpha$ CI is a function of the data, which covers the true β value with probability $1 - \alpha$, if the experiment and test are to be repeated infinite times. $1 - \alpha$ CI is not unique, but often that with the shortest length is unique. Under some situations the CI may consist of more than one interval. But we will not consider these situations here.

A.2   Fitting a regression model

To estimate β in the model

$$y_i = g(\mathbf{X}_i\beta) + \varepsilon_i \tag{A.3}$$

a commonly used approach is the least squares (LS) method, which finds β to minimizes the sum of squares of the difference between y_is and their model based prediction:

$$\sum_{i=1}^{n} (y_i - g(\mathbf{X}_i, \beta))^2. \tag{A.4}$$

The β minimizing it satisfies the following equation (estimating equation, as will be discussed later):

$$\sum_{i=1}^{n} \frac{\partial g(\mathbf{X}_i, \beta)}{\partial \beta} (y_i - g(\mathbf{X}_i, \beta)) = 0. \tag{A.5}$$

When $g(\mathbf{X}_i\boldsymbol{\beta}) = \mathbf{X}_i^T\boldsymbol{\beta}$, its solution is $\hat{\boldsymbol{\beta}}_{LS} = (\sum \mathbf{X}_i^T\mathbf{X}_i)^{-1}\sum \mathbf{X}_i^T y_i$. For non-linear models an LS algorithm is needed to calculate $\hat{\boldsymbol{\beta}}_{LS}$. But it is straightforward to specify the model and let software do the task. For example, using R function nls(.) we can fit an Emax model with

```
nls(resp ~ Emax*conc/(EC50 + conc), start=list(Emax=1, EC50 = 5))
```

where we specify starting values $Emax = 1, EC50 = 5$, as nonlinear LS algorithms typically need them for iteration. Estimating parameters using an estimating equation is a more general class of methods and will be discussed later.

A.3 Maximum likelihood estimation

The maximum likelihood estimate (MLE) approach is the most commonly used approach in statistics to fit a statistical model, estimate model parameters and test hypotheses regarding model parameters. The details can be found in standard textbooks. This appendix only serves as a brief introduction without detailed theory. Suppose we observe a number of samples $y_1, ..., y_n$ which have density function $f(y, \boldsymbol{\theta})$; the the likelihood function for $y_1, ..., y_n$ is defined as the product of individual density functions:

$$L(\boldsymbol{\theta}) = \prod_{i=1}^{n} f(y_i, \boldsymbol{\theta}). \tag{A.6}$$

For example, when $y_i \sim N(\mu, \sigma^2)$ with mean μ and variance σ^2, the likelihood function is

$$L(\mu, \sigma^2) = \prod_{i=1}^{n} \frac{1}{\sqrt{2\pi\sigma^2}} \exp(-\frac{(y_i - \mu)^2}{2\sigma^2}). \tag{A.7}$$

The MLE for μ and σ^2 are the values that maximize $L(y_1, ..., y_n, \mu, \sigma^2)$. To this end, we often use the log-likelihood function

$$l(\mu, \sigma^2) = log(L(\mu, \sigma^2)) = \sum_{i=1}^{n}(log(\sigma^2)/2 - \frac{(y_i - \mu)^2}{2\sigma^2}) + C. \tag{A.8}$$

where C is a constant. Taking derivatives with respect to μ and σ^2

$$\frac{\partial l(\mu, \sigma^2)}{\partial \mu} = \sum_{i=1}^{n} -(y_i - \mu)/(2\sigma^2)$$

$$\frac{\partial l(\mu, \sigma^2)}{\partial \sigma^2} = -\sum_{i=1}^{n}(\sigma^{-2} + (y_i - \mu)^2\sigma^{-4})/2 \tag{A.9}$$

$l(\mu, \sigma^2)$ reaches the maximum when the derivatives are equal to 0. The corresponding μ and σ^2 values

$$\hat{\mu} = \sum_{i=1}^{n} y_i / n$$

$$\hat{\sigma}^2 = \sum_{i=1}^{n} (y_i - \hat{\mu})^2 / n \tag{A.10}$$

are the MLE we are looking for. $\hat{\mu}$ is the same as the sample mean, while $\hat{\sigma}^2$ is slightly different from the commonly used sample variance estimate $\sum_{i=1}^{n} (y_i - \hat{\mu})^2 / (n-1)$.

For simplicity in dealing with general situations, we write data in vector format $\mathbf{y} = (y_1, ..., y_n)$ and $\boldsymbol{\theta}$ is a vector containing all parameters in the likelihood. In general, we can write the log-likelihood function as $l(\mathbf{y}, \boldsymbol{\theta})$. Its derivatives with respect to $\boldsymbol{\theta}$

$$S(\boldsymbol{\theta}) \equiv \frac{\partial l(\boldsymbol{\theta})}{\partial \boldsymbol{\theta}} \tag{A.11}$$

are called score functions, and the MLE $\hat{\boldsymbol{\theta}}$ is the solution to score equations $S(\boldsymbol{\theta}) = 0$. In the previous example, the score equations are (A.9). Under some technical conditions which we skip here, $\hat{\boldsymbol{\theta}}$ is consistent, i.e., $\hat{\boldsymbol{\theta}} \to \boldsymbol{\theta}$. A key condition for consistency of $\hat{\boldsymbol{\theta}}$ is $E(S(\boldsymbol{\theta})) = 0$, which is satisfied by the score function. The condition plays a central role in generalized estimating equation (GEE) in the next section. Also, under some technical conditions the MLE follows asymptotic normal distributions when n is sufficiently large

$$\sqrt{n}(\hat{\boldsymbol{\theta}} - \boldsymbol{\theta}) \sim N(0, nI^{-1}(\boldsymbol{\theta})) \tag{A.12}$$

where

$$I(\boldsymbol{\theta}) = E\left(\frac{\partial^2 l(\boldsymbol{\theta})}{\partial \boldsymbol{\theta} \partial \boldsymbol{\theta}^\top}\right) \tag{A.13}$$

is the Fisher information matrix.

For a number of models the score function has a simple form

$$S(\boldsymbol{\theta}) = \sum_{i=1}^{n} \mathbf{X}_i^T (y_i - \mu_i(\eta_i)) \tag{A.14}$$

with $E(y_i) = \mu_i(\eta_i)$ and $\eta_i = \mathbf{X}_i^T \boldsymbol{\theta}$. The models include $y_i = \mathbf{X}_i^T \boldsymbol{\theta} + \varepsilon_i$ assuming normality for y_i. Less obviously, they also include Poisson model $y_i \sim Poisson(\mu_i(\eta_i))$ with log-link $\eta_i = \log(E(y_i))$ and logistic model $y_i \sim Bin(\mu_i(\eta_i))$ with logit link $\eta_i = \text{logit}(E(y_i))$. For these two models the score function has a general form

$$S(\boldsymbol{\theta}) = \sum_{i=1}^{n} \mathbf{X}_i^T \frac{\partial \mu_i(\eta_i)}{\partial \eta_i} \text{var}(y_i)^{-1} (y_i - \mu_i(\eta_i)) \tag{A.15}$$

but with the link functions above $\partial \mu_i(\eta_i)/\partial \eta_i = \text{var}(y_i)$. Even with the general form (A.15), a weighted LS algorithm can be used to estimate θ, which is more stable than a general purpose maximization algorithm.

Our purpose in fitting a model is not only to estimate the parameters, but also to make statistical inferences for them. For example when θ_1 is our primary concern and we would like to test $H_0 : \theta_1 = \theta_{1r}$ with θ_{1r} given, there are three MLE based tests. The Wald test is based on (A.12) with the test statistic

$$z = (\hat{\theta}_1 - \theta_{1r})/se(\hat{\theta}_1) \qquad (A.16)$$

where $se(\hat{\theta}_1)$ is the standard error of $\hat{\theta}_1$ derived from $I(\theta)$. We will not give the details of derivation here as $se(\hat{\theta}_1)$ are provided by almost all statistical software. Under H_0 z follows the standard normal distribution asymptotically. Therefore, we reject H_0 when $|z| > u_0$ where u_0 can be found from the percentiles of the standard normal distribution.

The second test is the likelihood ratio test (LRT). To test $H_0 : \theta_1 = \theta_{1r}$, we fit two models, one set $\theta_1 = \theta_{1r}$ while the other allows θ_1 to be estimated. Writing the likelihood functions for the two models as $l(\theta|\theta_1 = \theta_{1r})$ and $l(\theta)$, the test statistic for the LRT is

$$\chi^2 = -2(l(\theta|\theta_1 = \theta_{1r}) - l(\theta)) \qquad (A.17)$$

which follows the χ^2 distribution with 1 degree of freedom. The test can be used for testing multiple parameters simultaneously and the degree of freedom in χ^2 equals the number of parameters under test.

The third test is the score test based on the score function. To test $H_0 : \theta_1 = \theta_{1r}$, we fit the model under H_0 and obtain parameter estimates for other parameters denoted as θ_2. The test statistic is

$$s = S_{\theta_1}(\theta_{1r}, \hat{\theta}_2)/\sqrt{\text{var}(S_{\theta_1}(\theta_{1r}, \hat{\theta}_2))} \qquad (A.18)$$

where $S_{\theta_1}(\theta)$ is the θ_1-component in $S(\theta)$ and $\text{var}(S_{\theta_1}(\theta))$ is its variance, which can be calculated from the information matrix. Asymptotically s follows the standard normal distribution, hence s can be calibrated by the standard normal percentiles. We reject H_0 when $|s| > u_0$. The major advantage of the score test is that only model fitting under H_0 is needed, which is often simpler than fitting the model under H_a. However, common software does not offer this test.

A.4 Large sample properties and approximation

Large sample properties and approximation is a large topic. Only the very basic concept and results are given here. Details can be found in specialized books such as Serfling (1980) and Barndorff-Nielsen and Cox (1994).

A.4.1   Large sample properties and convergence

Here we only give basic concept and omit all technical details. When a series random independent identical variables $y_1, ..., y_n$ satisfy some technical conditions, their mean $\bar{y}_n = \sum_{i=1}^{n} y_i/n \to E(y_i)$. It is called the large number theorem. Here $\bar{y}_n \to E(y_i)$ means that when n increases to ∞, the difference between $\bar{y}_n$ and $E(y_i)$ tends to zero, i.e., $\bar{y}_n$ converges to $E(y_i)$. However, since $\bar{y}_n$ is a random variable, we can only guarantee that the probability of the two being different tends to zero. This statement has a number of versions mathematically, with different assumptions on y_is, but we do not distinguish them in this book and write $X_n \to X$, where X can be a constant or a random variable, when X_n converges to X. Furthermore, $\sqrt{n}(\bar{y} - E(y_i)) \sim N(0, \text{var}(y_i))$, regardless the distribution of y_i, known as the central limit theorem, which also has a number of versions. The most common use of these properties is deriving the consistency and normal distribution of MLE, and that of the EE and GMM approaches.

The convergence sometimes can change order with operations. For example, when $g(X)$ is a continuous function and $X_n \to X$, the $g(X_n) \to g(X)$. Another commonly used property is the Slusky theorem: When $X_n \to X$ and $Y_n \to C$ where C is a constant, then $X_n Y_n \to CX$ and, if $C \neq 0$, $X_n/Y_n \to X/C$.

A.4.2   Approximation and delta approach

Recall that a function $f(c)$ can be expanded at a point c_0 as

$$f(c) = f(c_0) + \frac{df(c)}{dc}|_{c=c_0}(c - c_0) + o(c - c_0) \tag{A.19}$$

where $df(c)/dc$ is the derivative of $f(c)$ with respect to c and $o(\delta)$ is a term that goes to zero quicker than δ. In a number of situations, we have used an stochastic version of this expansion as an approximation to a complex function, including the unconditional and conditional first order expansion of a nonlinear function in NLMMs in Chapter 3. Suppose that c is a random variable with $E(c) = c_0$, then taking the mean and variance of (A.19) we get

$$
\begin{aligned}
E(f(c)) &\approx f(c_0) \\
\text{var}(f(c)) &\approx \frac{df(c)}{dc}^2 |_{c=c_0} \text{var}(c),
\end{aligned}
\tag{A.20}
$$

as long as $\text{var}(c)$ is small. The approximations have a multivariate version for a vector **c**. In this book we have used these approximation for a number of times.

Often one needs to calculate the variance of a nonlinear function of a random variable. Although it may involve complex calculation even simulation or numerical integration, a large sample approximation commonly used is the delta approach. Recall that MLE or GEE estimates have the property

$$\sqrt{n}(\hat{\boldsymbol{\beta}} - \boldsymbol{\beta}) \sim N(0, n\mathbf{I}^{-1}(\boldsymbol{\beta})). \tag{A.21}$$

We may want to calculate $\mathrm{var}(g(\hat{\boldsymbol{\beta}}))$, where for function $g(\boldsymbol{\beta})$, $\partial g/\partial \boldsymbol{\beta}$ exists, using the second approximation above. In fact, we have

$$\sqrt{n}(g(\hat{\boldsymbol{\beta}}) - g(\boldsymbol{\beta})) \sim N(0, n\Sigma_g) \tag{A.22}$$

with $\Sigma_g = \partial g/\partial\boldsymbol{\beta}\Sigma(\partial g/\partial\boldsymbol{\beta})^T$. Informally, we can write $\mathrm{var}(g(\hat{\boldsymbol{\beta}})) \approx \partial g/\partial\boldsymbol{\beta}\mathrm{var}(\hat{\boldsymbol{\beta}})(\partial g/\partial\boldsymbol{\beta})^T$, proving $\mathrm{var}(\hat{\boldsymbol{\beta}})$ is small.

A.5   Profile likelihood

The profile likelihood approach (Barndorff-Nielsen and Cox, 1994) is a useful tool in exposure-response modeling, since it may provide an easy alternative for dealing with technical difficulties when an additional parameter is added in a model and convergence cannot be achieved.

Suppose a model contains parameters $\boldsymbol{\beta}$ and λ and one is interested in inference for λ. The profile likelihood method fits the model fixing λ at different values. Specifically, supposing that $L(\boldsymbol{\beta}, \lambda)$ is the likelihood function, for each given λ value, the profile likelihood finds the MLE for $\boldsymbol{\beta}$ and calculates

$$L^*(\lambda) = \max_{\boldsymbol{\beta}} L(\boldsymbol{\beta}, \lambda). \tag{A.23}$$

In the case of the QT model, for fixed $K(= \lambda)$, it is still a linear mixed model and can be fitted easily. $L^*(\lambda)$ may be treated as an approximation to the marginal likelihood function for λ. Again since the profile likelihood is derived from the full likelihood function, residual maximum likelihood (REML) should not be used when fitting the model for the profile likelihood approach. Plotting $L^*(\lambda)$ around its peak is a good way to assess the impact of λ in the model. One can also calculate an approximate CI by finding the intercepts of the curve and a horizontal line of corresponding significance level. Note that the approximation may be poor if λ and $\boldsymbol{\beta}$ have strong association (in terms of orthogonality between them measured by the relative size of $\partial^2 \log(L(\boldsymbol{\beta}, \lambda))/\partial\boldsymbol{\beta}\partial\lambda$.

A.6   Generalized estimating equation

Recall that the score function for a number of models has the general form (A.15). This leads to the idea of quasi-likelihood that first specifies the score function as (A.15) assuming it comes from a "qausi" likelihood function so that solving (A.15) is to maximize this likelihood. For example, in Chapter 2 we introduced over dispersion models with negative- and beta-binomial distribution. An alternative to the MLE is to use the quasi-likelihood approach by solving (A.15). It is in principle less efficient than the MLE, but often the efficiency loss is very small.

In fact, (A.15) does not need to have a distribution model behind it. For a wide range of situations equations not derived from likelihood functions may also be used in the same way as the score equations. The quasi-likelihood approach was further developed into generalized estimation equations (GEE). A set of GEEs is equations

$$\mathbf{S}(\boldsymbol{\theta}) = 0 \qquad (A.24)$$

based on the model and observed data, e.g., as (A.15), and $S(\boldsymbol{\theta})$ is know as estimating equation, which is the score function in the MLE case.. A GEE estimate $\hat{\boldsymbol{\theta}}_{GEE}$ is the solution to the equations. For common GEEs $\hat{\boldsymbol{\theta}}_{GEE}$ can be obtained from software such as SAS proc GENMOD and R function glm(.); both allow the user to specify the mean, link and variance functions. Some software for Poisson or logistic regressions in fact allows y_i to be noninteger. Although a warning may be given, e.g., by R function glm(.), the estimate and SE are correct.

Under some technical conditions, $\hat{\boldsymbol{\theta}}_{GEE}$ is consistent and asymptotically normally distributed. A key condition for is that, letting $\boldsymbol{\theta}_0$ be the true values,

$$E(S(\boldsymbol{\theta}_0)) = 0 \qquad (A.25)$$

and $E(S(\boldsymbol{\theta})) \neq 0$ for any $\boldsymbol{\theta} \neq \boldsymbol{\theta}_0$. We call an EE unbiased if $E(S(\boldsymbol{\theta}_0)) = 0$. Furthermore, expanding the GEE at its estimate $\hat{\boldsymbol{\theta}}$

$$0 = \mathbf{S}(\hat{\boldsymbol{\theta}}) + \frac{\partial \mathbf{S}(\hat{\boldsymbol{\theta}})}{\partial \boldsymbol{\theta}}(\boldsymbol{\theta} - \hat{\boldsymbol{\theta}}) + o(\boldsymbol{\theta} - \hat{\boldsymbol{\theta}}), \qquad (A.26)$$

where $o(a)$ is a higher order term of a, we can derive

$$\sqrt{n}(\hat{\boldsymbol{\theta}} - \boldsymbol{\theta}) \sim N(0, n\mathbf{F}^{-1}\mathrm{var}(\mathbf{S}(\boldsymbol{\theta}))\mathbf{F}^{-1}) \qquad (A.27)$$

with $\mathbf{F} = E(\partial \mathbf{S}(\hat{\boldsymbol{\theta}})/\partial \boldsymbol{\theta})$. $\mathbf{F}^{-1}\mathrm{var}(\mathbf{S}(\boldsymbol{\theta}))\mathbf{F}^{-1}$ is known as the sandwich variance estimate. When $\mathbf{S}(\boldsymbol{\theta})$ has the form

$$S(\boldsymbol{\theta}) = \sum_{i=1}^{n} \frac{\partial \mu_i(\boldsymbol{\theta})}{\partial \boldsymbol{\theta}} \mathbf{W}_i (\mathbf{y}_i - \boldsymbol{\mu}_i(\boldsymbol{\theta}))/n \qquad (A.28)$$

we can derive

$$\mathbf{F} = \sum_{i=1}^{n} \frac{\partial \mu_i(\boldsymbol{\theta})}{\partial \boldsymbol{\theta}} \mathbf{W}_i \frac{\partial \mu_i(\boldsymbol{\theta})}{\partial \boldsymbol{\theta}}^T \qquad (A.29)$$

and

$$\mathrm{var}(\mathbf{S}(\boldsymbol{\theta})) = \sum_{i=1}^{n} \frac{\partial \mu_i(\boldsymbol{\theta})}{\partial \boldsymbol{\theta}} \mathbf{W}_i \mathrm{var}(\mathbf{y}_i) \mathbf{W}_i \frac{\partial \mu_i(\boldsymbol{\theta})}{\partial \boldsymbol{\theta}}^T . \qquad (A.30)$$

In practice, we can replace var($\mathbf{y}_i$) by its empirical estimate. The sandwich estimate is valid as long as the mean model μ_i is correctly specified. In summary, although the GEE approach is less efficient than MLE, it is more robust.

To compare the GEE and ML approaches, recall the EE

$$S(\boldsymbol{\theta}) = \sum_{i=1}^{n} \mathbf{X}_i^T (y_i - \mu(\boldsymbol{\theta})) \tag{A.31}$$

with $E(y_i) = \mu_i(\boldsymbol{\theta})$. The equation in the LS approach has this form, so it is the score equation for the MLE assuming normality for y_i. Less obviously, it is also the score equation when $y_i \sim Poisson(\mu_i(\boldsymbol{\theta})), \mu_i(\boldsymbol{\theta}) = \exp(\mathbf{X}_i^T \boldsymbol{\theta})$ and when $y_i \sim Bin(\mu_i(\boldsymbol{\theta})), \text{logit}(\mu_i(\boldsymbol{\theta})) = \mathbf{X}_i^T \boldsymbol{\theta}$. Furthermore, it is the optimal EE as long as $E(y_i) = \text{var}(y_i) = \exp(\mathbf{X}_i^T \boldsymbol{\theta})$ or $E(y_i) = \text{logit}(\mu_i(\boldsymbol{\theta})) = \mathbf{X}_i^T \boldsymbol{\theta}$ and $\text{var}(y_i) = E(y_i)(1 - E(y_i))$.

A.7 Model misspecification

Owning to the famous quote "All models are wrong, but some are useful" (George Box, exact source unknown), knowing how a model behaves when it is misspecified (i.e., different from the unknown true model that generates the data) is crucial. Suppose that the data are generated from a quadratic curve $y_i = a + bx_i^2 + \varepsilon_i$ but a linear model $y_i = a + bx_i + \varepsilon_i$ is fitted to the data. Intuitively, the linear model would act as the best linear approximation to the quadratic curve. In fact, if the data are generated from model $y_i = g(x_i, \boldsymbol{\beta}) + \varepsilon_i$ but are fitted by a (misspecified) model $y_i = g^*(x_i, \boldsymbol{\beta}) + \varepsilon_i$ with x_i taking K different values $x_1, ..., x_K$ with sampling proportions $w_1, ..., w_K$, then the LS estimate $\hat{\boldsymbol{\beta}}_{LS} \to \boldsymbol{\beta}^*$ and $\boldsymbol{\beta}^*$ minimizes

$$\sum_{k=1}^{K} w_k (g^*(x_k, \boldsymbol{\beta}^*) - g(x_k, \boldsymbol{\beta}))^2 \tag{A.32}$$

(White, 1981). Note that it can be written as $\sum_{i=1}^{n} (g^*(x_i, \boldsymbol{\beta}^*) - g(x_i, \boldsymbol{\beta}))^2$, in which there are w_k identical terms for each unique value x_k. Furthermore, asymptotically, $\hat{\boldsymbol{\beta}}_{LS} \sim N(\boldsymbol{\beta}^*, V^*)$, where V^* has a sandwich form, as the variance of the EE estimate. This property shows a very important fact that the limit of the parameter estimate for a misspecified model not only depends on the two models, but also on the distribution of x_i.

The property given by (A.32) can be extended in a number of directions. For longitudinal models fitted with a GEE approach, $\hat{\boldsymbol{\beta}}_{EE} \to \boldsymbol{\beta}^*$ and $\boldsymbol{\beta}^*$ is the solution to

$$\sum_{i=1}^{n} \frac{\partial g^*(x_i, \boldsymbol{\beta}^*)}{\partial \boldsymbol{\beta}^*} \mathbf{W}_i (g^*(x_i, \boldsymbol{\beta}^*) - g(x_i, \boldsymbol{\beta})) \tag{A.33}$$

and asymptotic normality of $\hat{\beta}_{EE}$ also holds and a sandwich form variance is also available.

The approach can also be extended to general EEs. Consider

$$S(\beta) = \sum_{i=1}^{n} S_i(\beta) \tag{A.34}$$

and suppose that a unique β^* exists such that $E(S(\beta^*)) = 0$. With a correctly specified model and confounding correctly dealt with, $\beta^* = \beta$, but we concentrate on the case of $\beta^* \neq \beta$. It can be under appropriate technical conditions, the solution to (A.34): $\hat{\beta} \to \beta^*$. The general results cover the situation when confounding factors exist but are ignored in the model fitting. For example, when the model is $y_i = \beta c_i + u_i + \varepsilon_i$ and u_i is a confounder, the LS estimate corresponds to $S_i = c_i(y_i - \beta c_i)$, and β^* is the solution of $\sum_{i=1}^{n} E(c_i(\beta c_i + u_i - \beta^* c_i)) = \sum_{i=1}^{n} E(c_i(\beta - \beta^*) + u_i) = 0$.

For MLE with a misspecified likelihood function, a classical work is White (1982), who showed that the MLE makes the misspecified likelihood function the best approximation to the true one. Letting $f(y_i|x_i, \beta)$ and $f^*(y_i|x_i, \beta)$ be the true and misspecified log-likelihood functions, then $\hat{\beta}_{MLE} \to \beta^*$, which minimizes

$$\int_{x_i} D(f(y_i|x_i, \beta), f^*(y_i|x_i, \beta^*)) dF(x_i) \tag{A.35}$$

where $D(f_1, f_2)$ is the KullbackLeibler divergency. Again the asymptotic normality of $\hat{\beta}_{MLE}$ and its sandwich form variance are ready to use.

In practice, one often does not know the true model. However, the true model might be estimated by a complex model that may not be easy to use. Therefore, the misspecified model results can be used to assess the sensitivity of using a simplified model to model misspecification. The true model may also come as a hypothetical model under some scenarios and one may want to know the performance of a given model under these scenarios.

A.8   Bootstrap and Bayesian bootstrap

Bootstrap is a procedure to reconstruct the distribution of data using a set of observations (Davison and Hinkley, 1997). The idea behind bootstrap is to take samples from the observations so that the samples behave like they were sampled from the population the observation came from. Suppose we have observed $y_1, ..., y_n$ from $y_i \sim F(y)$ with unknown distribution $F(y)$ and $E(y_i)$ is the parameter we are interested in. To obtain the distribution of its estimate $y. = \sum_{i=1}^{n} y_i/n$, we sample with replacement, i.e., we replace y_i in the set of data after it has been sampled, so each time any observation has $1/n$ probability to be selected, regardless whether it has been sampled or not. After taking samples $y_1^b, ..., y_n^b$, we can calculate a bootstrap sample $y_.^b = \sum_{i=1}^n y_i^b / n$. These steps can repeat B times so that we have $y_.^1, ..., y_.^B$. It can be found that $y_.^b$ could be considered as samples of $y_.$, had we repeatedly sample $y_1, ..., y_n$ from $y_i \sim F(y)$.

Using bootstrap in regression is also straightforward. Suppose that c_i and $y_i, i = 1, ..., n$ follow model

$$y_i = \beta_0 + \beta c_i + \varepsilon_i \tag{A.36}$$

and we are interested in statistical inference for β. To use bootstrap, we sample from c_is and y_is in pairs to form c_i^b and $y_i^b, i = 1, ..., n$, then fit them to the model and obtain estimates β^b. After repeating these steps B times, one can either use samples $\beta^1, ..., \beta^B$ to construct confidence intervals or calculating variance for $\hat{\beta}$, the LS estimate for β by fitting the model to the original dataset. There are other resampling methods in this case. One way is to fit the model to the original data and obtain the residuals $\hat{\varepsilon}_i = y_i - \hat{\beta}_0 - \hat{\beta} c_i$. Then the residuals are sampled with replacement to obtain $\hat{\varepsilon}_i^b, i = 1, ..., n$. Then bootstrap samples for y_i can be constructed as $y_i^b = \hat{\beta}_0 - \hat{\beta} c_i + \hat{\varepsilon}_i^b$. When the model is correct, this approach is more accurate than the direct sampling scheme.

A Bayesian counterpart of bootstrap was proposed by Rubin (1981). Without giving the details, we briefly describe the implementation of this approach. Taking the model (A.36) as an example, for Bayesian bootstrap, instead of taking samples from c_i and y_i, we weight them by a Dirichlet distribution $D(1, 1, ..., 1)$. In practice, the weight can be easily generated from exponentially distributed samples $-\log(U_i)$, where U_i is sampled from the uniform distribution. Then the bootstrap samples β^b are obtained by using weighted LS fitting the original data, but the weights change for each b. There are a number of interpretations for the Bayesian bootstrap. It can be considered a weighted likelihood approach (Newton and Raftery, 1994), or an empirical likelihood approach. Since the ordinary bootstrap is equivalent to weighting the pairs with integer (rather than exponential) weights, it can also be considered a smoothed version the ordinary bootstrap. In practice, it may also be easier to implement than the ordinary bootstrap.

Bibliography

[1] Aalen OO, Borgan Q and Gjessing HK (2008), *Survival and Event History Analysis: A Process Point of View.* Springer, New York.

[2] Abrahamowicz M, Bartlett G, Tamblyn R and du Berger R (2006) Modeling cumulative dose and exposure duration provided insights regarding the associations between benzodiazepines and injuries. *Journal of Clinical Epidemiology* **59**:393-403.

[3] Akaike H (1974) A new look at the statistical model identification, *I.E.E.E. Transactions on Automatic Control* **AC 19**:716–723.

[4] Andersen PK and Keiding N (2002) Multi-state models for event history analysis. *Statistical Methods for Medical Research* **11**:91-115.

[5] Atkinson AC and Donev AN (1992) *Optimum Experimental Designs.* Oxford University Press, Oxford.

[6] Austin PC (2012), Generating survival times to simulate Cox proportional hazards models with time-varying covariates. *Statistics in Medicine* **31**:3946-3958. DOI: 10.1002/sim.5452.

[7] Barndorff-Nielsen OE and Cox DR (1994) *Inference and Asymptotics.* Chapman and Hall, London.

[8] Bates DM and Watts DG (1988) *Nonlinear Regression Analysis and Its Applications.* John Wiley & Sons.

[9] Beal SL and Sheiner LB (1982) Estimating population kinetics. *CRC Crit. Rev. Biomed. Eng.,* **8**:195–222.

[10] Beal S, Sheiner LB and Boeckmann A (eds.) (2006) *NONMEM User's Guides.* Icon Development Solutions, Ellicott City.

[11] Bender R, Augustin T and Blettner M (2005) Generating survival times to simulate Cox proportional hazards models. *Statistics in Medicine* **24**:1713–1723.

[12] Bennett J and Wakefield J (2001) Errors-in-variables in joint population pharmacokinetic/pharmacodynamic modeling. *Biometrics* **57**:803–812.

[13] Bonate PL (2006) *Pharmacokinetic-Pharmacodynamic Modeling and Simulation.* Springer, New York.

[14] Bonate PL (2011) Modeling tumor growth in oncology. In *Pharmacokinetics in Drug Development.* Bonate PL and Howard DR (eds.) Springer, New York.

[15] Bonate PL (2013) Effect of assay measurement error on parameter estimation in concentration-QTc interval modeling. *Pharmaceut. Statist.* **12**:156–164.

[16] Brumback B, Greenland S, Redman M, Kiviat N and Diehr P (2003) The intensity score approach for adjusting for confounding. *Biometrics* **59**:274-285.

[17] Buckley J and James I (1979) Linear regression with censored data. *Biometrika* **66**:429-436.

[18] Burman CF and Senn S (2003) Examples of option values in drug development. *Pharmaceut. Statist*, **2**:113-125.

[19] Cameron AC and Trivedi PK (1998) *Regression Analysis of Count Data.* Cambridge University Press, New York.

[20] Cameron AC and Trivedi PK (2005) *Microeconometrics: Methods and Applications.* Cambridge University Press, New York.

[21] Carroll RJ and Ruppert D (1988) *Transformation and Weighting in Regression.* Chapman & Hall, New York.

[22] Carroll RJ and Ruppert D and Stefanski LA (2006) *Measurement Error in Nonlinear Models.* 2nd Ed. Chapman & Hall, New York.

[23] Claret L, Girard P, Hoff PM, Van Custem E, Zuideveld KP, Jorga K, Fagerberg J and Bruno R (2009) Modelbased prediction of Phase III overall survival in colorectal cancer on the basis of Phase II tumor dynamics. *Journal of Clinical Oncology* **27**:4103-4108.

[24] Clayton DG (1991) Models for the analysis of cohort and case-control studies with inaccurately measured exposures. In Dwyer JH, Feinleib MP, Lipsert P et al. (eds.), *Statistical Models for Longitudinal Studies of Health.* Oxford University Press, New York.

[25] Committee for Medicinal Products for Human Use (CHMP) (2007) *Guideline on the role of pharmacokinetics in the development of medicinal products in the paediatric population.* London.

[26] Cook RJ and Lawless JF (2007) *The Statistical Analysis of Recurrent Events.* Springer-Verlag, New York.

[27] Cox DR (1972) Regression models and life-tables (with discussion). *Journal of the Royal Statistical Society: Series B.* **34**:187-220.

[28] Cotterill A, Lorand D, Wang J and Jaki T (2015) A practical design for a dual-agent dose-escalation trial that incorporates pharmacokinetic data. *Statistics in Medicine* doi: 10.1002/sim.6482.

[29] Crowther MJ and Lambert PC (2013) Simulating biologically plausible complex survival data. *Statistics in Medicine* **32**:4118-4134.

[30] Csajka C and Verotta D (2006) Pharmacokinetic and Pharmacodynamic Modelling: History and Perspectives. *J Pharmacokin Pharmacodyn* **33**:227–279.

[31] Davidian M and Giltinan D (1995) *Nonlinear Models for Repeated Measurement Data.* Chapman and Hall, London.

[32] Davison AC and Hinkley DV (1997) *Bootstrap Methods and their Application.* Cambridge University Press, Cambridge.

[33] de Boor C (1978) *A Practical Guide to Splines.* Springer-Verlag, New York.

[34] Diaz FJ, Rivera TE, Josiassen RC, de Leon J (2007) Individualizing drug dosage by using a random intercept linear model. *Statistics in Medicine* **26**:2052-2073.

[35] Diggle P, Heagerty P, Liang KY Zeger S (2002) *Analysis of Longitudinal Data.* John Wiley & Sons.

[36] Dragalin V, Hsuan F, Padmanabhan SK (2007) Adaptive designs for dose-finding studies based on sigmoid Emax model. *J Biopharm Stat.* **17**:1051–1070

[37] Dragalin V, Fedorov V and Wu Y (2008) Two-stage design for dose-finding that accounts for both efficacy and toxicity. *Statistics in Medicine* **27**:5156–5176.

[38] Efron B (1989) Logistic regression, survival analysis and the Kaplan-Meier curve. *Journal of the American Statistical Association* **83**:414–425.

[39] Ette EI and Williams PJ (Eds) (2007) *Pharmacometrics: The Science of Quantitative Pharmacology.* John Wiley & Sons.

[40] Fayers P and Machin D (2007) *Quality of Life: The Assessment, Analysis and Interpretation of Patient-Reported Outcomes.* John Wiley & Sons, New York.

[41] FDA (2003) *Guidance for Industry: Exposure-Response Relationships-Study Design, Data Analysis, and Regulatory Applications.* FDA.

[42] Fedorov VV, Wu Y and Zhang R (2012) Optimal dose-finding designs with correlated continuous and discrete responses. *Statistics in Medicine* **31**:217-234.

[43] Fedorov VV and Leonov SL (2013) *Optimal Design for Nonlinear Response Models*. Chapman and Hall, London.

[44] Feng P, Zhou XH, Zou QM, Fan MY and Li XS (2012) Generalized propensity score for estimating the average treatment effect of multiple treatments. *Statistics in Medicine* **31**:681-697.

[45] Fitzmaurice G, Davidian M, Verbeke G and Molenberghs G (Eds.) (2006) *Longitudinal Data Analysis*. Chapman and Hall, London.

[46] Fleming TR and Harrington DP (1991) *Counting Processes and Survival Analysis*. New York, John Wiley & Sons.

[47] Florian JA, Torne CW, Brundage R, Parekh A, Garnett CE. (2011) Population pharmacokinetic and concentration–QTc models for moxifloxacin: pooled analysis of 20 thorough QT studies. *J Clin Pharmacol.* **51**:1152–62

[48] Fu H, Price KL, Nilsson ME and Ruberg SJ (2013) Identifying potential adverse events dose-response relationships via Bayesian indirect and mixed treatment comparison models. *J Biopharm Stat.* **23**:26–42.

[49] Gelman A, Carlin JB, Stern HS and Rubin DB (1995) *Bayesian Data Analysis*. Chapman & Hall, London.

[50] Gibaldi and Perrier B (1982) *Pharmacokinetics*. Dekker, New York.

[51] Hernan MA and Robins JM (2015) *Causal Inference* Chapman & Hall/CRC.

[52] Henderson R, Ansell P, and Alshibani D (2010) Regret-regression for optimal dynamic treatment regimes. *Biometrics* **66**:1192-1201.

[53] Hendry DJ (2013) Data generation for the Cox proportional hazards model with time-dependent covariates: A method for medical researchers. *Statistics in Medicine* **33**:436–454.

[54] Higgins J, Davidian M, Chew G and Burge H (1998) The effect of serial dilution error on calibration inference in immunoassay. *Biometrics* **54**:19–32.

[55] Hindmarsh AC (1983) ODEPACK, A Systematized collection of ODE solvers; in p.5564 of Stepleman RW et al.[eds.] (1983) *Scientific Computing*. North-Holland, Amsterdam.

[56] Hoeting J, Madigan D, Raftery A and Volinsky C (1999) Bayesian model averaging: A tutorial. *Statistical Science* **14**:382–401.

[57] Holford N, Black P, Couch R, Kennedy J, Briant R (1993a) Theophylline target concentration in severe airways obstruction-10 or 20 mg/L? A randomised concentration-controlled trial. *Clinical Pharmacokin* **25**:495–505.

[58] Holford N, Hashimoto Y, Sheiner LB (1993b) Time and theophylline concentration help explain the recovery of peak flow following acute airways obstruction. Population analysis of a randomised concentration controlled trial. *Clin Pharmacokin* **25**:506–515.

[59] Horn AR and Johnson CR (1985) *Matrix Analysis*. Cambridge University Press, Cambridge.

[60] Houede N, Thall PF, Nguyen H, Paoletti X and Kramar A (2010) Utility-based optimization of combination therapy using ordinal toxicity and efficacy in phase I/II trials. *Biometrics* **66**:532–540.

[61] Houk BE, Bello CL, Poland B, Rosen LS, Demetri GD and Motzer RJ (2009) Relationship between exposure to sunitinib and efficacy and tolerability endpoints in patients with cancer: Results of a pharmacokineticpharmacodynamic meta-analysis. *Cancer Chemotherapy and Pharmacology* **66**:357-371.

[62] Hsiao C (2003) *Analysis of Panel Data*. Cambridge University Press, Cambridge.

[63] Imai K and van Dyk DA (2004) Causal inference with general treatment regimes: generalizing the propensity score. *Journal of the American Statistical Association* **99**:854–866.

[64] Imbens G (2000) The role of the propensity score in estimating dose-response functions. *Biometrika* **87**:706–710.

[65] Kalbfleisch JD and Prentice RL (2002) *The Statistical Analysis of Failure Time Data* (2nd ed.). John Wiley & Sons, Hoboken.

[66] Karlsson KE, Grahnen A, Karlsson MO, Jonsson EN. (2007) Randomized exposure-controlled trials; impact of randomization and analysis strategies. *Br J Clin Pharmacol.* **64**:266-77.

[67] Klein JP (1992) Semiparametric estimation of random effects using the Cox model based on the EM algorithm. *Biometrics* **48**:795–806.

[68] Klein JP and Moeschberger ML (2003) *Survival Analysis. Techniques for Censored and Truncated Data* (2nd ed.) Springer-Verlag, New York.

[69] Ko H and Davidian M (2000) Correcting for measurement error in individual-level covariates in nonlinear mixed effects models. *Biometrics* **56**:368–375.

[70] Lacroix BD, Friberg LE, Karlsson MO (2012) Evaluation of IPPSE, an alternative method for sequential population PKPD analysis. *J Pharmacokin Pharmacodyn.* **39**:177-193.

[71] Lawless JF and Nadeau C (1995) Some simple robust methods for the analysis of recurrent events. *Technometrics.* **37**:158–168.

[72] Li F, Zaslavsky AM and Landrum MB (2013) Propensity score weighting with multilevel data. *Statistics in Medicine* **32**:3373-3387.

[73] Lindsey JK (1997) *Applying Generalized Linear Models.* Springer.

[74] Lindstrom MJ and Bates DM (1990) Nonlinear mixed-effects models for repeated measures data. *Biometrics* **46**:673-687.

[75] Littell RC, Milliken GA, Stroup WW, Wolfinger RD and Schabenberger O (2006) *SAS for Mixed Models,* Second Edition. SAS Institute Inc, Cary.

[76] Littell RC, Stroup WW, and Fruend RJ (2002) *SAS for Linear Models.* John Wiley & Sons.

[77] Lu G and Ades AE (2006) Assessing evidence inconsistency in mixed treatment comparisons. *Journal of American Statistical Association* 2006, **101**:447–459.

[78] Longford NT (1993) *Random Coefficient Models.* Oxford University Press, New York.

[79] Lumley, T (2002), Network meta-analysis for indirect treatment comparisons. *Statistics in Medicine* **21**:2313-2324.

[80] Mandel M (2010) Estimating disease progression using panel data. *Biostatistics.* **11**:304–16.

[81] McCullagh P and Nelder JA (1989) *Generalized Linear Models.* Chapman and Hall, London.

[82] Martinussen T and Scheike TH (2006) *Dynamic Regression Models for Survival Data.* Springer.

[83] McCulloch CE, Searle SR and Neuhaus JM (2008), *Generalized, Linear, and Mixed Models* 2nd Edition. John Wiley & Sons.

[84] Morita S, Thall PF, Mller P (2008) Determining the effective sample size of a parametric prior. *Biometrics* **64**:595–602.

[85] Mullahy J (1997) Instrumental variable estimation of Poisson regression models: application to models of cigarette smoking behavior. *Review of Economics and Statistics* **79**:586–593.

[86] Muller P, Berry DA, Grieve AP, Smith M and Krams M (2007) Simulation based sequential Bayesian design. *Journal of Statistical Planning and Inference* **137**:3140-3150.

[87] Murphy SA (2003) Optimal dynamic treatment regimes. *Journal of the Royal Statistical Society, Series B* **65**:331-366.

[88] Nelsen RB (1999) *An Introduction to Copulas.* Springer, New York.

[89] Nelson K, Lipsitz S, Fitzmaurice G, Ibrahim J, Parzen M and Strawderman R (2006) Use of the probability integral transformation to fit nonlinear mixed-effects models with non-normal random effects. *Journal of Computational and Graphical Statistics* **15**:39–57.

[90] Newton MA and Raftery AE (1994) Approximate Bayesian inference with the weighted likelihood bootstrap. *Journal of the Royal Statistical Society. Series B* **56**:3–48.

[91] O'Malley RE (1991) *Singular Perturbation Methods for Ordinary Differential Equations.* Springer-Verlag, New York.

[92] O'Quigley J, Hughes MD, Fenton T and Pei L (2010) Dynamic calibration of pharmacokinetic parameters in dose-finding studies. *Biostatistics* **11**:537-545.

[93] Overgaard RV, Jonsson N, Torne CW, Madsen H (2005) Non-linear mixed-effects models with stochastic differential equations: Implementation of an estimation algorithm. *J Pharmacokin Pharmacodyn.* **32**:85–107.

[94] Parmigiani G, Inoue LYT and Lopes HF (2009) *Decision Theory Principles and Approaches.* John Wiley & Sons.

[95] Parmigiani G (2002) *Modeling in Medical Decision Making: A Bayesian Approach.* John Wiley & Sons.

[96] Pearl J (1995) Causal diagrams for empirical research. *Biometrika* **82**:669–688.

[97] Pinheiro J and Bates D (2000) *Mixed-Effect Models in S and Splus.* Springer-Verlag: New York.

[98] Prentice RL (1982) Covariate measurement errors and parameter estimation in a failure time regression model. *Biometrika* **69**:331–342.

[99] Putter H, Fiocco M, Geskus RB (2007) Tutorial in biostatistics: Competing risks and multi-state models. *Statistics in Medicine* **26**:2389–430.

[100] Ouellet D, Patel N, Werth J, Feltner D, McCarthy B, Stone R, Mitchell D, Lalonde R (2006) The use of a clinical utility index in decision-making to select an insomnia compound. *Clinical Pharmacology & Therapeutics* **79**:277–282.

[101] Robins JM, Mark SD, Newey WK (1992) Estimating exposure effects by modelling the expectation of exposure conditional on confounders. *Biometrics* **48**:479–495.

[102] R Development Core Team (2008) *R: A Language and Environment for Statistical Computing*. R Foundation for Statistical Computing, Vienna. http://www.R-project.org

[103] Redyy PV (2013) Population pharmacokinetic-pharmacodynamic modeling of haloperidol in patients with schizophrenia using positive and negative syndrome rating scale. *J Clin Psychopharmacol.* **33**:731–9.

[104] Rizopoulos D, Verbeke G and Lesaffre E (2009) Fully exponential Laplace approximations for the joint modelling of survival and longitudinal data. *Journal of the Royal Statistical Society Series B* **71**:637–654.

[105] Rizopoulos D (2012) *Joint Models for Longitudinal and Time-to-Event Data: With Applications in R*. Chapman and Hall, London.

[106] Robert C (2001) *The Bayesian Choice*, second edition. Springer-Verlag, New York.

[107] Robins JM and Finkelstein DM (2000) Correcting for noncompliance and dependent censoring in an AIDS clinical trial with inverse probability of censoring weighted (IPCW) log-rank tests. *Biometrics* **56**:779-788.

[108] Rosenbaum PR and Rubin DB (1983) The central role of the propensity score in observational studies for causal effects. *Biometrika* **70**:41-55.

[109] Rubin DB (1981) The Bayesian bootstrap. *Annals of Statistics* **9**:130–134.

[110] Ruppert D, Wand MP and Carroll RJ (2003) *Semiparametric Regression*. Cambridge University Press.

[111] Sanathanan LP, and Peck CC (1991) The randomized concentration-controlled trial: An evaluation of its sample size efficiency. *Controlled Clinical Trials* **12**:780-94.

[112] SAS Institute Inc. (2011) *SAS/STAT 9.3 Users Guide*. SAS Institute Inc, Cary.

[113] Schwarz GE (1978) Estimating the dimension of a model. *Annals of Statistics* **6**:461-464.

[114] Schoenfeld DA (1983) Sample-size formula for the proportional-hazards regression model. *Biometrics* **39**:499-503.

[115] Senn S (2007) *Statistical Issues in Drug Development.* John Wiley & Sons.

[116] Serfling RJ (1980) *Approximation Theorems of Mathematical Statistics.* John Wiley & Sons.

[117] Snowden JM, Rose S, Mortimer KM (2011) Implementation of G-computation on a simulated data set: demonstration of a causal inference technique. *Am J Epidemiol* **173**:731–738.

[118] Stram DO and Lee JW (1994) Variance components testing in the longitudinal mixed effects model. *Biometrics* **50**:1171-1177.

[119] Sun J (2006) *The Statistical Analysis of Interval-Censored Failure Time Data.* Springer, New York.

[120] Sylvestre MP and Abrahamowicz M (2008) Comparison of algorithms to generate event times conditional on time-dependent covariates. *Statistics in Medicine* **27**:2618–2634

[121] Sylvestre MP and Abrahamowicz M (2009) Flexible modeling of the cumulative effects of time-dependent exposures on the hazard. *Statistics in Medicine* **28**:3437–53.

[122] Terza JV, Basu A, Rathouz PJ (2008) Two-stage residual inclusion estimation: Addressing endogeneity in health econometric modeling. *Journal of Health Economics* **27**:531–543.

[123] Therneau TM and Grambsch PM (2000) *Modeling Survival Data: Extending the Cox Model.* Springer-Verlag, New York.

[124] Ting N (Eds.) (2006) *Dose Finding in Drug Development.* Springer, New York.

[125] Tornoe CW et al. (2004a) Non-linear mixed-effects pharmacokinetic/pharmacodynamic modelling in NLME using differential equations. *Computer Methods and Programs in Biomedicine* **76**:31–40.

[126] Tosteson TD, Buonaccorsi JP, Demidenko E (1998) Covariate measurement error and the estimation of random effect parameters in a mixed model for longitudinal data. *Statistics in Medicine* **17**:1959–71.

[127] Verbeke G and Lesaffre E (1997) The effect of misspecifying the random-effects distribution in linear mixed models for longitudinal data. *Computational Statistics and Data Analysis* **53**:541-556.

[128] Wald A (1949) Statistical Decision Function. *Ann. Math. Statist.* **20**:165–205.

[129] Wang J (2001) Optimal design for linear interpolation of curves. *Statistics in Medicine* **20**:2467–77.

[130] Wang J, Donnan PD, Steinke D and MacDonald TM (2001) The multiple propensity score for analysis of dose-response relationships in drug safety studies. *J Pharmacoepidemiology and Drug Safety* **10**:105–111.

[131] Wang J, Donnan PT and MacDonald TM (2002) An approximate Bayesian risk analysis for the gastro-intestinal safety of ibuprofen. *J Pharmacoepidemiology and Drug Safety* **11**:695–701.

[132] Wang J (2005) A semi-parametric approach to fitting nonlinear mixed PK/PD model with effect compartment using SAS. *Pharmaceut. Statist.* **4**:59–69.

[133] Wang J (2006) Optimal parametric design with applications to pharmacokinetic and pharmacodynamic trials. *Journal of Applied Statistics* **33**:837–852.

[134] Wang J (2012) Semiparametric hazard function estimation in meta analysis for time to event data. *Research Synthesis Methods* **3**:240–249.

[135] Wang J (2012b) Dose as instrumental variable in exposure-safety analysis using count models. *J Biopharm Stat.* **22**:565-581.

[136] Wang J and Quartey G (2012c) Nonparametric estimation for cumulative duration of adverse events. *Biometrical Journal* **54**:61–74.

[137] Wang J and Quartey G (2013) A semi-parametric approach to analysis of event duration and prevalence. *Computational Statistics and Data Analysis* **68**:248–257.

[138] Wang J (2014) Determining causal effect in exposure-response relationship with randomized concentration controlled trials. *J Biopharm Stat.* **24**:874–892.

[139] Wang J (2014b) Causal effect estimation and dose adjustment in exposure-response relationship analysis. In *Developments in Statistical Evaluation in Clinical Trials.* van Montfort et al (eds.). Springer, New York.

[140] Wang J and Li W (2014c) Test hysteresis in phamacokinetic/pharmacodynamic relationship with mixed effect models: An instrumental model approach. *J Biopharm Stat.* **24**:326–343.

[141] Wang N and Davidian M (1996) A note on covariate measurement error in nonlinear mixed effects models. *Biometrika* **83**:801–812.

[142] Wang Y, Sung C, Dartois C, Ramchandani R, Booth BP, Rock E and Gobburu J (2009) Elucidation of relationship between tumor size and survival in non-small-cell lung cancer patients can aid early decision-making in clinical drug development. *Clinical Pharmacology and Therapeutics* **86**:167-174.

[143] Wei LJ, Lin DY and Weissfeld LO (1989) Regression analysis of multi-variate incomplete failure time data by modeling marginal distributions. *Journal of the American Statistical Association* **84**:1065-1073.

[144] White H (1981) Consequences and detection of misspecified nonlinear regression models. *Journal of the American Statistical Association* **76**:419-433.

[145] White H (1982) Maximum likelihood estimation of misspecified models. *Econometrica* **50**:1-25.

[146] Whitehead A (2002) *Meta-Analysis of Controlled Trials*, John Wiley & Sons.

[147] Whitehead J and Williamson D (1998) Bayesian decision procedures based on logistic regression models for dose-finding studies. *J Biopharm Stat.* **8**:445-467.

[148] Wooldridge JM (1997) On two stage least squares estimation of the average treatment effect in a random coefficient model. *Econ Lett.* **56**:129-133.

[149] Wu H et al. (2006) Pharmacodynamics of antiretroviral agents in hiv-1 infected patients: Using viral dynamic models that incorporate drug susceptibility and adherence. *J Pharmacokin Pharmacodyn* **33**:399-419.

[150] Wu L, Liu W, and Hu XJ (2010) Joint inference on HIV viral dynamics and immune suppression in presence of measurement errors. *Biometrics* **66**:327-335.

[151] Wu L, Hu XJ and Wu H (2008) Joint inference for nonlinear mixed-effects models and time to event at the presence of missing data. *Biostatistics* **9**:308-320.

[152] Yang W, Joffe MM, Hennessy S, and Feldman HI (2014) Covariance adjustment on propensity parameters for continuous treatment in linear models. *Statistics in Medicine* **33**:4577-4589.

[153] Zellner A (1986) On Assessing Prior Distributions and Bayesian Regression Analysis with g-Prior Distributions. in Goel PK and and Zellner A (eds.) *Bayesian Inference and Decision Techniques: Essays in Honor of Bruno de Finetti* North-Holland, Amsterdam.

[154] Zhang L, Beal SL and Sheiner LB (2003) Simultaneous vs. sequential analysis for population PK/PD Data I: Best case performance. *J Pharmacokin Pharmacodyn* **30**:387-403.

[155] Zhou M (2001) Understanding the Cox regression models with time-change covariates. *The American Statistician* **55**:153-155.

[156] Zohar S, Katsahian S and OQuigley J (2011) An approach to meta-analysis of dose-finding studies. *Statistics in Medicine* **30**:2109-2116.

Index

Printed in the United States
by Baker & Taylor Publisher Services